Loewy Decomposition of Linear Differential Equations

T0181355

Texts and Monographs in Symbolic Computation

A Series of the Research Institute
for Symbolic Computation,
Johannes Kepler University, Linz, Austria

For further volumes:
http://www.springer.com/series/3073

Fritz Schwarz

Loewy Decomposition of Linear Differential Equations

Fritz Schwarz
Institute SCAI
Fraunhofer Gesellschaft
Sankt Augustin
Germany

ISSN 0943-853X
ISBN 978-3-7091-1687-6   ISBN 978-3-7091-1286-1 (eBook)
DOI 10.1007/978-3-7091-1286-1
Springer Wien Heidelberg New York Dordrecht London

Printed on acid-free paper

Springer is part of Springer Science+Business Media (www.springer.com)

To Birgit

Preface

The aim of this book is to communicate some results on solving linear differential equations that have been achieved in the last two decades. The key concept is the factorization of a differential equation or the corresponding differential operator, and the resulting decomposition into unique objects of lower order. Although more than 100 years old, these results had been forgotten for almost a century before they were reawakened.

Several new developments have entailed novel interest in this subject. On the one hand, methods of differential algebra lead to a better understanding of the basic problems involved. Instead of dealing with individual equations, the corresponding differential operators are considered as elements of a suitable ring where they generate an ideal. This proceeding is absolutely necessary if partial differential equations and operators are investigated. In particular the concept of a Janet basis for the generators of an ideal and the Loewy decomposition of the ideal corresponding to the given equations are of fundamental importance. In order to apply these results for solving concrete problems, the availability of computer algebra software is indispensable due the enormous size of the calculations usually involved.

Proceeding along these lines, for large classes of linear differential equations – ordinary as well as partial – a fairly complete theory for obtaining its solutions in closed form has been achieved. Whenever feasible, constructive methods for algorithm design are given, and the possible limits of decidability are indicated. This proceeding may serve as a model for dealing with other problems in the area of differential equations.

I am grateful to Dima Grigoriev, Ziming Li, Michael Singer and Sergey Tsarev for numerous discussions and suggestions. Thanks are due to Hans-Heinrich Aumüller for carefully reading the final version of the manuscript, and to Winfried Neun at the Zuse Institut in Berlin for keeping the ALLTYPES website running. The ideal working environment at the Fraunhofer Institut SCAI is gratefully acknowledged.

Sankt Augustin, Germany   Fritz Schwarz
July 2012

Contents

Introduction

Factoring algebraic polynomials into irreducible components of lowest degree has been of basic importance for a long time. By default, a polynomial is called irreducible if it does not factor into components of lower order without enlarging its coefficient domain which is usually the rational number field $\mathbb{Q}$. For univariate polynomials, the relevance of this proceeding for determining the solutions of the corresponding algebraic equation is obvious. In the multivariate case, it is the basis for understanding the structure of the algebraic manifold associated with any such polynomial. Good introductions into the subject may be found, e.g., in the books by Cox et al. [13], Adams and Loustaunau [2] or Greuel and Pfister [20].

It turns out that factoring differential operators and the corresponding differential equations is of fundamental importance as well. Originally the problem of factoring linear *ordinary* differential equations (lode's) has been considered by Beke [3], Schlesinger [58] and Loewy [46]. Thereafter it has been forgotten for almost a century until it was reassumed by Grigoriev [21] and Schwarz [59], where essentially computational and complexity issues were considered; see also Bronstein [6] and van Hoeij [72]. A good survey of these results is given in the book by van der Put and Singer [71].

Due to the non-commutativity of differential operators, some new phenomena appear compared to algebraic polynomials. In the first place, left and right factors have to be distinguished. Correspondingly there are left and right least common multiples and left and right greatest common divisors. For the topics discussed in this monograph, the least common left multiple *Lclm* and the greatest common right divisor *Gcrd* are of paramount importance. The term *factor* without further specification means always *right factor*. An operator that does not have any right factor of a certain kind is called *irreducible*, otherwise it is called *reducible*. A more detailed specification of these terms may be found on page 26.

Furthermore, Loewy [46] realized that a new concept is needed beyond reducibility which he baptized *complete reducibility* of a given operator. For ordinary differential operators it may simply be expressed by saying that an operator, or the corresponding differential equation, is completely reducible if it may be represented as the least common left multiple of its irreducible right factors, the greatest common right divisor of which is trivial. This amounts to saying that the solution space of the originally given equation is the direct sum of the solution spaces of its right factors. Applying this concept repeatedly, Loewy obtained a unique decomposition of any ordinary differential operator into completely reducible components of highest possible order.

Starting from Loewy's results, it appears self-evident trying a similar approach for partial differential operators; the goal is to develop more systematic methods for solving linear partial differential equations than those available in the literature [14, 17, 18, 28, 31, 33]. It turns out that for a special class of problems, i.e. those systems of pde's that have a finite-dimensional solution space, Loewy's original theory for decomposing ordinary equations may be generalized almost straightforwardly as has been shown by Li et al. [43].

Factorization problems for general pde's were considered first in the dissertation of Blumberg [5]. He described a factorization of a third-order operator in two variables which appeared to preclude a generalization of Loewy's result beyond the ordinary case. Twenty years after that Miller [49] discussed several factorizations of partial operators in his dissertation. After a long period of inactivity, a lot of interest in this field has arisen in the last two decades, starting with some articles by Grigoriev and Schwarz [22–25]; see also [67]. At this point another analogy with the theory of algebraic polynomials comes into play. Beyond a certain point, further progress is only possible if the underlying scope is broadened and more abstract algebraic concepts are applied. In commutative algebra this means that polynomials are considered as elements of a ring generating certain ideals. Then the results known from ring theory may be applied to the problems involving these polynomials. Only in this way progress in this field has been possible, e.g. generating primary decompositions of ideals and above all the theory of Gröbner bases by Buchberger [7] and its ubiquitous applications.

In the differential case a similar development takes place. Although the solutions of differential equations are the ultimate objects of interest, it is advantageous to consider their left-hand sides as the result of applying a differential operator to a differential indeterminate. Solving an equation means to assign those elements from a suitable function space to the indeterminate such that the full expression vanishes. The differential operators are considered as elements of a non-commutative ring, or as modules over such a ring, so-called $\mathscr{D}$-modules. The extensive theory for these non-commutative rings or modules may then be applied to study their properties. It turns out that only in this way satisfactory results may be obtained. Good introductions into the underlying differential algebra are the books by Kolchin [37], Kaplanski [35] and Coutinho [12], and the article by Buium and Cassidy [8]. Those aspects that are relevant for this monograph may also be found in the book by van der Put and Singer [71].

It turns out that Loewy's original theory for factoring, and more generally decomposing, an ordinary differential operator uniquely into lower-order components may be extended to partial differential operators if this algebraic language is applied. The following observations are of particular relevance. In the first place, the left intersection of two partial differential operators is not necessarily principal;

in general it is an ideal in the ring determined by the originally given operators that may be generated by any number of elements. Secondly, right divisors of an operator need not be principal either; they are overideals of the ideal generated by the given operators that may be generated by any number of elements as well. Taking these observations into account, the objections concerning the generalization of Loewy's results do not apply.

This extended concept of factoring partial differential operators comprises Laplace's method for solving equations of the form $z_{xy} + az_x + bz_y + cz = 0$ as a special case. It consists of an iterative procedure which is described in detail in Chap. 5 of the second volume of Goursat's books on second order differential equations [18], or in Chap. 2 of the second volume of Darboux's series on surfaces [14]; see also [69] and Appendix C of this monograph. By proper substitutions, new equations are generated from the given one until eventually an equation is obtained that may be solved, from the solution of which a partial solution of the originally given equation may be constructed. An equivalent procedure which is also described in the books by Goursat and Darboux consists of generating an additional equation that is *in involution* with the given one. The operators corresponding to the two equations may be considered as generators of an ideal that contains the principal ideal generated by the operator corresponding to the originally given equation. In this way the apparent ad hoc nature of the method of Goursat and Darboux disappears.

These remarks may be summarized as follows. To any given system of linear differential equations, there corresponds the ideal or module generated by the differential operators at their left-hand sides. Factorization means finding a divisor of this ideal. In general there may be more than a single divisor. The complete answer consists of the sublattice in the lattice of left ideals or modules which has the given one as the lowest element. By analogy with Cohn's [11] lattice of factors, it is called the *divisor lattice*. The simplicity of this theory for ordinary differential operators, or for modules of partial differential operators corresponding to systems of pde's with finite dimensional solution space, originates from the fact that they form a sublattice in the respective ring or module.

This monograph deals with linear ordinary differential equations and linear partial differential equations in two independent variables of order not higher than three. Their decompositions are described along the lines described above.

It turns out that many calculations that occur when concrete problems are considered cannot be performed by pencil and paper due to their size, i.e. the complexity of the respective procedure is too large. Therefore computer algebra software has been developed for this purpose. It is available gratis on the website www.alltypes.de after registration. There is a special demo showing the functionality of this software for the applications relevant for this monograph. To start this demo, go to the ALLTYPES website and click on the button StartALLTYPES, the interactive alltypes window opens, then submit

Demo LoewyDecompopsitions;

The demo starts. If an example has been completed and the result has been displayed in a separate window, the system asks cont?; in order to continue with the next problem submit y.

However, it should be emphasized that it is by no means assured that any factorization problem may be solved algorithmically. On the contrary, there are severe indications that this may not be the case. If it could be proved that these problems are in general undecidable, it would be interesting to identify special classes of problems for which this is not true as it is the case, e.g. for diophantine equations.

The contents of the individual chapters may be summarized as follows:

Chapter 1.   Loewy's Results for Ordinary Differential Equations. The original results of Loewy for decomposing ordinary differential operators are presented. Its application to operators of order 2 and 3 is worked out in detail. It is shown how a nontrivial decomposition may be applied for solving the corresponding equation.

Chapter 2.   Rings of Partial Differential Operators. Basic properties of the ring $\mathbb{Q}(x, y)[\partial_x, \partial_y]$ and its left ideals are described, with particular emphasis on the sum and intersection of ideals and how they determine a lattice structure in this ring. This is the foundation for the decompositions described later on.

Chapter 3.   Equations with Finite-Dimensional Solution Space. Loewy's theory for linear ode's may be extended in a rather straightforward manner to systems of linear pde's with the property that their general solution involves only constants, i.e. if it is a finite-dimensional vector space. Systems of such equations of order 2 and 3 in two independent variables are discussed.

Chapter 4.   Decomposing Second-Order Operators. Principal ideals correspond to individual linear pde's. There is an extensive literature on such equations in the nineteenth and early twentieth century. These results may be obtained systematically, avoiding any ad hoc procedure, by applying the algebraic theory described in Chap. 2.

Chapter 5.   Solving Second-Order Equations. The results of the preceding chapter are applied for solving second-order homogeneous equations; at the end inhomogeneous equations are discussed because they are needed later on for solving certain third-order homogeneous equations.

Chapter 6.   Decomposing Third-Order Operators. Similar results as in Chap. 4 are obtained for third-order operators.

Chapter 7.   Solving Third-Order Equations. The results of the preceding chapter are applied for solving third-order homogeneous equations.

Chapter 8.   Conclusions and Summary. The results of this monograph are summarized and several possible extensions are outlined.

Appendix A.   Solutions to the exercises are given.

Appendix B.   Many algorithms in the main part of the book rely on the solution of Riccati equations and certain generalizations, e.g. Riccati equations of higher order and partial Riccati equations in two independent variables. Their solutions are described in detail. Furthermore, first integrals of first-order ode's are discussed.

Appendix C.   Laplace's solution procedure for certain second-order linear pde's in two variables is described.

Appendix D.   Lie developed a solution theory for certain second-order linear pde's in two variables based on its symmetries. These results are reviewed; its relation to decomposition properties is briefly discussed,

Appendix E.   On the website www.alltypes.de [62] userfunctions are provided that may be applied interactively for performing the voluminous calculations required for many problems in differential algebra. This appendix gives a short introduction to this website.

Chapter 1
Loewy's Results for Ordinary Differential Equations

Abstract The idea of factoring an ordinary differential operator, or the corresponding linear ordinary differential equation (ode), into components of lower order originated from the analogous problem for algebraic polynomials. In the latter case, it is the first step when the solutions of the corresponding algebraic equation are to be determined. It turns out that a similar strategy is the key for understanding the structure of the solution space of a linear ode. Good introductions into the subject are the books by Ince [29] or Kamke [32, 34], and the book by van der Put and Singer [71].

1.1 Basic Facts for Linear ODE's

Let $D \equiv \frac{d}{dx}$ denote the derivative w.r.t. the variable x. A differential operator of order n is a polynomial of the form

$$L \equiv D^n + a_1 D^{n-1} + \ldots + a_{n-1} D + a_n \tag{1.1}$$

where the coefficients a_i, $i = 1, \ldots, n$ are from some function field, the *base field* of L. Usually it is the rational function field of the variable x, i.e. $a_i \in \mathbb{Q}(x)$. If y is a differential indeterminate with $\frac{dy}{dx} \neq 0$, Ly becomes a differential polynomial, and $Ly = 0$ is the differential equation corresponding to L. Sometimes the notation $y' \equiv \frac{dy}{dx}$ is applied.

The equation $Ly = 0$ allows the *trivial solution* $y \equiv 0$. The general solution contains n constants $C_1, \ldots, C_n$. Due to its linearity, these constants appear in the form $y = C_1 y_1 + \ldots + C_n y_n$. The y_k are linearly independent over the field of constants; they form a so-called *fundamental system* and generate a n-dimensional vector space.

The coefficients a_k of (1.1) may be rationally expressed in terms of a fundamental system and its derivatives. To this end the *Wronskian*

F. Schwarz, *Loewy Decomposition of Linear Differential Equations*, Texts & Monographs in Symbolic Computation, DOI 10.1007/978-3-7091-1286-1_1,
© Springer-Verlag/Wien 2012

$$W^{(n)}(y_1, \ldots, y_n) \equiv \begin{vmatrix} y_1 & y_2 & \cdots y_n \\ y_1' & y_2' & \cdots y_n' \\ \cdots & \cdots & \cdots \\ y_1^{(n-1)} & y_2^{(n-1)} & \cdots y_n^{(n-1)} \end{vmatrix} \qquad (1.2)$$

is defined. It is different from zero if the $y_k(x)$ are independent over the constants.
More generally, the determinants

$$W_k^{(n)}(y_1, \ldots, y_n) \equiv \frac{\partial}{\partial y^{(n-k)}} \begin{vmatrix} y & y_1 & y_2 & \cdots y_n \\ y' & y_1' & y_2' & \cdots y_n' \\ \cdots & \cdots & \cdots & \cdots \\ y^{(n)} & y_1^{(n)} & y_2^{(n)} & \cdots y_n^{(n)} \end{vmatrix} \qquad (1.3)$$

are defined for $k = 0, 1, \ldots, n$. There are the obvious relations $W_0^{(n)} = (-1)^{n-1} W^{(n)}$ and $\frac{dW^{(n)}}{dx} = (-1)^{n-1} W_1^{(n)}$. The determinant at the right hand side of (1.3) vanishes for $y = y_k$. Therefore, given a fundamental system $\{y_1, \ldots, y_n\}$, (1.3) is another way of writing the left hand side of $Ly = 0$. This yields the representation

$$a_k = (-1)^k \frac{W_k^{(n)}(y_1, \ldots, y_n)}{W^{(n)}(y_1, \ldots, y_n)} \qquad (1.4)$$

for the coefficients a_k, $k = 1, \ldots, n$ in terms of a fundamental system; these expressions show that $Ly = 0$ is uniquely determined by its solution space. For $k = 1$ there follows

$$W_1^{(n)} + a_1 W^{(n)} = W^{(n)\prime} + a_1 W^{(n)} = 0, \qquad (1.5)$$

i.e. $W^{(n)} = \exp\left(-\int a_1 dx\right)$. Equation (1.5) is known as *Liouville's equation*. The cases $n = 2$ and $n = 3$ are discussed in more detail in Exercise 1.1.

Liouville's equation may be generalized for determinants that are generated by subsets $\{y_1, \ldots, y_m\}$ with $m < n$. Consider the matrix

$$\begin{pmatrix} y_1 & y_2 & \cdots & y_m \\ y_1' & y_2' & \cdots & y_m' \\ \cdots & \cdots & \cdots & \cdots \\ y_1^{(n-1)} & y_2^{(n-1)} & \cdots & y_m^{(n-1)} \end{pmatrix} ; \qquad (1.6)$$

for any fixed value of m there are $\binom{n}{m}$ square $m \times m$ matrices that may be formed out of its rows. Their determinants may be considered as new functions. This set of functions is closed under differentiations if the original differential equation $Ly = 0$ is used to substitute derivatives of order higher than $n - 1$.

By suitable differentiations and elimination, for each of these functions an $\binom{n}{m}$th order linear differential equation may be obtained. These equations are called *associated equations* for the original equation $Ly = 0$. For $n = 3$ and $m = 2$ they are shown next.

Example 1.1. Consider the third order linear ode

$$y''' + a_1 y'' + a_2 y' + a_3 y = 0; \tag{1.7}$$

let y_1 and y_2 be two members of a fundamental system and define

$$z_1 \equiv \begin{vmatrix} y_1 & y_2 \\ y_1' & y_2' \end{vmatrix}, \quad z_2 \equiv \begin{vmatrix} y_1 & y_2 \\ y_1'' & y_2'' \end{vmatrix}, \quad z_3 \equiv \begin{vmatrix} y_1' & y_2' \\ y_1'' & y_2'' \end{vmatrix}.$$

They obey $z_1' = z_2$, $z_2' = z_3 - a_1 z_2 - a_2 z_1$ and $z_3' = a_3 z_1 - a_1 z_3$. This system of linear homogeneous ode's may be transformed into a Janet basis in lex term order with $z_3 \succ z_2 \succ z_1$ with the result

$$z_1''' + 2a_1 z_1'' + (a_1' + a_1^2 + a_2)z_1' + (a_2' + a_1 a_2 - a_3)z_1 = 0,$$
$$z_2 - z_1' = 0, \quad z_3 - z_1'' - a_1 z_1' - a_2 z_1 = 0. \tag{1.8}$$

The first member is the desired associated equation for z_1. The respective equations for z_2 and z_3 may be obtained by suitable changes of the term order. This is considered in Exercise 1.2. □

1.2 Factorization and Loewy Decomposition

An operator L of order n is called *reducible* if it may be represented as the product of two operators L_1 and L_2, both of order lower than n; then one writes $L = L_1 L_2$, i.e. juxtaposition means the operator product. The product is defined by the rule $Da_i = a_i D + a_i'$. It is non-commutative, i.e. in general $L_1 L_2 \neq L_2 L_1$. Thus, left and right factors have to be distinguished. If $L = L_1 L_2$, the left factor L_1 is called the *exact quotient* of L by L_2, and L_2 is said to divide L from the right; it is a *right divisor* or simply *divisor* of L.

By default, the coefficient domain of the factors is assumed to be the base field of L, possibly extended by some algebraic numbers. An operator or an equation is called *irreducible* if such a decomposition is not possible without enlarging the base field. If the coefficients of the factors may be from an extension of the base field it has to be specified explicitly. Very much like in commutative algebra, any operator L may be represented as product of first-order factors if coefficients from a universal field are admitted.

For any two operators L_1 and L_2 the *least common left multiple* $Lclm(L_1, L_2)$ is the operator of lowest order such that both L_1 and L_2 divide it from the right. The *greatest common right divisor* $Gcrd(L_1, L_2)$ is the operator of highest order that divides both L_1 and L_2 from the right. Two operators L_1 and L_2 are called *relatively prime* if there is no operator of positive order dividing both on the right.

Two operators L_1 and L_2 are called *of the same type* (*von derselben Art*) if there exist operators P and Q with coefficients from the base field such that $PL_1 = L_2Q$. An operator that may be represented as $Lclm$ of irreducible right factors is called *completely reducible*. By definition, an irreducible operator is completely reducible.

Due to the non-commutativity of the product of differential operators another new phenomenon occurs in comparison to algebraic polynomials, i.e. the factorization of differential operators into irreducible factors is not unique as the following example due to Landau [39] shows.

Example 1.2. Consider $D^2 - \frac{2}{x}D + \frac{2}{x^2}$. Two possible factorizations are

$$\left(D - \frac{1}{x}\right)\left(D - \frac{1}{x}\right) = \left(D - \frac{1}{x}\right)^2 \text{ and } \left(D - \frac{1}{x(1+x)}\right)\left(D - \frac{1+2x}{x(1+x)}\right).$$

More generally, the factorization

$$\left(D - \frac{1}{x(1+ax)}\right)\left(D - \frac{1+2ax}{x(1+ax)}\right)$$

is valid with a constant parameter a.   □

However, the various factorizations are not uncorrelated as Loewy [46] has shown where also the proof may be found.

Proposition 1.1. *If an operator L allows two factorizations into irreducible factors $L = L_1 \ldots L_m = \bar{L}_1 \ldots \bar{L}_{\bar{m}}$, then $m = \bar{m}$, and there exists a permutation π such that L_i and $L_{\pi(i)}$ are of the same type.*

A systematic scheme for obtaining a *unique* decomposition of a differential operator of any order into lower order components has been given by Loewy [46]; see also Ore [52]. At first the irreducible right factors beginning with lowest order are determined, e.g. by applying Lemma 1.1 below. The $Lclm$ of these factors is the completely reducible right factor of highest order; by construction it is uniquely determined by the given operator. If its order equals n, the order of the given differential operator, this operator is completely reducible and the procedure terminates. If this is not the case, the $Lclm$ is divided out and the same procedure is repeated with the exact quotient. Due to the lowering of order in each step, this proceeding terminates after a finite number of iterations and the desired decomposition is obtained. Based on these considerations, Loewy [46] obtained the following fundamental result.

Theorem 1.1 ([46]). *Let $D = \frac{d}{dx}$ be a derivative and $a_i \in \mathbb{Q}(x)$. A differential operator*

$$L \equiv D^n + a_1 D^{n-1} + \ldots + a_{n-1}D + a_n$$

of order n may be written uniquely as the product of completely reducible factors
$L_k^{(d_k)}$ *of maximal order d_k over $\mathbb{Q}(x)$ in the form*

$$L = L_m^{(d_m)} L_{m-1}^{(d_{m-1})} \ldots L_1^{(d_1)}$$

with $d_1 + \ldots + d_m = n$. The factors $L_k^{(d_k)}$ are unique.

The decomposition described in this theorem is called the *Loewy decomposition* of L. The factors $L_k^{(d_k)}$ are called the *Loewy factors* of L, the rightmost of them is simply called the *Loewy factor*.

Loewy's decomposition may be refined if the completely reducible components are split into irreducible factors as shown next.

Corollary 1.1. *Any factor $L_k^{(d_k)}$, $k = 1, \ldots, m$ in Theorem 1.1 may be written as*

$$L_k^{(d_k)} = Lclm(l_{j_1}^{(e_1)}, l_{j_2}^{(e_2)}, \ldots, l_{j_k}^{(e_k)})$$

with $e_1 + e_2 + \ldots + e_k = d_k$; $l_{j_i}^{(e_i)}$ for $i = 1, \ldots, k$ denotes an irreducible operator of order e_i over $\mathbb{Q}(x)$.

Loewy's fundamental result described in Theorem 1.1 and the preceding corollary enable a detailed description of the function spaces containing the solution of a reducible linear differential equation as will be shown later in this chapter. More general field extensions associated to irreducible equations are studied by differential Galois theory. Good introductions to the latter are the above quoted articles by Kolchin, Singer and the lecture by Magid [47].

Next the question comes up how to obtain a factorization for any given equation or operator; for order 2 and 3 the answer is as follows.

Lemma 1.1. *Determining the right irreducible factors of an ordinary operator up to order 3 with rational function coefficients amounts to finding rational solutions of Riccati equations; as usual $D \equiv \dfrac{d}{dx} = {}'$.*

(i) *A second order operator $D^2 + AD + B$, $A, B \in \mathbb{Q}(x)$ has a right factor $D + a$ with $a \in \mathbb{Q}(x)$ if a is a rational solution of*

$$a' - a^2 + Aa - B = 0.$$

(ii) *A third order operator $D^3 + AD^2 + BD + C$, $A, B, C \in \mathbb{Q}(x)$ has a right factor $D + a$ with $a \in \mathbb{Q}(x)$ if a is a rational solution of*

$$a'' - 3aa' + a^3 + A(a' - a^2) + Ba - C = 0.$$

It has a right factor $D^2 + bD + c$, $b, c \in \mathbb{Q}(x)$, if b is a rational solution of

$$b'' - 3bb' + b^3 + 2A(b' - b^2) + (A' + A^2 + B)b - B' - AB + C = 0;$$

$$\text{then} \quad c = -(b' - b^2 + Ab - B).$$

Proof. Dividing the given second-order operator by $D + a$ and requiring that this division be exact yields immediately the given constraint. The same is true if the given third order operator is divided by $D + a$. Dividing the given third-order operator by $D + bD + c$ yields a system comprising two equations that may easily be simplified to the above conditions. □

This lemma reduces the problem of determining right factors in the base field for second- and third-order operators to finding rational solutions of Riccati equations. This latter problem is the subject of Appendix B. Because these Riccati equations have always solutions if the admitted function field is properly enlarged, any such equation factors into first-order factors in a universal field.

For equations of any order there is a more general scheme based on the associated equations [3, 59] introduced above. The basic principle behind it may be described as follows. Let the nth order equation

$$Ly \equiv y^{(n)} + a_1 y^{(n-1)} + \ldots + a_{n-1} y' + a_n y = 0 \tag{1.9}$$

be given with a fundamental system $S_n = \{y_1, \ldots, y_n\}$. The following question is asked: Does there exist an mth order right factor of L with $m < n$ such that the corresponding equation

$$y^{(m)} + b_1 y^{(m-1)} + \ldots + b_{m-1} y' + b_m y = 0 \tag{1.10}$$

has a fundamental system $S_m = \{y_1, \ldots, y_m\} \subset S_n$? On the one hand, the coefficients b_i may be expressed through certain determinants formed from the elements of S_m according to (1.4). On the other hand, these determinants obey the associated equations of (1.9). This provides the connection between the given coefficients a_k of (1.9) and the coefficients b_j of the possible factor (1.10). The following example of factoring a third-order operator illustrates these steps.

Example 1.3. Consider the equation

$$y''' + \left(1 - \frac{1}{x}\right) y'' + \left(x - \frac{2}{x}\right) y' + \frac{2}{x^2} y = 0. \tag{1.11}$$

Because by Lemma 1.1, case (i), a first-order right factor does not exist, ask for a second-order right factor $y'' + b_1 y' + b_2 y = 0$. According to Example 1.1 the associated equation for z_1 is

$$z_1''' + \left(2 - \frac{2}{x}\right) z_1'' + \left(x + 1 - \frac{4}{x} + \frac{2}{x^2}\right) z_1' + \left(x - \frac{2}{x} + \frac{2}{x^2}\right) z_1 = 0.$$

It has a single solution $z_1 = e^{-x}$ with rational logarithmic derivative. From system (1.8) there follows $z_2 = -e^{-x}$ and $z_3 = \left(1 - \frac{1}{x}\right)e^{-x}$. Thus, $b_1 = \frac{z_1'}{z_1}$ and $b_2 = -\frac{z_1''}{z_1} - a_1 \frac{z_1'}{z_1} - a_2$; they yield the following factorization

$$\left(D - \frac{1}{x}\right)\left(D^2 + D + x - \frac{1}{x}\right)y = 0$$

of the given Eq. (1.11).   □

In Exercise 1.8 the Loewy decomposition of Eq. (1.11) is determined by applying Lemma 1.1.

In Exercise 1.3 second-order factors of fourth-order equations are considered. It turns out that the complexity for factoring equations of order higher than four grows tremendously. Furthermore, when it comes to solving an equation, first- and second-order factors are most interesting; they correspond to Liouvillian or possibly special function solutions.

For operators of fixed order the possible decompositions, differing by the number and the order of factors, may be listed explicitly; some of the factors may contain parameters. Each alternative is called a *type of Loewy decomposition*. The complete answers for the most important cases $n = 2$ and $n = 3$ are detailed in the following corollaries to the above theorem.

Corollary 1.2. *Let L be a second-order operator. Its possible Loewy decompositions may be described as follows; $l^{(i)}$ and $l_j^{(i)}$ are irreducible operators of order i; C is a constant;*

$$\mathscr{L}_1^2 : L = l_2^{(1)}l_1^{(1)}; \quad \mathscr{L}_2^2 : L = Lclm(l_2^{(1)}, l_1^{(1)}); \quad \mathscr{L}_3^2 : L = Lclm(l^{(1)}(C)).$$

An irreducible second-order operator is defined to have decomposition type $\mathscr{L}_0^2$. The decompositions $\mathscr{L}_0^2$, $\mathscr{L}_2^2$ and $\mathscr{L}_3^2$ are completely reducible; for decomposition type $\mathscr{L}_1^2$ the unique first-order factors are the Loewy factors.

Proof. According to Lemma 1.1, case (i), the possible factors are determined by the solutions of a Riccati equation; they are described in Appendix B. The general solution may be rational; there may be two non-equivalent or a single rational solution; or there may be no rational solution at all; these alternatives correspond to decomposition types $\mathscr{L}_3^2$, $\mathscr{L}_2^2$, $\mathscr{L}_1^2$ or $\mathscr{L}_0^2$ respectively.   □

By definition, a Loewy factor comprises *all* irreducible right factors, either corresponding to special rational solutions or those involving a constant. As a consequence, the decomposition type $\mathscr{L}_1^2$ implies that a decomposition of type $\mathscr{L}_2^2$ or $\mathscr{L}_3^2$ does *not* exist, i.e. there is only a single first-order right factor that may not be obtained by specialization of C in a type $\mathscr{L}_3^2$ decomposition. By similar arguments, decomposition type $\mathscr{L}_2^2$ excludes type $\mathscr{L}_3^2$.

A factor containing a parameter corresponds to a factorization that is not unique; any special value for C generates a special irreducible factor. Because the originally given operator is of second order, two different special values may be chosen in order to represent it in the form $L = Lclm\left(l^{(1)}(C_1), l^{(1)}(C_2)\right)$, $C_1 \neq C_2$. The following corollary shows that not more than two such operators may be chosen.

Corollary 1.3. *The left-intersection of two operators $l^{(1)}(C_1)$ and $l^{(1)}(C_2)$ is a second-order operator; intersecting it with another operator $l^{(1)}(C_3)$ does not change it.*

The proof is considered in Exercise 1.5.

The next examples show that all possible decomposition types of a second-order equation actually do exist.

Example 1.4. The examples given below for the various types of Loewy decompositions of second order equations are taken from Kamke's collection [34], Chap. 2. In Kamke's enumeration they correspond to Eqs. (2.162), (2.136), (2.201) and (2.146) respectively.

$$\mathcal{L}_0^2: \; y'' + \frac{1}{x}y' + \left(1 - \frac{v^2}{x^2}\right)y = \left(D^2 + \frac{1}{x}D + 1 - \frac{v^2}{x^2}\right)y = 0,$$

$$\mathcal{L}_1^2: \; y'' + \frac{1}{x}y' - \left(\tfrac{1}{4} + \tfrac{1}{2x}\right)y = \left(D + \tfrac{1}{2} + \tfrac{1}{x}\right)\left(D - \tfrac{1}{2}\right)y = 0,$$

$$\mathcal{L}_2^2: \; y'' + (2 + \tfrac{1}{x})y' - \frac{4}{x^2}y =$$

$$Lclm\left(D + \frac{2}{x} - \frac{2x-2}{x^2 - 2x + \frac{3}{2}}, D + 2 + \frac{2}{x} - \frac{1}{x + \frac{3}{2}}\right)y = 0,$$

$$\mathcal{L}_3^2: \; y'' - \frac{6}{x^2}y = Lclm\left(D - \frac{5x^4}{x^5 + C} + \frac{2}{x}\right)y = 0. \qquad \square$$

For operators of order 3 twelve different types of Loewy decompositions are distinguished.

Corollary 1.4. *Let L be a third-order operator. Its possible Loewy decompositions may be described as follows; $l_j^{(i)}$ is an irreducible operator of order i; $L_j^{(i)}$ is a Loewy factor of order i as defined above; C, C_1 and C_2 are constants;*

$$\mathcal{L}_1^3 : L = l^{(2)}l^{(1)}; \quad \mathcal{L}_2^3 : L = l_1^{(1)}l_2^{(1)}l_3^{(1)}; \quad \mathcal{L}_3^3 : L = Lclm(l_1^{(1)}, l_2^{(1)})l_3^{(1)};$$

$$\mathcal{L}_4^3 : L = Lclm(l_1^{(1)}(C))l_2^{(1)}; \quad \mathcal{L}_5^3 : L = l^{(1)}l^{(2)};$$

$$\mathcal{L}_6^3 : L = l_1^{(1)}Lclm(l_2^{(1)}, l_3^{(1)});$$

$$\mathcal{L}_7^3 : L = l_1^{(1)}Lclm(l_2^{(1)}(C)); \quad \mathcal{L}_8^3 : L = Lclm(l_1^{(1)}, l_2^{(1)}, l_3^{(1)});$$

$$\mathcal{L}_9^3 : L = Lclm(l^{(2)}, l^{(1)});$$

$$\mathcal{L}_{10}^3 : L = Lclm(l_1^{(1)}(C), l_2^{(1)}); \quad \mathcal{L}_{11}^3 : L = Lclm(l^{(1)}(C_1, C_2)).$$

An irreducible third-order operator is defined to have decomposition type $\mathcal{L}_0^3$. The decompositions $\mathcal{L}_0^3$, and $\mathcal{L}_8^3$ through $\mathcal{L}_{11}^3$ are completely reducible; the decomposition types $\mathcal{L}_1^3$, $\mathcal{L}_3^3$ and $\mathcal{L}_4^3$ have the structure $L = L_2^{(2)}L_1^{(1)}$; the decomposition types $\mathcal{L}_5^3$, $\mathcal{L}_6^3$ and $\mathcal{L}_7^3$ have the structure $L = L_2^{(1)}L_1^{(2)}$.

Proof. According to Lemma 1.1, case (ii), the possible factors are determined by the solutions of a second-order Riccati equation. They are completely classified in Appendix B. The general solution may be rational involving two constants; there may be a rational solution involving a single constant and in addition a special rational solution; there may be only a rational solution involving a single constant; or there may be three, two or a single special rational solution, or nor rational solution at all. These alternatives may easily be correlated with the various decomposition types given above. □

Similar remarks apply as those following Corollary 1.2, i.e. the Loewy factors comprise *all* irreducible right factors. As a consequence, for example decomposition type $\mathcal{L}_6^3$ excludes decomposition types $\mathcal{L}_7^3$ and $\mathcal{L}_{11}^3$; or decomposition type $\mathcal{L}_1^3$ excludes types $\mathcal{L}_6^3$, $\mathcal{L}_8^3$, $\mathcal{L}_9^3$, $\mathcal{L}_{10}^3$ and $\mathcal{L}_{11}^3$. Whenever a parameter C appears in a factor, two different special values may be chosen such that the corresponding $Lclm$ generates a second-order operator. Similarly, for the two parameters C_1 and C_2 in the last decomposition type $\mathcal{L}_{11}^3$, three different pairs of special values may be chosen such that the $Lclm$ generates a third-order operator.

Example 1.5. Most of the examples of the various types of Loewy decompositions of third order equations are again taken from Kamke's collection, Chap. 3. The first three of them correspond to Eqs. (3.6), (3.76) and (3.73) respectively.

$$\mathcal{L}_0^3 : y''' + 2xy' + y = 0,$$

$$\mathcal{L}_1^3 : y''' + \frac{1}{x^4}y'' - \frac{2}{x^6}y = \left(D^2 + \left(\frac{2}{x} + \frac{1}{x^4}\right)D + \frac{2}{x^5}\right)\left(D - \frac{2}{x}\right)y = 0,$$

$$\mathcal{L}_2^3 : y''' - \left(\frac{2}{x+1} + \frac{2}{x}\right)y'' + \left(\frac{6}{x} + \frac{4}{x^2} - \frac{6}{x+1}\right)y' + \left(\frac{8}{x} - \frac{8}{x^2} - \frac{4}{x^3} - \frac{8}{x+1}\right)y$$

$$= \left(D - \frac{2}{x+1} - \frac{1}{x}\right)\left(D - \frac{1}{x}\right)\left(D - \frac{2}{x}\right)y = 0.$$

There is no type $\mathcal{L}_3^3$ decomposition in Kamke's collection. An example is

$$\mathcal{L}_3^3 : y''' + \left(x - \frac{1}{x+1} - \frac{1}{x-1} + \frac{2}{x}\right) y'' + \left(1 + \frac{1}{x+1} - \frac{1}{x-1}\right) y'$$
$$= Lclm\left(D + x, D + \frac{1}{x}\right) Dy = 0.$$

The next four examples are Eqs. (3.58), (3.45), (3.37) and (3.74) respectively from Kamke's collection.

$$\mathcal{L}_4^3 : y''' + \left(\frac{1}{4} + \frac{7}{x} - \frac{\frac{1}{4}}{x^2}\right) y'' + \left(\frac{1}{x} + \frac{1}{x^2}\right) y' + \frac{\frac{1}{2}}{x^2} y =$$
$$Lclm\left(D - \frac{1}{x+C} + \frac{2}{x}\right) \left(D + \frac{1}{4} - \frac{\frac{1}{2}}{x} - \frac{\frac{1}{4}}{x^2}\right) y = 0,$$

$$\mathcal{L}_5^3 : y''' + \frac{4}{x} y'' + \left(1 + \frac{2}{x^2}\right) y' + \frac{3}{x} y = \left(D + \frac{3}{x}\right) \left(D^2 + \frac{1}{x} D + 1\right) y = 0,$$

$$\mathcal{L}_6^3 : y''' - y'' - \left(\frac{1}{x-2} - \frac{1}{x}\right) y' + \left(\frac{1}{x-2} - \frac{1}{x}\right) y =$$
$$\left(D + \frac{1}{x-2} + \frac{1}{x}\right) Lclm\left(D - 1, D - \frac{2}{x}\right) y = 0,$$

$$\mathcal{L}_7^3 : y''' - \frac{1}{x} y'' + \frac{1}{x^2} y' = DLclm\left(D - \frac{2x}{x^2 + C}\right) y = 0.$$

The decompositions of type $\mathcal{L}_8^3$ in Kamke's collection are fairly complicated. A simple example is

$$\mathcal{L}_8^3 : y''' + \left(x + 1 - \frac{2x}{x^2 - 2} - \frac{2x+1}{x^2 + x + 1}\right) y'' + \left(x - 2 - \frac{2x-1}{x^2 + x + 1}\right) y'$$
$$- \left(2 - \frac{2x}{x^2 - 2} - \frac{2}{x^2 + x + 1}\right) y = Lclm\left(D + 1, D + x, D - \frac{2}{x}\right) y = 0.$$

The next example is Kamke's equation (3.29).

$$\mathcal{L}_9^3 : y''' + \frac{3}{x} y'' + y = Lclm\left(D^2 - \left(1 - \frac{2}{x}\right) D + 1 - \frac{1}{x}, D + 1 + \frac{1}{x}\right) y = 0.$$

There is no type $\mathcal{L}_{10}^3$ decomposition in Kamke's collection. An example is

$$\mathcal{L}_{10}^3 : y''' - \left(1 - \frac{3}{x} + \frac{2x+1}{x^2 + x - 1}\right) y'' - \frac{x+3}{x^2 + x - 1} y'$$
$$- \left(\frac{3}{x} - \frac{3x+4}{x^2 + x - 1}\right) y = Lclm\left(D + \frac{1}{x} - \frac{2x}{x^2 + C}, D - 1\right) y = 0.$$

Finally, there is Eq. (3.71) from Kamke's collection with a decomposition involving two constants.

$$\mathcal{L}_{11}^3 : y''' - \left(\frac{1}{x+3} + \frac{2}{x}\right) y'' + \left(\frac{\frac{4}{3}}{x} + \frac{2}{x^2} - \frac{\frac{4}{3}}{x+3}\right) y' + \left(\frac{\frac{2}{3}}{x} - \frac{2}{x^2} - \frac{\frac{2}{3}}{x+3}\right) y$$

$$= Lclm \left(D - \frac{3x^2 + 2C_1 x + C_2}{x^3 + C_1 x^2 + C_2(x+1)}\right) y = 0. \ \square$$

Again, these examples show that each possible decomposition type actually does exist.

1.3 Solving Linear Homogeneous Ode's

There remains to be discussed how the solution procedure for a linear ode with a nontrivial Loewy decomposition is simplified. The general procedure that applies to reducible equations of any order is described first.

Proposition 1.2. *Let a linear differential operator P of order n factor into $P = QR$ with R of order m and Q of order $n - m$. Further let $y_1, \ldots, y_m$ be a fundamental system for $Ry = 0$, and $\bar{y}_1, \ldots, \bar{y}_{n-m}$ a fundamental system for $Qy = 0$. Then a fundamental system for $Py = 0$ is given by the union of $y_1, \ldots, y_m$ and special solutions of $Ry = \bar{y}_i$ for $i = 1, \ldots, n - m$.*

Proof. From $Ry_i = 0$ there follows $Py_i = QRy_i = 0$, i.e. y_i belongs to a fundamental system of $Py = 0$ if this is true for $Ry = 0$; this proves the first part. Furthermore, from $Ry_i = \bar{y}_i$ and $Q\bar{y}_i = 0$ there follows $Py_i = QRy_i = 0$; this proves the second part. $\square$

This proceeding will be applied for solving reducible second- and third order equations. The following two corollaries are obtained by straightforward application of the above Proposition 1.2. In some cases a solution of an inhomogeneous second-order linear ode is required; this problem is considered at the end of this section.

Corollary 1.5. *Let L be a second-order differential operator, $D \equiv \frac{d}{dx}$, y a differential indeterminate, and $a, a_i \in \mathbb{Q}(x)$. Define $\varepsilon_i(x) \equiv \exp\left(-\int a_i dx\right)$ for $i = 1, 2$ and $\varepsilon(x, C) \equiv \exp\left(-\int a(C) dx\right)$, C is a parameter; the barred quantities $\bar{C}$ and $\bar{\bar{C}}$ are numbers, $\bar{C} \neq \bar{\bar{C}}$. For the three nontrivial decompositions of Corollary 1.2 the following elements y_1 and y_2 of a fundamental system are obtained.*

$$\mathcal{L}_1^2: \quad Ly = (D + a_2)(D + a_1)y = 0; \quad y_1 = \varepsilon_1(x), \quad y_2 = \varepsilon_1(x) \int \frac{\varepsilon_2(x)}{\varepsilon_1(x)} dx.$$

$$\mathcal{L}_2^2: \quad Ly = Lclm(D + a_2, D + a_1)y = 0; \quad y_i = \varepsilon_i(x); \quad a_1 \text{ is not equivalent}$$
to a_2.

$$\mathcal{L}_3^2: \quad Ly = Lclm(D + a(C))y = 0; y_1 = \varepsilon(x, \bar{C}), y_2 = \varepsilon(x, \bar{\bar{C}}).$$

Here two rational functions $p, q \in \mathbb{Q}(x)$ are called *equivalent* if there exists another rational function $r \in \mathbb{Q}(x)$ such that $p - q = \frac{r'}{r}$ holds [41]. The proof of the above corollary is considered in Exercise 1.6.

There is an important difference between the last two cases. A fundamental system corresponding to a type $\mathcal{L}_2^2$ decomposition is linearly independent over the base field, whereas this is not true for the last decomposition type $\mathcal{L}_3^2$.

Example 1.6. The decompositions of Example 1.4 are considered again. The first equation is the well-known Bessel equation with a fundamental system $J_\nu(x)$ and $I_\nu(x)$ in terms of series expansions. For the type $\mathcal{L}_1^2$ decomposition with $a_1 = -\frac{1}{2}$ and $a_2 = \frac{1}{2} + \frac{1}{x}$, a fundamental system is $y_1 = \exp\left(\frac{1}{2}x\right)$, $y_2 = \exp\left(\frac{1}{2}x\right)\int \exp\left(-x\right)\frac{dx}{x}$. For the type $\mathcal{L}_2^2$ decomposition, $a_2 = \frac{2}{x} - \frac{2x-2}{x^2 - 2x + \frac{3}{2}}$, $a_1 = 2 + \frac{2}{x} - \frac{1}{x + \frac{3}{2}}$. The two independent integrations yield the fundamental system $y_1 = \frac{2}{3} - \frac{4}{3x} + \frac{1}{x^2}$ and $y_2 = \frac{2}{x} + \frac{3}{x^2}e^{-2x}$. Finally in the last case $a(C) = -\frac{5x^4}{x^5 + C} + \frac{2}{x}$, integration yields $y = (x^5 + C)\frac{1}{x^2}$ from which the fundamental system $y_1 = x^3$ and $y_2 = \frac{1}{x^2}$ corresponding to $C = 0$ and $C \to \infty$ follows.   □

For differential equations of order 3 there are four decomposition types into first-order factors with no constants involved. Fundamental systems for them may be obtained as follows.

Corollary 1.6. *Let L be a third-order differential operator, $D \equiv \frac{d}{dx}$, y a differential indeterminate, and $a_i \in \mathbb{Q}(x)$. Define $\varepsilon_i(x) \equiv \exp\left(-\int a_i\,dx\right)$ for $i = 1, 2, 3$. For the four nontrivial decompositions of Corollary 1.4 involving only first-order factors without parameters the following elements y_i, $i = 1, 2, 3$, of a fundamental system are obtained.*

$\mathcal{L}_2^3 : Ly = (D + a_3)(D + a_2)(D + a_1)y = 0; \quad y_1 = \varepsilon_1(x),$

$$y_2 = \varepsilon_1(x)\int \frac{\varepsilon_2(x)}{\varepsilon_1(x)}dx,$$

$$y_3 = \varepsilon_1(x)\left(\int \frac{\varepsilon_3(x)}{\varepsilon_2(x)}dx \int \frac{\varepsilon_2(x)}{\varepsilon_1(x)}dx - \int \frac{\varepsilon_3(x)}{\varepsilon_2(x)}\int \frac{\varepsilon_2(x)}{\varepsilon_1(x)}dxdx\right).$$

$\mathcal{L}_3^3 : Ly = Lclm(D + a_3, D + a_2)(D + a_1)y = 0;$

$$y_1 = \varepsilon_1(x), \quad y_2 = \varepsilon_1(x)\int \frac{\varepsilon_2(x)}{\varepsilon_1(x)}dx, \quad y_3 = \varepsilon_1(x)\int \frac{\varepsilon_3(x)}{\varepsilon_1(x)}dx.$$

$\mathscr{L}_6^3 :\ Ly = (D + a_3)Lclm(D + a_2, D + a_1)y = 0; \quad y_i = \varepsilon_i(x)\ for\ i = 1, 2;$

$$y_3 = \varepsilon_1(x) \int \frac{\varepsilon_3(x)}{\varepsilon_1(x)} \frac{dx}{a_2 - a_1} - \varepsilon_2(x) \int \frac{\varepsilon_3(x)}{\varepsilon_2(x)} \frac{dx}{a_2 - a_1}.$$

$\mathscr{L}_8^3 :\ Ly = Lclm(D + a_3, D + a_2, D + a_1)y = 0, \quad y_i = \varepsilon_i(x), \quad i = 1, 2, 3.$

The proof is a straightforward application of Proposition 1.2. In the following examples the above corollary is applied.

Example 1.7. The type $\mathscr{L}_2^3$ decomposition considered in Example 1.5 has coefficients $a_1 = -\frac{2}{x}, a_2 = -\frac{1}{x}$ and $a_3 = \frac{1}{x} - \frac{2}{x+1}$; they yield $\varepsilon_1 = x^2$, $\varepsilon_2 = x$ and $\varepsilon_3 = \frac{1}{x}(x + 1)^2$. If they are substituted in the above expressions the fundamental system

$$y_1 = x^2, \quad y_2 = x^2 \log x \quad \text{and} \quad y_3 = x + x^3 + x^2 \log (x)^2$$

is obtained. □

Example 1.8. The type $\mathscr{L}_6^3$ decomposition considered in Example 1.5 has coefficients $a_1 = -\frac{2}{x}, a_2 = -1$ and $a_3 = \frac{1}{x} + \frac{1}{x-2}$. If they are substituted in the above expressions, the fundamental system

$$y_1 = x^2, \quad y_2 = e^x, \quad \text{and} \quad y_3 = \frac{x(x^2 - 2)}{4(x - 2)} + \frac{x^2}{4} \log \frac{x - 2}{x} + e^x \int e^{-x} \frac{dx}{(x - 2)^2}$$

is obtained. □

In addition to the decompositions dealt with in the above corollary there are four decompositions into first-order factors involving one or two parameters. They are considered next.

Corollary 1.7. *Let L be a third-order differential operator,* $D \equiv \frac{d}{dx}$, *y a differential indeterminate, and* $a, a_i \in \mathbb{Q}(x)$. *Define* $\varepsilon_i(x) \equiv \exp(-\int a_i dx)$ *for* $i = 1, 2, 3,$ $\varepsilon(x, C) \equiv \exp(-\int a(C)dx)$ *and* $\varepsilon(x, C_1, C_2) = \exp(-\int a(C_1, C_2)dx);$ C, C_1 *and* C_2 *are parameters; the barred quantities* $\bar{C}, \bar{C}_i$ *etc are numbers; for each case they are pairwise different from each other. For the four decompositions of Corollary 1.4 involving first-order factors with parameters the following elements* $y_i, i = 1, 2, 3,$ *of a fundamental system are obtained.*

$\mathscr{L}_4^3 :\ Ly = Lclm\,(D + a(C))\,(D + a_1)y = 0;$

$$y_1 = \varepsilon_1(x), \ y_2 = \varepsilon_1(x) \int \frac{\varepsilon(x, \bar{C})}{\varepsilon_1(x)} dx, \ y_3 = \varepsilon_1(x) \int \frac{\varepsilon(x, \bar{\bar{C}})}{\varepsilon_1(x)} dx;$$

$\mathscr{L}_7^3 :\ Ly = (D + a_3)Lclm\,(D + a(C))\,y = 0; y_1 = \varepsilon(x, \bar{C}), y_2 = \varepsilon(x, \bar{\bar{C}}),$

$$y_3 = \varepsilon(\bar{C}, x) \int \frac{\varepsilon_3(x)}{\varepsilon(x, \bar{C})} \frac{dx}{a(\bar{\bar{C}}) - a(\bar{C})}$$

$$- \varepsilon(x, \bar{\bar{C}}) \int \frac{\varepsilon_3(x)}{\varepsilon(x, \bar{\bar{C}})} \frac{dx}{a(\bar{\bar{C}}) - a(\bar{C})}$$

$\mathcal{L}^3_{10} : Ly = Lclm(D + a(C), D + a_1)y = 0;$

$$y_1 = \varepsilon_1(x), \quad y_2 = \varepsilon(x, \bar{C}), \quad y_3 = \varepsilon(x, \bar{\bar{C}}).$$

$\mathcal{L}^3_{11} : Ly = Lclm(D + a(C_1, C_2)) y = 0,$

$$y_1 = \varepsilon(x, \bar{C}_1, \bar{C}_2), \quad y_2 = \varepsilon(x, \bar{\bar{C}}_1, \bar{\bar{C}}_2), \quad y_2 = \varepsilon(x, \bar{\bar{\bar{C}}}_1, \bar{\bar{\bar{C}}}_2).$$

Again the proof follows immediately from Proposition 1.2 and is omitted. The four alternatives are illustrated by the following examples.

Example 1.9. The equation with the type $\mathcal{L}^3_4$ decomposition

$$y''' + \frac{x^2 - 3}{x} y'' + 4y' + \frac{2}{x} y = Lclm \left(D + \frac{1}{x} - \frac{1}{x + C} \right) \left(D + x - \frac{5}{x} \right) y = 0$$

has the fundamental system

$$y_1 = x^2 \exp\left(-\tfrac{1}{2}x^2\right), \quad y_2 = x^3 + 2x - x^5 \exp\left(-\tfrac{1}{2}x^2\right) \int \exp\left(\tfrac{1}{2}x^2\right) \frac{dx}{x},$$

$$y_3 = x^5 \exp\left(-\tfrac{1}{2}x^2\right) \int \exp\left(\tfrac{1}{2}x^2\right) \frac{dx}{x^6};$$

the elements of this fundamental system are linearly independent over the base field.   □

Example 1.10. The equation with the type $\mathcal{L}^3_7$ decomposition

$$y''' - +\frac{3}{x} y'' - \frac{2}{x^2} y' + \frac{2}{x^3} y = \left(D + \frac{1}{x} \right) Lclm \left(D + \frac{2}{x} - \frac{3x^2}{x^3 + C} \right) y = 0$$

leads to $\varepsilon(C, x) = x^3 + \frac{C}{x^2}$ and $\varepsilon_3(x) = \frac{1}{x^3}$. It yields the fundamental system $y_1 = x, y_2 = \frac{1}{x^2}$ and $y_3 = x \log x$; y_1 and y_2 are linearly dependent over the base field, they obey $y_1 - x^3 y_2 = 0$.   □

Example 1.11. An equation with type $\mathscr{L}_{10}^3$ decomposition is

$$y''' - \frac{x^3 - 3x + 3}{x(x^2 + x - 1)} - \frac{x + 3}{x^2 + x - 1}y' + \frac{x + 3}{x(x^2 + x - 1)}y$$

$$= Lclm\left(D + \frac{1}{x} - \frac{2x}{x^2 + C}, D - 1\right)y = 0$$

with fundamental system $y_1 = e^x$, $y_2 = x$ and $y_3 = \frac{1}{x}$. There follows $y_2 - x^2 y_3 = 0$ whereas y_1 is linearly independent of y_2 and y_3 over the base field. □

Example 1.12. An equation with type $\mathscr{L}_{11}^3$ decomposition is

$$y''' - \frac{3(x + 2)}{x(x + 3)}y'' + \frac{6(x + 1)}{x^2(x + 3)}y' - \frac{6}{x^2(x + 3)}y$$

$$= Lclm\left(D - \frac{3x^2 + 2C_2x + C_1}{x^3 + C_2x^2 + C_1x + C_1}\right)y = 0.$$

Its fundamental system $y_1 = x^3$, $y_2 = x^2$ and $y_3 = x + 1$ is linearly dependent over the base field; there holds $y_1 + y_2 - x^3 y_3 = 0$. □

Finally there are two decomposition types involving second-order irreducible factors.

Corollary 1.8. *Let L be a third-order differential operator,* $D = \frac{d}{dx}$, *y a differential indeterminate, and* $a_i \in \mathbb{Q}(x)$. *Define* $\varepsilon_i(x) \equiv \exp(-\int a_i dx)$ *for* $i = 1, 3$. *For the two decompositions of Theorem 1.4 involving second-order factors and parameters the elements* y_i, $i = 1, 2, 3$, *of a fundamental system may be described as follows.*

$\mathscr{L}_1^3$: $(D^2 + a_3 D + a_2)(D + a_1)y = 0$. *Let* $\bar{y}_2$ *and* $\bar{y}_3$ *be a fundamental system of the left factor. Then*

$$y_1 = \varepsilon_1(x), \quad y_2 = y_1 \int \frac{\bar{y}_2}{y_1} dx, \quad y_3 = y_1 \int \frac{\bar{y}_3}{y_1} dx.$$

$\mathscr{L}_5^3$: $(D + a_3)(D^2 + a_2 D + a_1)y = 0$. *Let* y_1 *and* y_2 *be a fundamental system of the right factor,* $W = y_1' y_2 - y_2' y_1$ *its Wronskian. Then*

$$y_3 = y_1 \int \varepsilon_3(x) \frac{y_2}{W} dx - y_2 \int \varepsilon_3(x) \frac{y_1}{W} dx.$$

The decompositions described in the preceding corollary are interesting because they may lead to certain special function solutions as shown in the following examples.

Example 1.13. Consider the following equation with the type $\mathcal{L}_1^3$ decomposition.

$$y''' + \frac{x^2+1}{x}y'' + \frac{4x^2-4}{x^2}y' + \frac{x^2-3}{x}y = \left(D^2 + \frac{1}{x}D + \frac{x^2-4}{x^2}\right)(D+x)y = 0.$$

The left factor corresponds to the Bessel equation for $n = 2$ with the fundamental system $J_2(x)$ and $Y_2(x)$. Applying the above formulas, the fundamental system

$$y_1 = \exp\left(-\tfrac{1}{2}x^2\right), \ \ y_2 = y_1 \int J_2(x)\exp\left(\tfrac{1}{2}x^2\right)dx, \ \ y_3 = y_1 \int Y_2(x)\exp\left(\tfrac{1}{2}x^2\right)dx$$

in terms of integrals over Bessel functions and exponentials is obtained.   □

Example 1.14. The following equation with the type $\mathcal{L}_5^3$ decomposition

$$y''' + \frac{x^2+1}{x}y'' + \frac{4x^2-4}{x^2}y' + \frac{x^2-3}{x}y = (\partial_x+x)\left(D^2 + \frac{1}{x}D + \frac{x^2-4}{x^2}\right)y = 0$$

differs from the preceding example by the order of its factors. A fundamental system now is $y_1 = J_2(x)$, $y_2 = Y_2(x)$ with Wronskian $W = \frac{2}{\pi x}$ and

$$y_3 = xJ_2(x)\int \exp\left(-\tfrac{1}{2}x^2\right)Y_2(x)dx - xY_2(x)\int \exp\left(-\tfrac{1}{2}x^2\right)J_2(x)dx. \quad □$$

The website `alltypes.de` provides functions for decomposing linear ode's of order 2 or 3, and to generate a fundamental system from it. Furthermore, a data file contains the equations listed in Kamke's collection; they may be used for testing the software and for studying the behaviour of equations under variations of the input, e.g. by assigning different values for parameters that may be contained in any given equation.

1.4   Solving Second-Order Inhomogeneous Ode's

The rest of this section deals with the solution of *inhomogeneous linear* ode's of the form

$$Ly \equiv (D^n + q_1 D^{n-1} + q_2 D^{n-2} + \ldots + q_{n-1}D + q_n)y = R \qquad (1.12)$$

where $D = \frac{d}{dx}$. As usual, the coefficients $q_1, \ldots, q_n$ determine the base field. The right hand side R is not necessarily contained in it. The smallest field extension containing both the coefficients q_i and R is called the *extended base field* of (1.12). The general solution of (1.12) is the sum $C_1y_1 + \ldots + C_ny_n + y_0$ where $\{y_1, \ldots, y_n\}$ is a basis for the solution space of the homogeneous equation.

The *special solution* y_0 satisfies $Ly_0 = R$, i.e. L applied to y_0 generates the inhomogeneity R. However, y_0 is by no means unique. Any linear combination of the y_i with numerical coefficients may be added to it.

The classical method for solving an inhomogeneous linear ode is the method of *variation of constants*, see Ince [29], Sect. 5.23, page 122; originally it is due to Lagrange. According to this method, solving a linear inhomogeneous ode (1.12) may be traced back to the corresponding homogeneous problem with $R = 0$ and integrations. To this end, the C_i in the general solution $y = C_1 y_1 + \ldots + C_n y_n$ of the homogeneous equation are considered as functions of x. Substituting this expression into (1.12) and imposing the $n - 1$ constraints

$$C_1' y_1^{(k)} + C_2' y_2^{(k)} + \ldots + C_n' y_n^{(k)} = 0 \tag{1.13}$$

for $k = 0, \ldots, n - 2$ yields the additional condition

$$C_1' y_1^{(n-1)} + C_2' y_2^{(n-1)} + \ldots C_n' y_n^{(n-1)} = R. \tag{1.14}$$

The linear algebraic system (1.13) and (1.14) comprising n equations for the C_k' has always a solution due to the non-vanishing determinant of its coefficient matrix which is the Wronskian $W^{(n)}$. Consequently, solving the inhomogeneous Eq. (1.12) requires only integrations if a fundamental system for the corresponding homogeneous problem is known; details may be found in the book by Ince quoted above.

For second-order equations $y'' + q_1 y' + q_2 y = R$ with fundamental system $\{y_1, y_2\}$ a special solution y_0 may be written explicitly as

$$y_0 = y_2 \int \frac{R y_1}{W} dx - y_1 \int \frac{R y_2}{W} dx; \tag{1.15}$$

$W = y_1 y_2' - y_1' y_2$ is its Wronskian. As mentioned above, any linear combination of a fundamental system with numerical coefficients may be added to y_0. Therefore the difference of any two expressions for y_0 is a solution of the homogeneous equation.

Example 1.15. Let the equation

$$y'' - \frac{4x}{2x - 1} y' + \frac{4}{2x - 1} y = \frac{e^{-2x}}{2x - 1}$$

be given. A fundamental system for the homogeneous equation is $y_1 = x$, $y_2 = e^{2x}$. Furthermore $R = \frac{e^{-2x}}{2x - 1}$ and $W = (2x - 1)e^{2x}$. Substituting these expressions into (1.15) leads to

$$y_0 = e^{2x} \int \frac{x e^{-4x}}{(2x - 1)^2} dx - x \int \frac{e^{-2x}}{(2x - 1)^2} dx. \qquad \square$$

Although the variation of constants is perfectly suited for solving inhomogeneous linear ode's, it does not seem to be possible to generalize it for linear pde's.

Therefore for second order linear ode's with a nontrivial Loewy decomposition, a different procedure is described next. To this end the following result will be needed.

Lemma 1.2. *Let* $L \equiv D^2 + A_1 D + A_2$ *with* $D \equiv \dfrac{d}{dx}$ *and* $A_1, A_2 \in \mathbb{Q}(x)$ *be a reducible second-order differential operator. Its coefficients* A_1 *and* A_2 *may be represented as follows.*

(i) *If* $L = (D + a_2)(D + a_1)$ *then*

$$A_1 = a_1 + a_2 \quad \text{and} \quad A_2 = a_1' + a_1 a_2. \tag{1.16}$$

(ii) *If* $L = Lclm(D + a_2, D + a_1)$ *then*

$$A_1 = a_1 + a_2 - \frac{a_1' - a_2'}{a_1 - a_2}, \quad A_2 = a_1 a_2 - \frac{a_1' a_2 - a_2' a_1}{a_1 - a_2}. \tag{1.17}$$

Proof. In case (i), multiplication yields (1.16). In case (ii), reduction of L w.r.t. $D + a_1$ and $D + a_2$ leads to $A_2 - a_1 A_1 = a_1' - a_1^2$ and $A_2 - a_2 A_1 = a_1' - a_2^2$. Solving for A_1 and A_2 yields the expressions (1.17).   □

The implications of commutativity of $D + a_1$ and $D + a_2$ for the representation (1.17) are investigated in Exercise 1.10.

If a second-order operator L is reducible, a special solution of an inhomogeneous equation $Ly = R$ may be determined from its factors, avoiding the method of variation of constants. This proceeding has the advantage that it may be generalized for partial differential equations as will be seen in later chapters.

Proposition 1.3. *Let* $Ly = R$ *be a reducible linear second-order ode. A fundamental system of the corresponding homogeneous equation has been determined in Corollary 1.5. A special solution* y_0 *may be obtained as follows. Define* $\varepsilon_i(x) \equiv \exp\left(-\int a_i dx\right)$ *for* $i = 1, 2$.

(i) *If* $L = (D + a_2)(D + a_1)$, $a_1 \neq a_2$, *then*

$$y_0 = \varepsilon_1(x) \int \frac{\varepsilon_2(x)}{\varepsilon_1(x)} \int \frac{R(x)}{\varepsilon_2(x)} dx dx. \tag{1.18}$$

(ii) *If* $L = Lclm(D + a_2, D + a_1)$, $a_1 \neq a_2$, *then*

$$y_0 = \varepsilon_1(x) \int \frac{R(x)}{\varepsilon_1(x)} \frac{dx}{a_2 - a_1} - \varepsilon_2(x) \int \frac{R(x)}{\varepsilon_2(x)} \frac{dx}{a_2 - a_1}. \tag{1.19}$$

Proof. In case (i), let $y' + a_1 y = r$ be the equation corresponding to the right factor; $r(x)$ is an undetermined function of x. Substitution into $Ly = R$ yields $r' + a_2 r = R$ with special solution $r = \varepsilon_2(x) \int \dfrac{R(x)}{\varepsilon_2(x)} dx$. Solving the above first-order equation with this right hand side yields the special solution (1.18).

In case (ii), let $y_1' + a_1 y_1 = r_1$ and $y_2' + a_2 y_2 = r_2$ be the equations corresponding to the two arguments of the $Lclm$; its right hand sides r_1 and r_2 are undetermined functions of x. Substituting $y = y_1 + y_2$ into $Ly = R$, reducing the result w.r.t. these first-order equations and applying Lemma 1.2 yields

$$r_1' + r_2' + \left(a_2 - \frac{a_1' - a_2'}{a_1 - a_2} \right) r_1 + \left(a_1 - \frac{a_1' - a_2'}{a_1 - a_2} \right) r_2 = R. \qquad (1.20)$$

Choosing $r_1 = r$ and $r_2 = -r$ leads to $r = \frac{R}{a_2 - a_1}$. Substituting this value into the above first-order equations and taking the sum of its solution yields the special solution (1.19).   □

The correctness of expressions (1.18) and (1.19) may be proved by substitution into Ly, taking into account Lemma 1.2.

The two representations for the special solution y_0 in the above proposition show clearly the simplification due to the existence of two first-order right factors in case (ii) where the operator is completely reducible. Whereas in case (i), Eq. (1.18), two successive integrations are necessary, Eq. (1.19) requires only a single integration. As a consequence, for case (i) the function field containing the special solution in general will be larger. In any case, y_0 is Liouvillian over the extended base field.

Example 1.16. Let the equation

$$y'' - \frac{3}{x} y' + \frac{4}{x^2} y = \left(D - \frac{1}{x} \right) \left(D - \frac{2}{x} \right) y = x + 1$$

be given. A fundamental system is $y_1 = x^2$ and $y_2 = x^2 \log x$. With the notation of Proposition 1.3, it follows that $a_1 = -\frac{2}{x}$, $a_2 = -\frac{1}{x}$, $R = x + 1$, $\varepsilon_1(x) = x^2$ and $\varepsilon_2(x) = x$. According to (1.18) there follows $y_0 = \frac{1}{2} x^2 \log x^2 + x^3$.   □

Example 1.17. Consider the equation

$$y'' - \frac{3}{x} y' - \frac{5}{x^2} y = Lclm \left(y' + \frac{1}{x} y, y' - \frac{5}{x} y \right) = \frac{1}{x} \log x$$

with fundamental system $y_1 = \frac{1}{x}$ and $y_2 = x^5$. There follows $a_1 = -\frac{5}{x}$, $a_2 = \frac{1}{x}$, $R = \frac{1}{x} \log x$, $\varepsilon_1(x) = x^5$ and $\varepsilon_2(x) = \frac{1}{x}$. Substitution into Eq. (1.19) yields $y_0 = -\frac{1}{8} x \log x + \frac{1}{32} x$.   □

Example 1.18. The equation

$$Ly \equiv y'' - \frac{5}{x} y' + \frac{8}{x^2} y = Lclm \left(D - \frac{2}{x} - \frac{2x}{x^2 + C} \right) y = x + 1$$

allows a Loewy decomposition of type $\mathscr{L}_3^2$. Choosing two special values for C, case (ii) of Proposition 1.3 applies. With $C = \pm 1$ there follows

$$a_1 = -\frac{2}{x} - \frac{2x}{x^2 - 1}, \quad a_2 = -\frac{2}{x} - \frac{2x}{x^2 + 1}, \quad R = x + 1,$$

$$\varepsilon_1(x) = x(x^2 - 1) \quad \text{and} \quad \varepsilon_2(x) = x(x^2 + 1).$$

Substitution into (1.19) yields $y_0 = -\frac{1}{2}x^2 \log x - x^2 - \frac{1}{4}x^2$.   □

1.5 Exercises

Exercise 1.1. Express the coefficients of $y'' + a_1 y' + a_2 y = 0$ explicitly in terms of the fundamental system $\{y_1, y_2\}$. The same for $y''' + a_1 y'' + a_2 y' + a_3 y = 0$ and the fundamental system $\{y_1, y_2, y_3\}$.

Exercise 1.2. Generate the associated equations for z_2 and z_3 as defined in (1.8).

Exercise 1.3. Determine the associated equations necessary for finding a second-order factor for a fourth-order equation.

Exercise 1.4. Any second-order linear ode of the form $y'' + py' + qy = 0$ may be transformed into an equation $z'' + rz = 0$. Determine a transformation of the form $y = \phi(x)z$ with this property and the corresponding value of r.

Exercise 1.5. Prove Corollary 1.3.

Exercise 1.6. Prove Corollary 1.5.

Exercise 1.7. Discuss the possible factorizations and the corresponding solutions for a second-order equation $Ly \equiv (D^2 + AD + B)y = 0$ if A and B are constant.

Exercise 1.8. Determine the Loewy decomposition of the operator

$$L \equiv D^3 + \left(1 - \frac{1}{x}\right)D^2 + \left(x - \frac{2}{x}\right)D + \frac{2}{x^2}$$

by applying Lemma 1.1.

Exercise 1.9. Generalize the expression (1.15) for the solutions of an inhomogeneous third-order equation $y''' + q_1 y'' + q_2 y' + q_3 y = R$; consider also the special case $q_1 = q_2 = q_3 = 0$.

Exercise 1.10. Determine the condition for two operators $l_1 = D + a_1$ and $l_2 = D + a_2$ to commute. What does this mean for the representation (1.17) in Lemma 1.2?

Chapter 2
Rings of Partial Differential Operators

Abstract In the ring of ordinary differential operators all ideals are principal. Consequently, the relation between an individual operator and the ideal that is generated by it is straightforward. The situation is different in rings of partial differential operators where in general ideals may have any number of generators, and only a Janet basis provides a unique representation. Therefore a more algebraic language is appropriate for dealing with partial differential operators and the ideals or modules they generate. It is introduced in the first section of this chapter. Subsequently it is applied for discussing certain properties of ideals in those rings of partial differential operators that are applied in later parts of this monograph. General references for this chapter are the books by Kolchin [37] or van der Put and Singer [71], or the article by Buium and Cassidy [8].

2.1 Basic Differential Algebra

This section summarizes some basic terminology from differential algebra that is used throughout this monograph. In order to study partial differential equations, rings of partial differential operators and modules over such rings are the proper concepts which are defined first.

A field $\mathscr{F}$ is called a *differential field* if it is equipped with a *derivation operator*. An operator δ on a field $\mathscr{F}$ is called a derivation operator if $\delta(a + b) = \delta(a) + \delta(b)$ and $\delta(ab) = \delta(a)b + a\delta(b)$ for all elements $a, b \in \mathscr{F}$. A field with a single derivation operator is called an *ordinary differential field*; if there is a finite set Δ containing several commuting derivation operators the field is called a *partial differential field*.

In this monograph rings of differential operators with derivative operators $\partial_x = \frac{\partial}{\partial x}$ and $\partial_y = \frac{\partial}{\partial y}$ with coefficients from some differential field are considered. Its elements have the form $\sum_{i,j} r_{i,j}(x, y)\partial_x^i \partial_y^j$; almost all coefficients $r_{i,j}$ are zero. The coefficient field is called the *base field*. If constructive and algorithmic methods

F. Schwarz, *Loewy Decomposition of Linear Differential Equations*, Texts & Monographs in Symbolic Computation, DOI 10.1007/978-3-7091-1286-1_2,
© Springer-Verlag/Wien 2012

are the main issue it is $\mathbb{Q}(x, y)$. However, in some places this is too restrictive and a suitable extension $\mathscr{F}$ of it may be allowed. The respective ring of differential operators is denoted by $\mathscr{D} = \mathbb{Q}(x, y)[\partial_x, \partial_y]$ or $\mathscr{D} = \mathscr{F}[\partial_x, \partial_y]$; if not mentioned explicitly, the exact meaning will be clear from the context.

The ring $\mathscr{D}$ is non-commutative, in general $\partial_x a = a\partial_x + \frac{\partial a}{\partial x}$ and similarly for the other variables; a is from the base field.

The following lemma is a simple consequence of this definition; it will be useful for understanding several details of the lattice structure in $\mathbb{Q}(x, y)[\partial_x, \partial_y]$. Its proof is considered in Exercise 2.1.

Lemma 2.1. *Two cases are distinguished for the commutativity of two first-order operators in the plane with coordinates x and y.*

(i) *Two operators $l_1 \equiv \partial_x + a_1\partial_y + b_1$ and $l_2 \equiv \partial_x + a_2\partial_y + b_2$ commute if*

$$(a_1 - a_2)_x + a_{1,y}a_2 - a_{2,y}a_1 = 0, \quad (b_1 - b_2)_x + b_{1,y}a_2 - b_{2,y}a_1 = 0.$$

(ii) *Two operators $l \equiv \partial_x + a\partial_y + b$ and $k \equiv \partial_y + c$ commute if*

$$(a + b)_y - c_x - ac_y = 0.$$

For an operator $L = \sum_{i+j \leq n} r_{i,j}(x, y)\partial_x^i\partial_y^j$ of order n the *symbol of L* is the homogeneous algebraic polynomial $symb(L) \equiv \sum_{i+j=n} r_{i,j}(x, y)X^i Y^j$, X and Y algebraic indeterminates.

Let I be a left ideal which is generated by elements $l_i \in \mathscr{D}$, $i = 1, \ldots, p$. Then one writes $I = \langle l_1, \ldots, l_p \rangle$. Because right ideals are not considered in this monograph, sometimes I is simply called an ideal.

A m-dimensional left *vector module* $\mathscr{D}^m$ over $\mathscr{D}$ has elements $(l_1, \ldots, l_m)$, $l_i \in \mathscr{D}$ for all i. The sum of two elements of $\mathscr{D}^m$ is defined by componentwise addition; multiplication with a ring element l by $l(l_1, \ldots, l_m) = (ll_1, \ldots, ll_m)$.

The relation between left ideals in $\mathscr{D}$ or submodules of $\mathscr{D}^m$ on the one hand, and systems of linear pde's on the other is established as follows. Let $(z_1, \ldots, z_m)^T$ be an m-dimensional column vector of differential indeterminates such that $\partial_x z_i \neq 0$ and $\partial_y z_i \neq 0$. Then the product

$$(l_1, \ldots, l_m)(z_1, \ldots, z_m)^T = l_1z_1 + l_2z_2 + \ldots + l_mz_m \tag{2.1}$$

defines a linear differential polynomial in the z_i that may be considered as the left hand side of a partial differential equation; $z_1, \ldots, z_m$ are called the *dependent variables* or *functions*, depending on the *independent variables* x and y.

A $N \times m$ matrix $\{c_{i,j}\}$, $i = 1, \ldots, N$, $j = 1, \ldots, m$, $c_{i,j} \in \mathscr{D}$, defines a system of N linear homogeneous pde's

$$c_{i,1}z_1 + \ldots + c_{i,m}z_m = 0, \quad i = 1, \ldots, N. \tag{2.2}$$

The $i - th$ equation of (2.2) corresponds to the vector

$$(c_{i,1}, c_{i,2}, \ldots, c_{i,m}) \in \mathscr{D}^m \text{ for } i = 1, \ldots, N. \qquad (2.3)$$

This correspondence between the elements of $\mathscr{D}^m$, the differential polynomials (2.1), and its corresponding pde's (2.2) allows to turn from one representation to the other whenever it is appropriate.

For $m = 1$ this relation becomes more direct. The elements $l_i \in \mathscr{D}$ are simply applied to a single differential indeterminate z. In this way the ideal $I = \langle l_1, l_2, \ldots \rangle$ corresponds to the system of pde's $l_1 z = 0$, $l_2 z = 0$, ... for the single function z. Sometimes the abbreviated notation $I z = 0$ is applied for the latter.

2.2 Janet Bases of Ideals and Modules

The generators of an ideal are highly non-unique; its members may be transformed in infinitely many ways by taking linear combinations of them or its derivatives over the base field without changing the ideal. This ambiguity makes it difficult to decide membership in an ideal or to recognize whether two sets of generators represent the same ideal. Furthermore, it is not clear what the solutions of the corresponding system of pde's are, if there are any. The same remarks apply to the vector-modules introduced above.

This was the starting point for Maurice Janet [30] early in the twentieth century to introduce a normal form for systems of linear pde's that has been baptized *Janet basis* in [60]. They are the differential analog to Gröbner bases of commutative algebra, originally introduced by Bruno Buchberger [7]; therefore they are also called *differential Gröbner bases*. Good introductions to the subject may be found in the articles by Oaku [50], Castro-Jiménez and Moreno-Frías [9], Plesken and Robertz [54] or Chap. 2 of Schwarz [61].

In order to generate a Janet basis, a ranking of derivatives must be defined. It is a total ordering such that for any derivatives δ, δ_1 and δ_2, and any derivation operator θ there holds $\delta \preceq \theta\delta$, and $\delta_1 \preceq \delta_2 \rightarrow \delta\delta_1 \preceq \delta\delta_2$. In this monograph lexicographic term orderings lex and graded lexicographic term orderings $grlex$ are applied. For partial derivatives of a single function their definition is analogous to the monomial orderings in commutative algebra. If $x \succ y$ is defined, derivatives up to order 3 in lex order are arranged like

$$\partial_{xxx} \succ \partial_{xxy} \succ \partial_{xx} \succ \partial_{xyy} \succ \partial_{xy} \succ \partial_x \succ \partial_{yyy} \succ \partial_{yy} \succ \partial_y \succ 1, \qquad (2.4)$$

and in $grlex$ ordering

$$\partial_{xxx} \succ \partial_{xxy} \succ \partial_{xyy} \succ \partial_{yyy} \succ \partial_{xx} \succ \partial_{xy} \succ \partial_{yy} \succ \partial_x \succ \partial_y \succ 1. \qquad (2.5)$$

For modules these orderings have to be generalized appropriately, e.g. the orderings TOP or POT of Adams and Loustaunau [2] may be applied.

The following convention will always be obeyed. In an individual operator or differential polynomial the terms are arranged decreasingly from left to right, i.e. the first term contains the highest derivative. A collection of such objects like the generators of an ideal or a module is arranged such that the leading terms do not increase. In particular, if the leading terms are pairwise different they will decrease from left to right, and from top to bottom. If the term order is not explicitly given it is assumed to be $grlex$ with $x \succ y$.

The most distinctive feature of a Janet basis is the fact that it contains all algebraic consequences for the derivatives in the ideal generated by its members explicitly. This is achieved by two basic operations, *reductions* and adding *integrability conditions*; the latter correspond to the S-pairs in commutative algebra.

An operator l_1 may be reduced w.r.t. another operator l_2 if the leading derivative of l_2 or a derivative thereof occurs in l_1. If this is true, its occurrence in l_1 may be removed by replacing it by the negative reductum of l_2 or its appropriate derivative. This process may be repeated until no further reduction is possible. This process will always terminate because in each step the derivatives in l_1 are lowered. The following example shows a single-step reduction of two operators.

Example 2.1. Let two operators l_1 and l_2 be given.

$$l_1 \equiv \partial_{xy} - \frac{x^2}{y^2}\partial_x - \frac{x-y}{y^2}, \quad l_2 \equiv \partial_x + \frac{1}{y}\partial_y + x.$$

The derivatives ∂_{xy} and ∂_x may be removed from l_1 with the result

$$Reduce(l_1, l_2) = -\frac{1}{y}\partial_{yy} + \frac{1}{y^2}\partial_y - x\partial_y + \frac{x^2}{y^2}\left(\frac{1}{y}\partial_y + x\right) - \frac{x-y}{y^2}$$

$$= -\frac{1}{y}\left(\partial_{yy} + \frac{1}{y^2}(xy^3 - x^2 - y)\partial_y - \frac{1}{y}(x^3 - x + y)\right).$$

There are no further reductions possible.   □

If a system of operators or differential polynomials is given, various reductions may be possible between pairs of its members. If all of them have been performed such that no further reduction is possible, the system is called *autoreduced*.

For an autoreduced system the integrability conditions have to be investigated. They arise if the same leading derivative occurs in two different members of the system or its derivatives. Upon subtraction, possibly after multiplication with suitable factors from the base field, the difference does not contain it any more. If it does not vanish after reduction w.r.t. the remaining members of the system, it is called an integrability condition that has to be added to the system. The following example shows this process.

Example 2.2. Consider the ideal

$$I = \left\langle l_1 \equiv \partial_{xx} - \frac{1}{x}\partial_x - \frac{y}{x(x+y)}\partial_y, \ l_2 \equiv \partial_{xy} + \frac{1}{x+y}\partial_y, \ l_3 \equiv \partial_{yy} + \frac{1}{x+y}\partial_y \right\rangle$$

in *grlex* term order with $x \succ y$. Its generators are autoreduced. If the integrability condition

$$l_{1,y} - l_{2,x} = \partial_{xy} + \frac{y}{2x+y}\partial_{yy}$$

is reduced w.r.t. to I, the new generator ∂_y is obtained. Adding it to the generators and performing all reductions, the given ideal is represented as $I = \langle \partial_{xx} - \frac{1}{x}\partial_x, \partial_y \rangle$. Its generators are autoreduced and the single integrability condition is satisfied. $\square$

It may be shown that for any given system of operators or differential polynomials and a fixed ranking autoreduction and adding integrability conditions always terminates with a unique result; the proof may be found in the above quoted literature. Due to its fundamental importance a special term is introduced for it.

Definition 2.1 (Janet basis). For a given ranking an autoreduced system of differential operators is called a Janet basis if all integrability conditions reduce to zero.

If a system of operators or differential polynomials forms a Janet basis, it is a unique representation for the ideal or module it generates.

Due to its importance the following notation will be applied from now on. If the generators of an ideal or module are assured to be a Janet basis they are enclosed by a pair of $\langle\!\langle \ldots \rangle\!\rangle$. In general, if the Janet basis property is not known, the usual notation $\langle \ldots \rangle$ will be applied. According to this convention, in the preceding example the result may be written as $I = \langle\!\langle \partial_{xx} - \frac{1}{x}\partial_x, \partial_y \rangle\!\rangle$. By definition, a single element l is a Janet basis, i.e. there holds always $\langle l \rangle = \langle\!\langle l \rangle\!\rangle$. A system of operators or pde's with the property that all integrability conditions are satisfied is called *coherent*.

2.3 General Properties of Ideals and Modules

Just like in commutative algebra, the generators of an ideal in a ring of differential operators obey certain relations which are known as *syzygies*. Let a set of generators be $f = \{f_1, \ldots, f_p\}$ where $f_i \in \mathcal{D}$ for all i. Syzygies of f are relations of the form

$$d_{k,1}f_1 + \ldots + d_{k,p}f_p = 0$$

where $d_{k,i} \in \mathcal{D}$, $i = 1, \ldots p$, $k = 1, 2, \ldots$. The $(d_{k,1}, \ldots, d_{k,p})$ may be considered as elements of the module $\mathcal{D}^p$. The totality of syzygies generates a submodule.

Example 2.3. Consider the ideal $\langle f_1 \equiv \partial_x + a, f_2 \equiv \partial_y + b \rangle$ with the constraint $a_y = b_x$. The coherence condition for $\partial_x - a - f_1 = 0$ and $\partial_y + b - f_2 = 0$ yields $a\partial_y + a_y - f_{1,y} - b\partial_x - b_x + f_{2,x} = 0$. Reduction w.r.t. to the given generators and some simplification yields the single syzygy $(\partial_y + b)f_1 - (\partial_x + a)f_2 = 0$. $\square$

Example 2.4. Consider the ideal

$$\left\langle\!\!\left\langle f_1 \equiv \partial_{xx} + \frac{4}{x}\partial_x + \frac{2}{x^2}, \ f_2 \equiv \partial_{xy} + \frac{1}{x}\partial_y, \ f_3 \equiv \partial_{yy} + \frac{1}{y}\partial_y - \frac{x}{y^2}\partial_x - \frac{2}{y^2} \right\rangle\!\!\right\rangle.$$

The integrability condition for $\partial_{xx} + \frac{4}{x}\partial_x + \frac{2}{x^2} - f_1 = 0$ and $\partial_{xy} + \frac{1}{x}\partial_y - f_2 = 0$ yields upon reduction and simplification $f_{1,y} + f_{2,x} - \frac{3}{x}f_2 = 0$. Similarly from the last two elements $f_1 - \frac{y^2}{x}f_{2,y} - \frac{y}{x}f_2 + \frac{y^2}{x}f_{3,x} + \frac{y^2}{x^2}f_3 = 0$ is obtained. Autoreduction of these two equations yields the following two syzygies as the final answer

$$\left(\partial_{yy} + \frac{3}{y}\partial_y - \frac{x}{y^2}\partial_x - \frac{2}{y^2}\right)f_2 - \left(\partial_{xy} + \frac{1}{x}\partial_y + \frac{2}{y}\partial_x + \frac{2}{xy}\right)f_3 = 0,$$

$$f_1 - \left(\frac{y^2}{x}\partial_y + \frac{y}{x}\right)f_2 + \left(\frac{y^2}{x}\partial_x + \frac{y^2}{x^2}\right)f_3 = 0. \qquad \square$$

Given any ideal I it may occur that it is properly contained in some larger ideal J with coefficients in the base field of I; then J is called a *divisor* of I. If the divisor J has the same differential type as I the latter is called *reducible*; if such a divisor does not exist it is called *irreducible*. If a divisor ideal of the same differential type does not exist even if a universal differential field is allowed for its coefficients, I is called *absolutely irreducible*. According to this definition an ideal may be irreducible, yet it may have divisors of lower differential type as the following example shows.

Example 2.5. Consider the operator L defined by $L \equiv \partial_{xx} + \frac{2}{x}\partial_x + \frac{y}{x^2}\partial_y - \frac{1}{x^2}$ of differential dimension $(1, 2)$ (see Definition 2.2 below), i.e. its differential type is 1. The principal ideal $\langle L \rangle$ has the two divisors $l_1 = \langle\!\langle \partial_x + \frac{1}{x}, \partial_y - \frac{1}{y} \rangle\!\rangle$ and $l_2 = \langle\!\langle \partial_x, \partial_y - \frac{1}{y} \rangle\!\rangle$, both of differential type 0; $l_1 z = 0$ has the solution $z = \frac{y}{x}$, $l_2 z = 0$ has the solution $z = y$. Both are also solutions of $Lz = 0$. In Chap. 4 it will be shown that L is irreducible according to the above definition, and even absolutely irreducible. $\qquad \square$

The *greatest common right divisor* or *sum* of two ideals I and J is denoted by $Gcrd(I, J) = I + J$; it is the smallest ideal with the property that both I and J are contained in it. If they have the representation

$$I \equiv \langle f_1, \ldots, f_p \rangle \quad \text{and} \quad J \equiv \langle g_1, \ldots, g_q \rangle,$$

$f_i, g_j \in \mathcal{D}$ for all i and j, the sum is generated by the union of the generators of I and J (Cox et al. [13], page 191). The solution space of the equations corresponding to $Gcrd(I, J)$ is the intersection of the solution spaces of its arguments.

The *least common left multiple* or *left intersection* of two ideals I and J denoted by $Lclm(I, J) = I \cap J$; it is the largest ideal with the property that it is contained both in I and J. The solution space of $Lclm(I, J)z = 0$ is the smallest space containing the solution spaces of its arguments. In the last two subsections of this chapter the properties of sum and intersection ideals will be discussed in some detail.[1]

---

[1] Some authors define it as the lowest element w.r.t. the term order in the ideal defined above [26].

Example 2.6. Consider the ideals $I = \langle\!\langle \partial_{yyy} + \frac{3}{y}\partial_{yy}, \ \partial_x + \frac{y}{x}\partial_y \rangle\!\rangle$ and

$$J = \left\langle\!\!\!\left\langle \partial_{xx} + \frac{1}{x}\partial_x - \frac{1}{x^2}, \ \partial_{xy} + \frac{1}{x}\partial_y + \frac{1}{y}\partial_x + \frac{1}{xy}, \ \partial_{yy} + \frac{1}{y}\partial_y - \frac{1}{y^2} \right\rangle\!\!\!\right\rangle.$$

Applying the above specification, the $Gcrd$ and the $Lclm$ are

$$Gcrd(I, J) = \left\langle\!\!\!\left\langle \partial_{yy} + \frac{1}{y}\partial_y - \frac{1}{y^2}, \ \partial_x + \frac{y}{x}\partial_y \right\rangle\!\!\!\right\rangle,$$

$$Lclm(I, J) = \left\langle\!\!\!\left\langle \partial_{yyy} + \frac{3}{y}\partial_{yy}, \ \partial_{xx} - \frac{y^2}{x^2}\partial_{yy} + \frac{1}{x}\partial_x - \frac{y}{x^2}\partial_y, \right.\right.$$

$$\left.\left. \partial_{xy} + \frac{y}{x}\partial_{yy} + \frac{1}{y}\partial_x + \frac{2}{x}\partial_y \right\rangle\!\!\!\right\rangle.$$

It follows that $lc(H_I) = lc(H_J) = 3$, $lc(H_{I+J}) = 2$ and $lc(H_{I\cap J}) = 4$ in accordance with Sit's relation (2.8). In terms of solution spaces this result may be understood as follows. For $Iz = 0$ a basis of the solution space is $\{1, \frac{x}{y}, \frac{y}{x}\}$, and for $Jz = 0$ a basis is $\{\frac{1}{xy}, \frac{x}{y}, \frac{y}{x}\}$. A basis for their two-dimensional intersection space is $\{\frac{x}{y}, \frac{y}{x}\}$, it is the solution space of $Gcrd(I, J)z = 0$. □

Example 2.7. Consider the two ideals

$$I = \left\langle\!\!\!\left\langle \partial_x + \frac{1}{x}, \partial_y + \frac{1}{y} \right\rangle\!\!\!\right\rangle \quad \text{and} \quad J = \left\langle\!\!\!\left\langle \partial_x + \frac{1}{x+y}, \partial_y + \frac{1}{x+y} \right\rangle\!\!\!\right\rangle.$$

Their one-dimensional solution spaces are generated by $\{\frac{1}{xy}\}$ and $\{\frac{1}{x+y}\}$ respectively. Then $Gcrd(I, J) = \langle 1 \rangle$ and

$$Lclm(I, J) = \left\langle\!\!\!\left\langle \partial_{yy} + \frac{2}{y}\frac{x+2y}{x+y}\partial_y + \frac{2}{y(x+y)}, \ \partial_x - \frac{y^2}{x^2}\partial_y + \frac{x-y}{x^2} \right\rangle\!\!\!\right\rangle.$$

A basis for the solution space of $Lclm(I, J)z = 0$ is $\{\frac{1}{xy}, \frac{1}{x+y}\}$; it will be determined algorithmically in Example 3.5. □

For ordinary differential operators the exact quotient has been defined on page 3. Because all ideals of ordinary differential operators are principal, it is obtained by the usual division scheme. This is different in rings of partial differential operators and a proper generalization of the exact quotient is required. Let $I \equiv \langle f_1, \ldots, f_p \rangle \in \mathcal{D}$ and $J \equiv \langle g_1, \ldots, g_q \rangle \in \mathcal{D}$ be such that $I \subseteq J$, i.e. J is a *divisor* of I. The *exact quotient* is generated by

$$\{(e_{i,1}, \ldots, e_{i,q}) \in \mathcal{D}^q | e_{i,1}g_1 + \ldots + e_{i,q}g_q = f_i, i = 1, \ldots, p\}.$$

The *exact quotient module* $Exquo(I, J)$ is generated by

$$\{h = (h_1, \ldots, h_q) \in \mathcal{D}^q | h_1g_1 + \ldots + h_qg_q \in I\}.$$

It generalizes the syzygy module of J; the latter is obtained for the special choice $I = 0$. If the elements of the exact quotient module are arranged as rows of a matrix with q columns, and the generators of J as elements of a q-dimensional vector, I may be represented as

$$I = Exquo(I, J)J. \tag{2.6}$$

This defines the juxtaposition of $Exquo(I, J)$ and J in terms of matrix multiplication; it generalizes the product representation $L = L_1 L_2$ of an operator L. In general, the result at the right hand side of (2.6) has to be transformed into a Janet basis in order to obtain the original generators $f_1, \ldots, f_p$.

Example 2.8. Consider again the ideals

$$I = \left\langle\!\!\left\langle \partial_{yy} + \frac{2}{y}\frac{x + 2y}{x + y}\partial_y + \frac{2}{y(x + y)}, \ \partial_x - \frac{y^2}{x^2}\partial_y + \frac{x - y}{x^2} \right\rangle\!\!\right\rangle$$

and $J = \left\langle\!\!\left\langle \partial_x + \frac{1}{x}, \partial_y + \frac{1}{y} \right\rangle\!\!\right\rangle$ of the preceding example; obviously $I \subset J$. Division yields $\left\{ \left(0, \partial_y + \frac{x + 3y}{y(x + y)} \right), \left(1, -\frac{y^2}{x^2} \right) \right\}$. There is a single syzygy $(\partial_y + \frac{1}{y}, -\partial_x - \frac{1}{x})$. The Janet basis representation for the exact quotient module is

$$Exquo(I, J) = \left\langle\!\!\left\langle \left(0, \partial_x + \frac{x - y}{x(x + y)} \right), \left(0, \partial_y + \frac{x + 3y}{y(x + y)} \right), \left(1, -\frac{y^2}{x^2} \right) \right\rangle\!\!\right\rangle \subset \mathcal{D}^2.$$

Continued in Example 3.5.   □

Example 2.9. Consider the ideal $I = \left\langle\!\!\left\langle \partial_{xx} + \frac{1}{x}\partial_x, \partial_{xy}, \partial_{yy} + \frac{1}{y}\partial_y \right\rangle\!\!\right\rangle$ in *grlex*, $x \succ y$ term order. The divisor $J = \left\langle\!\!\left\langle \partial_x, \partial_y \right\rangle\!\!\right\rangle$ yields the exact quotient module $Exquo(I, J) = \left\langle\!\!\left\langle \left(\partial_x + \frac{1}{x}, 0 \right), (\partial_y, 0), (0, \partial_x), \left(0, \partial_y + \frac{1}{y} \right) \right\rangle\!\!\right\rangle$. It may be represented as the intersection of two maximal modules of order 1, i.e.

$$Exquo(I, J) = Lclm\left(\left\langle\!\!\left\langle (1, 0), (0, \partial_x)\left(0, \partial_y + \frac{1}{y} \right) \right\rangle\!\!\right\rangle, \right.$$

$$\left. \left\langle\!\!\left\langle \left(\frac{x}{y}, 1 \right), \left(\partial_x + \frac{1}{x}, 0 \right), (\partial_y, 0) \right\rangle\!\!\right\rangle \right).$$ □

In the next chapter it will be shown how to obtain their decomposition algorithmically, and how the exact quotient module may be applied for solving systems of linear pde's that are not completely reducible.

Consider $I \subseteq \mathcal{D}$, and denote by I_n the intersection of I with the $\mathcal{F}$-linear space of all derivatives of order not higher than n. Then according to Kolchin [37], see also Buium and Cassidy [8], and Pankratiev et al. [38], the Hilbert-Kolchin polynomial of I is defined by

$$H_I(n) \equiv \binom{n + k}{k} - \dim I_n; \tag{2.7}$$

k is the number of variables. The first term equals the number of all derivatives of order not higher than n. Hence, for sufficiently large n the value of $H_I(n)$ counts the number of derivatives of order not higher than n which is not in the ideal generated by the leading derivatives of the generators of I. The degree $deg(H_I)$ of H_I is called the *differential type* of I ([37], page 130; [8], page 602). Its leading coefficient $lc(H_I)$ is called the *typical differential dimension* of I, ibid. If $I, J \subseteq \mathcal{D}$ are two ideals, Sit [65], Theorem 4.1, has shown the important equality

$$lc(H_{I+J}) + lc(H_{I \cap J}) = lc(H_I) + lc(H_J) \tag{2.8}$$

for its typical differential dimensions.

According to Kolchin [36], $deg(H_I)$ and $lc(H_I))$ are differential birational invariants; their importance justifies the introduction of a special term for these quantities.

Definition 2.2. The pair $(deg(H_I), lc(H_I))$ for an ideal I is called the differential dimension of I, denoted by d_I [23].

For the solutions of the differential equations attached to any ideal or module of differential operators, these quantities have an important meaning [8, 36].

Theorem 2.1. *The differential type denotes the largest number of arguments occurring in any undetermined function of the general solution. The typical differential dimension means the number of functions depending on this maximal number of arguments.*

Apparently the differential dimension describes somehow the *"size"* of the solution space. In this terminology the differential dimension $(0, m)$ corresponds to a system of pde's with a finite-dimensional solution space of dimension m over the field of constants. This discussion shows that the differential dimension is the proper generalization of the dimension of a solution space to general systems of linear pde's.

Example 2.10. For the ideal $I = \langle\!\langle \partial_{xx} - \frac{1}{x}\partial_x, \partial_y \rangle\!\rangle$ of the preceding example, only the two derivatives 1 and ∂_x are not contained in the ideal generated by the leading derivatives. Thus $H_I = 2$ and $d_I = (0, 2)$. □

Example 2.11. Let the principal ideal $I = \langle \partial_x + a\partial_y + b \rangle$ be given. There are $\frac{1}{2}(n^2 + n)$ derivatives of order not higher than n containing at least a single derivative ∂_x. Therefore $H_I = n + 1$ and $d_I = (1, 1)$. □

Example 2.12. Consider the ideal $I = \langle\!\langle \partial_{xxx}, \partial_{xxy} \rangle\!\rangle$. The number of derivatives which are multiples of either leading term is $\frac{1}{2}(n-2)(n+1)$. Therefore $H_I = 2n+2$ and $d_I = (1, 2)$. □

By analogy with the well known *Landau symbol* of asymptotic analysis, the following notation will frequently be applied. Whenever in an expression terms of order lower than some fixed term τ are not relevant, they are collectively denoted by $o(\tau)$. This will frequently occur in lex term orderings where τ denotes the highest term involving a particular variable.

Another short hand notation concerns the generators of ideals or modules of differential operators. If only the number of generators and its leading derivatives are of interest, the abbreviated notation $\langle \ldots \rangle_{LT}$ will be used. For example, if an ideal of differential operators is generated by two elements with leading derivatives ∂_{xx} and ∂_{xy}, it is denoted by $\langle \partial_{xx}, \partial_{xy} \rangle_{LT}$. A principal ideal that is generated by a single generator with highest derivative ∂_{xxx} is abbreviated by $\langle \partial_{xxx} \rangle_{LT}$.

2.4 Differential Type Zero Ideals in $\mathbb{Q}(x, y)[\partial_x, \partial_y]$

Understanding the ideal structure in the ring $\mathbb{Q}(x, y)[\partial_x, \partial_y]$ will be of utmost importance when dealing with linear pde's and their decompositions. In the first place this means to characterize the ideals that are relevant in a certain context as detailed as possible. Ultimately this comes down to a partial classification of such ideals. Secondly, the relationships between individual ideals must be understood in order to utilize them for the solution procedure. For any two ideals this means deciding pairwise inclusion, determining their sum and their intersection.

To begin with, the range of ideals to be covered has to be delimited by suitable constraints. In particular they are selected with regard to solving differential equations that are of interest for some application. This requirement confines the derivatives to order not higher than three. In addition the differential dimension of an ideal turns out to be an important distinctive feature that may be applied for classification. Furthermore, some kind of completeness in the mathematical sense will be desirable.

Taking these considerations into account the subsequent three propositions describe certain "large" ideals close to the top of the complete lattice. The *grlex* term order with $x \succ y$ is always applied. The first result concerns ideals of differential type zero; they correspond to systems of linear pde's with a finite-dimensional solution space. They will be denoted by $\mathbb{J}^{(0,k)}$ where $(0, k)$ means their differential dimension. This proceeding is only meaningful if the coherence conditions for the coefficients are explicitly known and it is assured that they are satisfied.

Proposition 2.1. *There are six types* $\mathbb{J}^{(0,k)}$ *of ideals with differential dimension* $(0, k)$ *and* $k \leq 3$. *The integrability conditions (IC's for short) are given for each case in terms of a Janet basis for the coefficients.*

$$\mathbb{J}^{(0,1)} : \langle \partial_x + a, \partial_y + b \rangle, \text{ integrability condition } a_y = b_x.$$

$$\mathbb{J}_1^{(0,2)} : \langle \partial_{yy} + a_1 \partial_y + a_2, \partial_x + b_1 \partial_y + b_2 \rangle,$$

integrability conditions

$$b_{2,yy} - 2b_{1,y} a_2 + b_{2,y} a_1 - a_{2,x} - a_{2,y} b_1 = 0,$$

$$b_{1,yy} - b_{1,y} a_1 + 2b_{2,y} - a_{1,x} - a_{1,y} b_1 = 0.$$

$$\mathbb{J}_2^{(0,2)} : \langle \partial_{xx} + a_1 \partial_x + a_2, \partial_y + b \rangle,$$

integrability conditions

$$b_x - \tfrac{1}{2}a_{1,y} = 0, \quad a_{2,y} - \tfrac{1}{2}a_{1,xy} - \tfrac{1}{2}a_{1,y}a_1 = 0.$$

$$\mathbb{J}_1^{(0,3)} : \ \langle \partial_{yyy} + a_1\partial_{yy} + a_2\partial_y + a_3, \partial_x + b_1\partial_y + b_2 \rangle,$$

integrability conditions

$$b_{2,yy} + \tfrac{1}{6}a_{1,xy} + \tfrac{1}{6}a_{1,yy}b_1 + b_{1,y}(\tfrac{1}{3}a_{1,y} + \tfrac{2}{9}a_1^2 - a_2) + \tfrac{1}{3}b_{2,y}a_1$$
$$+\tfrac{2}{9}a_{1,x}a_1 + \tfrac{2}{9}a_{1,y}b_1a_1 - \tfrac{1}{2}a_{2,x} - \tfrac{1}{2}a_{2,y}b_1 = 0,$$
$$b_{1,yy} - \tfrac{1}{3}b_{1,y}a_1 + b_{2,y} - \tfrac{1}{3}a_{1,x} - \tfrac{1}{3}a_{1,y}b_1 = 0,$$

$$a_{1,xyy} + a_{1,yyy}b_1 + 2a_{1,xy}a_1 + a_{1,yy}(3b_{1,y} + 2b_1a_1) - 3a_{2,xy} - 3a_{2,yy}b_1$$
$$+b_{1,y}(6a_{1,y}a_1 - 9a_{2,y} + \tfrac{4}{3}a_1^3 - 6a_1a_2 + 18a_3)$$
$$+a_{1,x}(2a_{1,y} + \tfrac{4}{3}a_1^2 - 2a_2) + 2a_{1,y}^2b_1 + a_{1,y}(\tfrac{4}{3}b_1a_1^2 - 2b_1a_2)$$
$$-2a_{2,x}a_1 - 2a_{2,y}b_1a_1 + 6a_{3,x} + 6a_{3,y}b_1 = 0.$$

$$\mathbb{J}_2^{(0,3)} : \ \langle \partial_{xx} + a_1\partial_x + a_2\partial_y + a_3, \partial_{xy} + b_1\partial_x + b_2\partial_y + b_3,$$
$$\partial_{yy} + c_1\partial_x + c_2\partial_y + c_3 \rangle,$$

integrability conditions

$$b_{3,y} - c_{3,x} + a_3c_1 - b_1b_3 - b_2c_3 + b_3c_2 = 0,$$
$$b_{2,y} - c_{2,x} + a_2c_1 - b_1b_2 + b_3 = 0,$$
$$b_{1,y} - c_{1,x} + a_1c_1 - b_1^2 + b_1c_2 - b_2c_1 - c_3 = 0,$$
$$a_{3,y} - b_{3,x} - a_1b_3 - a_2c_3 + a_3b_1 + b_2b_3 = 0,$$
$$a_{2,y} - b_{2,x} - a_1b_2 + a_2b_1 - a_2c_2 + a_3 + b_2^2 = 0,$$
$$a_{1,y} - b_{1,x} - a_2c_1 + b_1b_2 - b_3 = 0.$$

$$\mathbb{J}_3^{(0,3)} : \ \langle \partial_{xxx} + a_1\partial_{xx} + a_2\partial_x + a_3, \partial_y + b \rangle,$$

integrability conditions

$$b_x - \tfrac{1}{3}a_{1,y} = 0, \quad a_{1,xy} + \tfrac{2}{3}a_{1,y}a_1 - a_{2,y} = 0,$$
$$a_{2,xy} - a_{1,y}(\tfrac{2}{3}a_{1,x} + \tfrac{2}{9}a_1^2 - a_2) + \tfrac{1}{3}a_{2,y}a_1 - 3a_{3,y} = 0.$$

Proof. Any ideal of differential dimension $(0, k)$ must contain generators with leading terms ∂_{x^i} and ∂_{y^j} with $i, j \geq 1$. From the listing (2.5) and the constraints $i, j \leq 3$ the following choices of leading derivatives are possible.

$$(\partial_x, \partial_y), \ (\partial_{yy}, \partial_x), \ (\partial_{xx}, \partial_y), \ (\partial_{yyy}, \partial_x), \ (\partial_{xx}, \partial_{xy}, \partial_{yy}), \ (\partial_{xxx}, \partial_y).$$

They yield the six ideals given above. The integrability condition for $\mathbb{J}^{(0,1)}$ is obvious. For $\mathbb{J}_1^{(0,2)}$, differentiating $\partial_x + b_1\partial_y + b_2$ w.r.t. y and reducing the result w.r.t. $\partial_{yy} + a_1\partial_y + a_2$ yields

$$\partial_{xy} + (b_{1,y} + b_2 - a_1 b_1)\partial_y + b_{2,y} - a_1 b_1.$$

Deriving a second time w.r.t. y and performing all possible reductions leads to

$$\partial_{xyy} + (b_{1,yy} + 2b_{2,y} - a_{1,y}b_1 - 2a_1 b_{1,y} - a_1 b_2 - a_2 b_1 + a_1^2 b_1)\partial_y$$
$$+ b_{2,yy} - a_{2,y}b_1 - 2a_2 b_{1,y} - a_2 b_2 + a_1 a_2 b_1.$$

Differentiating now the first generator $\partial_{yy} + a_1\partial_y + a_2$ w.r.t. x and applying the above expression with leading term ∂_{xy} for reduction yields

$$\partial_{xyy} + (a_{1,x} - a_1 b_{1,y} - a_1 b_2 - a_2 b_1 + a_1^2 b_1)\partial_y + a_{2,x} - a_1 b_{2,y} - a_2 b_2 + a_1 a_2 b_1.$$

Equating the coefficients of the two operators with leading term ∂_{xyy} leads to the given IC's after some simplifications. The calculations for the other ideals are similar and are therefore omitted.   □

It should be noticed that the $IC's$ given in the above proposition are coherent. This is obvious due to the term order that has been chosen; they are linear in the leading derivatives with constant leading coefficient, and the leading derivatives contain pairwise different functions. The importance of the $IC's$ given in the above proposition will become clear later on, e.g. in proving the existence of certain divisors in Exercise 3.2.

2.5 Differential Type Zero Modules over $\mathbb{Q}(x, y)[\partial_x, \partial_y]$

In this subsection vector modules of differential dimension $(0, 1)$ and $(0, 2)$ in $\mathscr{D}^2$ are considered; they correspond to systems of linear pde's in two dependend variables.

Proposition 2.2. *There are seven types* $\mathbb{M}^{(0,k)}$ *of modules of differential dimension* $(0, k)$ *with* $k = 1$ *or* $k = 2$ *in* $\mathscr{D}^2$. *The integrability conditions are given for each case in terms of a Janet basis for its coefficients.*

$\mathbb{M}_1^{(0,1)} : \langle (1, a), (0, \partial_x + b), (0, \partial_y + c)\rangle$, *integrability condition* $b_y - c_x = 0$.

$\mathbb{M}_2^{(0,1)} : \langle (0, 1), (\partial_x + a, 0), (\partial_y + b, 0)\rangle$, *integrability condition* $a_y - b_x = 0$.

$\mathbb{M}_1^{(0,2)} : \langle (1, 0), (0, \partial_{yy} + a_1\partial_y + a_2), (0, \partial_x + b_1\partial_y + b_2)\rangle$

with the same integrability conditions as for ideal type $\mathbb{J}_1^{(0,2)}$.

$$\mathbb{M}_2^{(0,2)} : \langle (1,0), (0, \partial_{xx} + a_1\partial_x + a_2), (0, \partial_y + b)\rangle$$

with the same integrability conditions as for ideal type $\mathbb{J}_2^{(0,2)}$.

$$\mathbb{M}_3^{(0,2)} : \langle (\partial_x + a_1, a_2), (\partial_y + b_1, b_2), (c_1, \partial_x + c_2), (d_1, \partial_y + d_2)\rangle$$

with integrability conditions

$$a_{1,y} - b_{1,x} - a_2d_1 + b_2c_1 = 0,$$
$$a_{2,y} - b_{2,x} - a_1b_2 + a_2b_1 - a_2d_2 + b_2c_2 = 0,$$

$$c_{1,y} - d_{1,x} + a_1d_1 - b_1c_1 + c_1d_2 - c_2d_1 = 0,$$
$$c_{2,y} - d_{2,x} + a_2d_1 - b_2c_1 = 0.$$

$$\mathbb{M}_4^{(0,2)} : \quad \langle (1, c), (0, \partial_{yy} + a_1\partial_y + b_2), (0, \partial_x + b_1\partial_y + b_2)\rangle$$

with the same integrability conditions as for ideal type $\mathbb{J}_4^{(0,2)}$.

$$\mathbb{M}_5^{(0,2)} : \quad \langle (1, c), (0, \partial_{xx} + a_1\partial_x + a_2), (0, \partial_y + b)\rangle$$

with the same integrability conditions as for ideal type $\mathbb{J}_5^{(0,2)}$.

Proof. For any component the generators must contain leading derivatives ∂_{x^i} and ∂_{y^j} with $i, j \geq 0$. For type $\mathbb{M}^{(0,1)}$ this allows the combination of leading terms $\{(1,0), (0, \partial_x), (0, \partial_y)\}$ and $\{(0,1), (\partial_x, 0), \partial_y, 0)\}$. For type $\mathbb{M}^{(0,2)}$ the possible leading terms are

$$\{(1,0), (0, \partial_{xx}), (0, \partial_y)\}, \quad \{(1,0), (0, \partial_{yy}), (0, \partial_x)\},$$
$$\{(\partial_x, 0), (\partial_y, 0), (0, \partial_x), (0, \partial_y)\},$$
$$\{(0, 1), (\partial_{xx}, 0), (\partial_y, 0)\}, \quad \{(0, 1), (\partial_{yy}, 0), (\partial_x, 0)\}.$$

The integrability conditions are obtained as in the preceding proposition.   □

There is one more module of differential dimension $(0, 1)$ in $\mathcal{D}^3$ needed for the decompositions discussed in the next chapter; its type is defined to be $\mathbb{M}_3^{(0,1)}$ and is the subject of Exercise 2.8.

Example 2.13. The module

$$M \equiv \left\langle \left(\partial_x - \frac{x^2 - 3y}{x(x^2 - y)}, -\frac{1}{x^2 - y}\right), (\partial_y, 0),\right.$$

$$\left.\left(\frac{4y(x^2 + y)}{x^2(x^2 - y)}, \partial_x - \frac{2(x^2 + y)}{x(x^2 - y)}\right)\left(-\frac{2x}{x^2 - y}, \partial_y + \frac{1}{x^2 - y}\right)\right\rangle$$

is of type $M_3^{(0,2)}$; the coherence conditions may easily be tested by applying the explicit form as given in the above proposition without running a costly Janet basis algorithm. Continued in Example 3.4.   □

2.6 Laplace Divisors $\mathbb{L}_{x^m}(L)$ and $\mathbb{L}_{y^n}(L)$

The origin of these ideals goes back to Laplace who introduced an iterative solution scheme for equations with leading derivative z_{xy}; it is described in Appendix C. Later on it was realized that this procedure is essentially equivalent to determining a so called *involutive system*. In more modern language this comes down to constructing a Janet basis for an ideal that is generated by the operator corresponding to the originally given equation, and an ordinary operator of fixed order involving exclusively derivatives w.r.t. x or y. Thus this important concept may be generalized to large classes of operators with a mixed leading derivative. Due to its origin the following definition is suggested.

Definition 2.3. Let L be a partial differential operator in the plane; define

$$l_m \equiv \partial_{x^m} + a_{m-1}\partial_{x^{m-1}} + \ldots + a_1\partial_x + a_0 \tag{2.9}$$

and

$$\mathfrak{k}_n \equiv \partial_{y^n} + b_{n-1}\partial_{y^{n-1}} + \ldots + b_1\partial_y + b_0 \tag{2.10}$$

be ordinary differential operators w.r.t. x or y; $a_i, b_i \in \mathbb{Q}(x, y)$ for all i; m and n are natural numbers not less than 2. Assume the coefficients $a_i, i = 0, \ldots, m-1$ are such that L and l_m form a Janet basis. If m is the smallest integer with this property then $\mathbb{L}_{x^m}(L) \equiv \langle\!\langle L, l_m \rangle\!\rangle$ is called a Laplace divisor of L. Similarly, if b_j, $j = 0, \ldots, n-1$ are such that L and $\mathfrak{k}_n$ form a Janet basis and n is minimal, then $\mathbb{L}_{y^n}(L) \equiv \langle\!\langle L, \mathfrak{k}_n \rangle\!\rangle$ is called a Laplace divisor of L. Both Laplace divisors have differential dimension $(1, 1)$.

The possible existence of a Laplace divisor for operators of order 2 or 3 is investigated next.

Proposition 2.3. *Let the second-order partial differential operator*

$$L \equiv \partial_{xy} + A_1\partial_x + A_2\partial_y + A_3 \tag{2.11}$$

be given with $A_i \in \mathbb{Q}(x, y)$ for all i; m and n are natural numbers not less than 2.

(i)   *If A_1, A_2 and A_3 satisfy the differential polynomial constructed in the proof below, there exists a Laplace divisor $\mathbb{L}_{x^m}(L) = \langle\!\langle L, l_m \rangle\!\rangle$.*

(ii)   *If A_1, A_2 and A_3 satisfy the differential polynomial constructed in the proof below there exists a Laplace divisor $\mathbb{L}_{y^n}(L) = \langle\!\langle L, \mathfrak{k}_n \rangle\!\rangle$.*

(iii) *If there are two Laplace divisors $\mathbb{L}_{x^m}(L)$ and $\mathbb{L}_{y^n}(L)$, the operator L is completely reducible, $\langle L \rangle = Lclm\big(\mathbb{L}_{x^m}(L), \mathbb{L}_{y^n}(L)\big)$.*

Proof. The proof will be given for case (ii). If the operator (2.11) is derived repeatedly w.r.t. y, and the reductum is reduced in each step w.r.t. (2.11), $n - 2$ expressions of the form

$$\partial_{xy^i} + p_i\partial_x + p_{i,i}\partial_{y^i} + p_{i,i-1}\partial_{y^{i-1}} + \ldots + p_{i,0} \qquad (2.12)$$

for $2 \leq i \leq n - 1$ are obtained. All coefficients p_i and $p_{i,j}$ of (2.12) are differential polynomials in the ring $\mathbb{Q}\{A_1, A_2, A_3\}$ with derivatives ∂_x and ∂_y. There is no reduction w.r.t. (2.10) possible. Deriving the last expression (2.12) corresponding to $i = n - 1$ once more w.r.t. y and reducing the reductum w.r.t. both (2.11) and (2.10) yields

$$\partial_{xy^n} + p_n\partial_x + (p_{n,n-1} - p_{n,n}b_{n-1})\partial_{y^{n-1}}$$

$$+(p_{n,n-2} - p_{n,n}b_{n-2})\partial_{y^{n-2}} + \ldots \qquad (2.13)$$

$$+(p_{n,1} - p_{n,n}b_1)\partial_y + p_{n,0} - p_{n,n}b_0.$$

In the first derivative of (2.10) w.r.t. x

$$\partial_{xy^n} + b_{n-1,x}\partial_{y^{n-1}} + b_{n-2,x}\partial_{y^{n-2}} + \ldots + b_{1,x}\partial_y + b_{0,x}$$

$$+b_{n-1}\partial_{xy^{n-1}} + b_{n-2}\partial_{xy^{n-2}} + \ldots + b_1\partial_{xy} + b_0\partial_x$$

the terms containing derivatives of the form ∂_{xy^k} for $1 \leq k \leq n - 1$ may be reduced w.r.t. (2.11) or (2.12) with the result

$$\partial_{xy^n} + (b_{n-1,x} - p_{n-1,n-1}b_{n-1})\partial_{y^{n-1}}$$

$$+(b_{n-2,x} - p_{n-1,n-2}b_{n-1} - p_{n-2,n-2}b_{n-2})\partial_{y^{n-2}}$$

$$\vdots \qquad\qquad \vdots$$

$$+(b_{1,x} - p_{n-1,1}b_{n-1} - p_{n-2,1}b_{n-2} \ldots - p_{2,1}b_2 - A_2b_1)\partial_y \qquad (2.14)$$

$$+b_{0,x} - p_{n-1,0}b_{n-1} - p_{n-2,0}b_{n-2} - \ldots - p_{2,0}b_2 - A_3b_1$$

$$+(b_0 - p_1b_{n-1} - p_{n-2}b_{n-2} - \ldots - p_2b_2 - A_1b_1)\partial_x.$$

If this expression is subtracted from (2.13), the coefficients of the derivatives must vanish in order that (2.11) and (2.10) combined form a Janet basis. The resulting system of equations is

$$b_{n-1,x} + (p_{n,n} - p_{n-1,n-1})b_{n-1} - p_{n,n-1} = 0,$$

$$b_{n-2,x} - p_{n-1,n-2}b_{n-1} + (p_{n,n} - p_{n-2,n-2})b_{n-2} - p_{n,n-2} = 0,$$

$$\vdots \qquad\qquad \vdots \qquad (2.15)$$

$$b_{1,x} - p_{n-1,1}b_{n-1} - \ldots + (p_{n,n} - A_2)b_1 - p_{n,1} = 0,$$

$$b_{0,x} - p_{n-1,0}b_{n-1} - \ldots - A_3b_1 + p_{n,n}b_0 - p_{n,0} = 0,$$

$$b_0 - p_{n-1}b_{n-1} - p_{n-2}b_{n-2} - \ldots - p_2b_2 - A_1b_1 = 0.$$

The last equation may be solved for b_0. Substituting it into the equation with leading term $b_{0,x}$, and eliminating the first derivatives $b_{j,x}$ for $j = 1,\ldots,n-1$ by means of the preceding equations, it may be solved for b_1. Proceeding in this way, due to the triangular structure, finally b_{n-1} is obtained from the equation with leading term $b_{n-2,x}$. Backsubstituting these results, all b_k are explicitly known. By construction they are all rational in the A_i and its derivatives, i.e. they are contained in the base field of L. Substituting them into the first equation a constraint for the coefficients A_i is obtained; it is the condition for the existence of the Laplace divisor.

The proof for case (i) is similar and is therefore omitted. In case (iii), the intersection ideal must divide both Laplace divisors. Due to the Janet basis property of the latter, its generators are linear combinations of L and $\mathfrak{l}_m$ as well as L and $\mathfrak{k}_n$. Transformation into a Janet basis yields the result given above.   □

For low values of n and m the conditions for A_1, A_2 and A_3 for the existence of a Laplace divisor of the operator (2.11) may be given explicitly as shown next.

Corollary 2.1. *Let the second-order operator $L \equiv \partial_{xy} + A_1\partial_x + A_2\partial_y + A_3$ be given and define the quantities*

$$H_0 \equiv A_{1,x} + A_1 A_2 - A_3, \quad K_0 = A_{2,y} + A_1 A_2 - A_3,$$

$$H_1 \equiv H_{0,xy} H_0 - H_{0,x} H_{0,y} - H_0^2(2H_0 - K_0),$$

$$K_1 \equiv K_{0,xy} K_0 - K_{0,x} K_{0,y} - K_0^2(2K_0 - H_0),$$

$$H_2 \equiv H_{1,xy} H_1 - H_{1,x} H_{1,y} - H_1^2(3H_0 - 2K_0),$$

$$K_2 \equiv K_{1,xy} K_1 - K_{1,x} K_{1,y} - K_1^2(3K_0 - 2H_0).$$

Operators $\mathfrak{l}_m$ and $\mathfrak{k}_n$ for $m,n = 2,3$ may be constructed as follows. A divisor $\langle\!\langle L, \partial_{xx} + a_1\partial_x + a_0\rangle\!\rangle$ exists if $K_0 \neq 0$, $K_1 = 0$; then

$$a_1 = 2A_2 - \frac{K_{0,x}}{K_0}, \quad a_0 = A_{2,x} + A_2^2 - A_2\frac{K_{0,x}}{K_0}.$$

A divisor $\langle\!\langle L, \partial_{yy} + b_1\partial_y + b_0\rangle\!\rangle$ exists if $H_0 \neq 0$, $H_1 = 0$; then

$$b_1 = 2A_1 - \frac{H_{0,y}}{H_0}, \quad b_0 = A_{1,y} + A_1^2 - A_1\frac{H_{0,y}}{H_0}.$$

A divisor $\langle\!\langle L, \partial_{xxx} + a_2\partial_{xx} + a_1\partial_x + a_0\rangle\!\rangle$ exists if $K_0 \neq 0$, $K_1 \neq 0$, $K_2 = 0$; then

$$a_2 = 3A_2 - \frac{K_{1,x}}{K_1}, \quad a_1 = 3(A_{2,x} + A_2^2) - 2A_2\frac{K_{1,x}}{K_1} - \frac{K_{0,xx}}{K_0} + \frac{K_{0,x}}{K_0}\frac{K_{1,x}}{K_1},$$

$$a_0 = A_{2,xx} + 3A_2 A_{2,x} + A_2^3 - (A_{2,x} + A_2^2)\frac{K_{1,x}}{K_1} - A_2\left(\frac{K_{0,xx}}{K_0} - \frac{K_{0,x}}{K_0}\frac{K_{1,x}}{K_1}\right).$$

A divisor $\langle\!\langle L, \partial_{yyy} + b_2\partial_{yy} + b_1\partial_y + b_0 \rangle\!\rangle$ exists if $H_0 \neq 0$, $H_1 \neq 0$, $H_2 = 0$; then

$$b_2 = 3A_1 - \frac{H_{1,y}}{H_1}, \quad b_1 = 3(A_{1,y} + A_1^2) - 2A_1\frac{H_{1,y}}{H_1} - \frac{H_{0,yy}}{H_0} + \frac{H_{0,y}}{H_0}\frac{H_{1,y}}{H_1},$$

$$b_0 = A_{1,yy} + 3A_1A_{1,y} + A_1^3 - (A_{1,y} + A_1^2)\frac{H_{1,y}}{H_1} - A_1\left(\frac{H_{0,yy}}{H_0} - \frac{H_{0,y}}{H_0}\frac{H_{1,y}}{H_1}\right).$$

This corollary and Proposition 2.3 show that the existence of a Laplace divisor may be decided for any fixed value of m or n. If the answer is affirmative, it may be constructed explicitly. However, in general there is no upper bound for m and n known. Only in special cases as in Example 2.14 below a general answer may be obtained.

The relation of this problem to the number of Laplace's iterations, and the relation between the quantities H_i and K_j to the Laplace invariants h_i and k_j is discussed in Appendix C.

The operators $\mathfrak{l}_m$ and $\mathfrak{k}_n$ given in the preceding corollary are reducible, they allow a first order right factor as shown next.

Corollary 2.2. *The operator $\mathfrak{l}_m$ defined by (2.9) allows the right factor $\partial_x + A_2$; the operator $\mathfrak{k}_n$ defined by (2.10) allows the right factor $\partial_y + A_1$.*

Proof. The proof will be given for the second case. Differentiating the operator $\partial_y + A_1$ $(i-1)$-times and reducing the reductum of the result w.r.t. to it yields

$$\partial_{y^i} + p_i(A_{1,y^{i-1}}, \ldots, A_{1,y}, A_1);$$

p_i is a differential polynomial of A_1. One more differentiation leads to

$$\partial_{y^{i+1}} + \left(\frac{dp_i}{dy} - A_1p_i\right) = \partial_{y^{i+1}} + p_{i+1}(A_{1,y^i}, \ldots, A_{1,y}, A_1).$$

From this there follows the recursion

$$p_{i+1} = \frac{dp_i}{dy} - A_1p_i \quad \text{for } i \geq 1, \quad p_1 = A_1. \tag{2.16}$$

The same recursion with the same initial condition is obtained for the coefficients p_i of Proposition 2.3 if (2.12) is differentiated once more w.r.t. y, i.e. the p_i are the same in either case.

If $\mathfrak{k}_n$ has the factor $\partial_y + A_1$ it may be reduced to zero w.r.t. it, i.e. it vanishes upon substitution of the above expressions $\partial_y^i + p_i$. This yields the condition

$$b_0 - p_{n-1}b_{n-1} - p_{n-2}b_{n-2} - \ldots - p_1b_1 = 0$$

for the existence of the factor $\partial_y + A_1 = 0$ of $\mathfrak{k}_n$. Due to the above mentioned property of the p_i it is identical to the last equation of (2.15) which is obeyed due to the Janet basis property. □

Example 2.14. The operator $L \equiv \partial_{xy} + xy\partial_x - 2y$ has been considered by Imschenetzky [28]. The invariants are

$$H_0 = 3y, \; H_1 = -36y^3, \; H_2 = -6,480y^7 \text{ and } K_0 = 2y, \; K_1 = -4y^3, \; K_2 = 0.$$

According to Corollary 2.1 a divisor $\mathbb{L}_{y^k}(L)$ for $k \leq 3$ does not exist. There follows $a_0 = a_1 = a_2 = 0$, hence there is the divisor $\mathbb{L}_{x^3}(L) = \langle\!\langle \partial_{xxx}, \partial_{xy} + xy\partial_x - 2y \rangle\!\rangle$.

In Exercise C.2 it will be shown that a divisor $\mathbb{L}_{y^k}(L)$ does not exist for any k by applying Laplace's iteration to the given operator.   □

The syzygies for the generators L and $\mathfrak{l}_m$ or $\mathfrak{k}_n$ respectively of the Laplace divisors described in Proposition 2.3 are given next; they will be applied in later chapters for solving equations $Lz = 0$ allowing a Laplace divisor.

Corollary 2.3. *Let* $L \equiv \partial_{xy} + A_1\partial_x + A_2\partial_y + A_3$ *be such that a Laplace divisor* $\mathbb{L}_{x^m}(L) = \langle\!\langle L, \mathfrak{l}_m \rangle\!\rangle$ *exists. For* $m \leq 4$ *there are the following syzygies.*

$$m = 2 : L_x - (A_2 - a_1)L = \mathfrak{l}_{2,y} + A_1\mathfrak{l}_2,$$

$$m = 3 : L_{xx} - (A_2 - a_2)L_x - (2A_{2,x} - A_2^2 + A_2a_2 - a_1)L = \mathfrak{l}_{3,y} + A_1\mathfrak{l}_3,$$

$$m = 4 : L_{xxx} - (A_2 - a_3)L_{xx} - (3A_{2,x} - A_2^2 + A_2a_3 - a_2)L_x$$
$$-(3A_{2,xx} - 5A_{2,x}A_2 + 2A_{2,x}a_3 + A_2^3 - A_2^2a_3 + A_2a_2 - a_1)L = \mathfrak{l}_{4,y} + A_1\mathfrak{l}_4.$$

For any value of m the right hand side has the form $\mathfrak{l}_{m,y} + A_1\mathfrak{l}_m$.
For Laplace divisors $\mathbb{L}_{y^n}(L) = \langle\!\langle L, \mathfrak{k}_n \rangle\!\rangle$ *and* $n \leq 4$ *the syzygies are as follows.*

$$n = 2 : L_y - (A_1 - b_1)L = \mathfrak{k}_{2,x} + A_2\mathfrak{k}_2,$$

$$n = 3 : L_{yy} - (A_1 - b_2)L_y - (2A_{1,y} - A_1^2 + A_1b_2 - b_1)L = \mathfrak{k}_{3,x} + A_2\mathfrak{k}_3$$

$$n = 4 : L_{yyy} - (A_1 - b_3)L_{yy} - (3A_{1,y} - A_1^2 + A_1b_3 - b_2)L_y$$
$$-(3A_{1,yy} - 5A_{1,y}A_1 + 2A_{1,y}b_3 + A_1^3 - A_1^2b_3 + A_1b_2 - b_1)L = \mathfrak{k}_{4,x} + A_2\mathfrak{k}_4.$$

For any value of n the right hand side has the form $\mathfrak{k}_{n,y} + A_2\mathfrak{k}_n$.

Proof. The proof is given for the divisor $\mathbb{L}_{x^m}(L)$. As usual the term order *grlex*, $x \succ y$ is applied; the symbol $o(\tau)$, τ any derivative, is defined on page 29. Deriving L w.r.t. x up to order $m - 1$, using the definition of L for substituting the non-leading mixed derivatives in each step, and the definition of $\mathfrak{l}_m$ for substituting the derivative ∂_{x^m}, the expression

$$\partial_{x^m y} + A_1\mathfrak{l}_m + o(\partial_{x^{m-1}}) = \sum_{i=0}^{m-1} p_i L_{x^i}$$

is obtained; the p_i are differential polynomials in A_2. Deriving $\mathfrak{l}_m$ once w.r.t. y and substituting all non-leading mixed derivatives by derivatives of L, an expression of the form

$$\partial_{x^m y} - \mathfrak{l}_{m,y} + O(\partial_{x^{m-1}}) = \sum_{i=0}^{m-1} q_i L_{x^i}$$

follows; the q_i are differential polynomials in A_1, A_2, A_3 and the a_j. Subtracting these expressions from each other leads to

$$\mathfrak{l}_{m,y} + A_1 \mathfrak{l}_m = \sum_{i=1}^{m-1} r_i L_{x^i} .$$

This is the desired syzygy; r_i is a differential polynomial in the coeffcients of L and $\mathfrak{l}_m$; due to the Janet basis property of the Laplace divisor all terms subsumed under the Landau symbol vanish. The proof for $\mathbb{L}_{y^n}(L)$ is similar. For low values of m and n the above syzygies are obtained explicitly. $\qquad\square$

The ideals constructed in Proposition 2.3 may be generalized for higher order operators. The following proposition deals with Laplace Divisors for operators with leading derivative ∂_{xxy}.

Proposition 2.4. *Let an operator*

$$L \equiv \partial_{xxy} + A_1 \partial_{xyy} + A_2 \partial_{xx} + A_3 \partial_{xy} + A_4 \partial_{yy} + A_5 \partial_x + A_6 \partial_y + A_7 \quad (2.17)$$

be given; $\mathfrak{l}_m$ and $\mathfrak{k}_n$ are defined by (2.9) and (2.10) respectively). The following divisors may be constructed.

(i) *If $m \geq 3$ is a natural number and the coefficients $A_1, \ldots, A_6$ satisfy two differential polynomials constructed below in the proof, there exists a Laplace divisor $\mathbb{L}_{x^m}(L) = \langle\!\langle L, \mathfrak{l}_m \rangle\!\rangle$.*

(ii) *If $n \geq 2$ is a natural number, $A_1 = A_4 = 0$ and the coefficients $A_1, \ldots, A_7$ satisfy in addition the differential polynomials constructed below in the proof, there exists a Laplace divisor $\mathbb{L}_{y^n}(L) = \langle\!\langle L, \mathfrak{k}_n \rangle\!\rangle$.*

(iii) *If there are two Laplace divisors $\mathbb{L}_{x^m}(L)$ and $\mathbb{L}_{y^n}(L)$, the operator L is completely reducible; then $\langle L \rangle = Lclm\big(\mathbb{L}_{x^m}(L), \mathbb{L}_{y^n}(L)\big)$.*

Proof. It is given for case (i). If the operator L is derived k times w.r.t. x for $0 \leq k \leq m-2$, and in each step the reductum is reduced w.r.t. (2.17), the operator

$$\partial_{x^{k+2}y} + \sum_{j=0}^{k+2} P_{j,k} \partial_{x^j} + Q_k \partial_{xy} + R_k \partial_y \quad (2.18)$$

is obtained; $P_{j,k}$, Q_k and R_k are differential polynomials in the coefficients $A_1, \ldots, A_7$. There is no reduction w.r.t. (2.9) possible. For $k \leq 0 \leq m-3$, there is no reduction w.r.t. (2.9) possible either. However, if $k = m-2$ the highest derivative w.r.t. x in the reductum may be eliminated with the help of (2.9); the result is

$$\partial_{x^m y} + \sum_{j=0}^{m} P_{j,m-2}\partial_{x^j} + Q_{m-2}\partial_{xy} + R_{m-2}\partial_y$$

$$= \partial_{x^m y} + P_{m,m-2}\partial_{x^m} + \sum_{j=0}^{m-1} P_{j,m-2}\partial_{x^j} + Q_{m-2}\partial_{xy} + R_{m-2}\partial_y$$

$$= \partial_{x^m y} + \sum_{j=0}^{m-1}(P_{j,m-2} - P_{m,m-2}a_{m-j})\partial_{x^j} + Q_{m-2}\partial_{xy} + R_{m-2}\partial_y.$$
(2.19)

Deriving $\mathfrak{l}_m$ as defined in (2.9) once w.r.t. y and assuming $a_0 = 1$ yields

$$\left(\sum_{j=0}^{m} a_{m-j}\partial_{x^j} \right)_y = \sum_{j=0}^{m} a_{m-j}\partial_{x^j y} + \sum_{j=0}^{m-1} a_{m-j,y}\partial_{x^j}$$

$$= \partial_{x^m y} + \sum_{j=2}^{m-1} a_{m-j}\partial_{x^j y} + a_{m-1}\partial_{xy} + a_m\partial_y + \sum_{j=0}^{m-1} a_{m-j,y}\partial_{x^j}.$$

The terms in the first sum may be reduced w.r.t. (2.18) with the result

$$\partial_{x^m y} - \sum_{j=2}^{m-1} a_{m-j}\left(\sum_{i=0}^{j} P_{i,j-2}\partial_{x^i} + Q_{j-2}\partial_{xy} + R_{j-2}\partial_y \right)$$

$$+ a_{m-1}\partial_{xy} + a_m\partial_y + \sum_{j=0}^{m-1} a_{m-j,y}\partial_{x^j}. \qquad (2.20)$$

Equating the coefficients of this expression and the expression in the last line of (2.19) leads to the following system of equations.

$$a_{m-i,y} - \sum_{j=max(i,2)}^{m-1} a_{m-j} P_{i,j-2} + a_{m-i} P_{m,m-2} - P_{i,m-2} = 0, \qquad (2.21)$$

$$a_{m-1} - \sum_{j=2}^{m-1} a_{m-j} Q_{j-2} - Q_{m-2} = 0,$$

$$a_m - \sum_{j=2}^{m-1} a_{m-j} R_{j-2} - R_{m-2} = 0.$$

The last two equations may be used to express a_m and a_{m-1} in terms of the variables with lower indices. Substituting these values into the system (2.21), and autoreducing it leads to a linear algebraic system comprising m equations for $a_1, a_2, \ldots, a_{m-2}$; they are rational in the coefficients of L. In order for a nontrivial solution to exist, two differential polynomials in the coefficients must be satisfied.

The calculations for case (ii) are similar and are therefore omitted. Case (iii) follows by the same reasoning as in Proposition 2.3. $\square$

For later applications the syzygies for the above Laplace divisors are given next; the proof is similar as for Corollary 2.3 and is therefore omitted.

Corollary 2.4. *Let L be defined by*

$$L \equiv \partial_{xxy} + A_1\partial_{xyy} + A_2\partial_{xx} + A_3\partial_{xy} + A_4\partial_{yy} + A_5\partial_x + A_6\partial_y + A_7$$

and allow a Laplace divisor $\mathbb{L}_{x^m}(L) = \langle\!\langle L, \mathfrak{l}_m \rangle\!\rangle$. *For* $m \leq 4$ *there are the following syzygies.*

$$m = 3 : L_x + (A_3 - a_2)L = \mathfrak{l}_{3,y} + A_2\mathfrak{l}_3,$$
$$m = 4 : L_{xx} - A_1L_{xy} + (A_1A_2 - A_3 + a_3)L_x + A_1^2L_{yy}$$
$$-(2A_{1,x} - A_{1,y}A_1 + 2A_1^2A_2 - 2A_1A_3 + A_1a_3 + A_4)L_y$$
$$+(2A_{1,x}A_2 - A_{1,y}A_1A_2 + A_{2,x}A_1 - 2A_{2,y}A_1^2 - 2A_{3,x} + A_{3,y}A_1$$
$$+2A_1^2A_2^2 - 3A_1A_2A_3 + A_1A_2a_3 + A_1A_5 + A_2A_4$$
$$+A_3^2 - A_3a_3 - A_6 + a_2)L = \mathfrak{l}_{4,y} + A_2\mathfrak{l}_4.$$

For Laplace divisors $\mathbb{L}_{y^n}(L) = \langle\!\langle L, \mathfrak{k}_n \rangle\!\rangle$ *and* $n \leq 3$ *the syzygies are as follows.*

$$n = 2 : L_y - (A_2 - b_1)L = \mathfrak{k}_{2,xx} + A_1\mathfrak{k}_{2,xy}$$
$$+(A_{1,y} - A_1A_2 + A_3)\mathfrak{k}_{2,x} + A_4\mathfrak{k}_{2,y} + (A_{4,y} - A_2A_4 + A_6)\mathfrak{k}_2,$$
$$n = 3 : L_{yy} - (A_2 - b_2)L_y - (2A_{2,y} - A_2^2 + A_2b_2 - b_1)L = \mathfrak{k}_{3,xx} + A_1\mathfrak{k}_{3,xy}$$
$$+(2A_{1,y} - A_1A_2 + A_3)\mathfrak{k}_{3,x} + A_4\mathfrak{k}_{3,y} + (2A_{4,y} - b_{2,x}A_1 - A_2A_4 + A_6)\mathfrak{k}_3.$$

Finally, Laplace divisors for an operator with leading derivative ∂_{xyy} are considered.

Proposition 2.5. *Let an operator*

$$L \equiv \partial_{xyy} + A_1\partial_{xx} + A_2\partial_{xy} + A_3\partial_{yy} + A_4\partial_x + A_5\partial_y + A_6 \qquad (2.22)$$

be given. The following divisors may be constructed.

(i)   *If* $n \geq 3$ *is a natural number and the coefficients* $A_1, \ldots, A_6$ *satisfy the differential polynomials constructed below in the proof, there exists a Laplace divisor* $\mathbb{L}_{y^n}(L) = \langle\!\langle L, \mathfrak{k}_n \rangle\!\rangle$.

(ii)   *If $m \geq 2$ is a natural number and the coefficients $A_1, \ldots A_6$ satisfy the
        differential polynomials constructed below in the proof, there exists a Laplace
        divisor $\mathbb{L}_{x^m}(L) = \langle\!\langle L, \mathfrak{l}_m \rangle\!\rangle$.*
(iii)   *If there are two Laplace divisors $\mathbb{L}_{x^m}(L)$ and $\mathbb{L}_{y^n}(L)$, the operator L is
        completely reducible; there holds $\langle L \rangle = Lclm(\mathbb{L}_{x^m}(L), \mathbb{L}_{y^n}(L))$.*

Proof. At first case (i) is considered. If (2.22) is derived repeatedly w.r.t. y, and the
reductum is reduced in each step w.r.t. (2.22), $n - 3$ expressions of the form

$$\partial_{xy^k} + R_{k,1}\partial_{xy} + R_{k,0}\partial_x + P_{k,k}\partial_{y^k} + P_{k,k-1}\partial_{y^{k-1}} + \ldots + P_{k,0} \qquad (2.23)$$

for $3 \leq k \leq n - 1$ are obtained. All coefficients $R_{k,l}$ and $P_{i,j}$ are differential
polynomials in the ring $\mathbb{Q}\{A_1, \ldots, A_6\}$. There is no reduction w.r.t. (2.10) possible.
Deriving the last expression once more w.r.t. y and reducing the reductum w.r.t. both
(2.22) and (2.10) yields

$$\partial_{xy^n} + R_{n,1}\partial_{xy} + R_{n,0}\partial_x + (P_{n,n-1} - P_{n-1,n-1}b_1)\partial_{y^{n-1}}$$
$$+(P_{n,n-2} - P_{n-1,n-1}b_2)\partial_{y^{n-2}} + \ldots \qquad (2.24)$$
$$+(P_{n,1} - P_{n-1,n-1}b_{n-1})\partial_y + P_{n,0} - P_{n-1,n-1}b_n.$$

In the first derivative of (2.10) w.r.t. x

$$\partial_{xy^n} + b_{1,x}\partial_{y^{n-1}} + b_{2,x}\partial_{y^{n-2}} + \ldots + b_{n-1,x}\partial_y + b_{n,x}$$
$$+b_1\partial_{xy^{n-1}} + b_2\partial_{xy^{n-2}} + \ldots + b_{n-1}\partial_{xy} + b_n\partial_x$$

the terms containing derivatives of the form ∂_{xy^k} may be reduced w.r.t. (2.23) or
(2.22) with the result

$$\partial_{xy^n} + (b_{1,x} - P_{n-1,n-1}b_1)\partial_{y^{n-1}}$$
$$+(b_{2,x} - P_{n-1,n-2}b_1 - P_{n-2,n-2}b_2)\partial_{y^{n-2}}$$
$$\vdots \qquad\qquad \vdots$$
$$+(b_{n-1,x} - P_{n-1,1}b_1 - P_{n-2,1}b_2 \ldots - A_4 b_{n-2})\partial_y$$
$$+b_{n,x} - P_{n-1,0}b_1 - P_{n-2,0}b_2 \ldots - A_5 b_{n-2}$$
$$+(b_{n-1} - R_{n-1,1}b_1 - R_{n-2,1}b_2 - \ldots - A_1 b_{n-2})\partial_{xy}$$
$$+(b_n - R_{n-1,0}b_1 - R_{n-2,0}b_2 - \ldots - A_3 b_{n-2})\partial_x$$

where $Q_k, Q_{i,j} \in \mathbb{Q}\{A_1, \ldots, A_6\}$. If this expression is subtracted from (2.24), the
coefficients of the derivatives must vanish in order that (2.22) and (2.10) form a

Janet basis. The resulting system of equations is

$$b_{1,x} + (P_{n-1,n-2} - P_{n-1,n-1})b_1 - P_{n,n-1} = 0,$$

$$b_{2,x} - P_{n-1,n-2}b_1 + (P_{n-1,n-1} - P_{n-2,n-2})b_2 - P_{n,n-2} = 0,$$

$$\vdots \qquad \qquad \vdots$$

$$b_{n-1,x} - P_{n-1,1}b_1 - \ldots - A_4 b_{n-2} + P_{n-1,n-1}b_{n-1} - P_{n,1} = 0,$$

$$b_{n,x} - P_{n-1,0}b_1 - P_{n-2,0}b_2 - \ldots - A_5 b_{n-2} + P_{n-1,n-1}b_n - P_{n,0} = 0,$$

$$b_{n-1} - R_{n-1,1}b_1 - R_{n-2,1}b_2 - \ldots - A_1 b_{n-2} - R_{n,1} = 0,$$

$$b_n - R_{n-1,0}b_1 - R_{n-2,0}b_2 - \ldots - A_3 b_{n-2} - R_{n,0}.$$

The last two equations may be solved for b_n and b_{n-1}. Substituting these values into the equations with leading terms $b_{n,x}$ and $b_{n-1,x}$, and eliminating the first derivatives $b_{j,x}$ for $j = 1, \ldots, n - 2$ by means of the preceding equations, they may be solved for b_{n-2} and b_{n-3}. Proceeding in this way, due to the triangular structure, all b_j are obtained in terms of b_i with $i < j$. Backsubstituting these results, all b_k are explicitly known; due to their linear occurrence they are rational in the coefficients of L. Substituting them into the remaining equation, a constraint for the coefficients $A_1, \ldots, A_6$ expressing the condition for the existence of a Janet basis comprising (2.22) and (2.10), is obtained.

For case (ii) expression (2.22) is derived repeatedly w.r.t. x, and the reductum is reduced in each step w.r.t. (2.22). In this way $m - 2$ expressions of the form

$$\partial_{x^k yy} + P_{k,k} \partial_{x^k y} + P_{k,k-1} \partial_{x^{k-1}y} + \ldots + P_{k,2} \partial_{x^2 y}$$

$$+ Q_{k,k} \partial_{x^k} + Q_{k,k-1} \partial_{x^{k-1}} + \ldots + Q_{k,2} \partial_{x^2} \qquad (2.25)$$

$$+ R_{k,1} \partial_{xy} + R_{k,2} \partial_{yy} + R_{k,3} \partial_x + R_{k,4} \partial_y + R_{k,5}$$

for $2 \le k \le m - 1$ may be obtained. All coefficients $P_{i,j}$, $Q_{i,j}$ and $R_{i,j}$ are differential polynomials in the ring $\mathbb{Q}\{A_1, \ldots, A_6\}$. There is no reduction w.r.t. (2.9) possible. Deriving the last expression once more w.r.t. x, and reducing the reductum w.r.t. (2.22), (2.9) and its first derivative w.r.t. y yields a completely reduced expression with leading term $\partial_{x^m yy}$. Similarly, deriving (2.9) twice w.r.t. y, and performing all possible reductions w.r.t. (2.25) leads to another completely reduced expression with leading term $\partial_{x^m yy}$. Equating the coefficients of the lower order terms of these two expressions yields the following equations for the desired coefficients a_k, $1 \le k \le m$.

$$2a_{1,y} + (P_{m,m} - P_{m-1,m-1})a_1 - P_{m,m-1} = 0,$$

$$2a_{2,y} + (P_{m,m} - P_{m-1,m-2})a_1 - P_{m-2,m-2}a_2 - P_{m,m-2} = 0,$$

$$\vdots \qquad\qquad \vdots$$

$$2a_{m-2,y} - P_{m-1,2}a_1 - P_{m-2,2}a_2 - \ldots + (P_{m,m} - P_{2,2})a_{m-2} - P_{m,2} = 0, \quad (2.26)$$

$$2a_{m-1,y} + (P_{m,m} - A_1)a_{m-1} - \ldots - R_{m-2,1}a_2 - R_{m-1,1}a_1 - R_{m,1} = 0,$$

$$2a_{m,y} + P_{m,m}a_m - A_4a_{m-1} - \ldots - R_{m-2,4}a_2 - R_{m-1,4}a_1 - R_{m,4} = 0,$$

$$a_m - A_2a_{m-1} - \ldots - R_{m-2,2}a_2 - R_{m-1,2}a_1 - R_{m,2} = 0.$$

In addition there are m equations of the general structure

$$a_{k,yy} + pa_{k,y} + q_ka_k + q_{k-1}a_{k-1} + \ldots + q_2a_2 + q_1a_1 = 0 \qquad (2.27)$$

for $1 \leq k \leq m$. The system (2.26) has the same form as in case (i), therefore the coefficients a_k may be determined in a similar way, beginning with the highest coefficient a_m, then a_{m-1} and so on until a_1 is reached; due to the linearity of these equations the a_k are rational in the coefficients of L. Substituting them into the remaining equations (2.27) a set of constraints for $A_1, \ldots, A_6$ is obtained. If they are satisfied, (2.22) and (2.9) form a Janet basis. Case (iii) follows by the same reasoning as in Proposition 2.3.   □

The syzygies for the above Laplace divisors are given next; the proof is again omitted because it is similar as for Corollary 2.3.

Corollary 2.5. *Let L be defined by (2.22) and allow a Laplace divisor $\mathbb{L}_{x^m}(L) = \langle\!\langle L, \mathfrak{l}_m \rangle\!\rangle$. For $m \leq 3$ there are the following syzygies.*

$$m = 2 : L_x + (a_1 - A_3)L = \mathfrak{l}_{2,yy} + A_1\mathfrak{l}_{2,x} + A_2\mathfrak{l}_{2,y} + (A_{1,x} - A_1A_3 + A_4)\mathfrak{l}_2,$$

$$m = 3 : L_{xx} + (a_2 - A_3)L_x + (a_1 - a_2A_3 - 2A_{3,x} + A_3^2)L$$

$$= \mathfrak{l}_{3,yy} + A_1\mathfrak{l}_{3,x} + A_2\mathfrak{l}_{3,y} + (2A_{1,x} - A_1A_3 + A_4)\mathfrak{l}_3.$$

For Laplace divisors $\mathbb{L}_{y^n}(L) = \langle\!\langle L, \mathfrak{k}_n \rangle\!\rangle$ and $n \leq 4$ the syzygies are as follows.

$$n = 3 : L_y - (A_2 - b_2)L = \mathfrak{k}_{3,x} + A_3\mathfrak{k}_3,$$

$$n = 4 : L_{yy} - A_1L_x + (b_3 - A_2)L_y$$

$$-(2A_{2,y} + b_3A_2 - b_2 - A_1A_3 - A_2^2 + A_4)L = \mathfrak{k}_{4,x} + A_3\mathfrak{k}_4.$$

2.7 The Ideals $\mathbb{J}_{xxx}$ and $\mathbb{J}_{xxy}$

Among the ideals involving only derivatives of order not higher than three those ideals that occur as generic intersection ideals of first-order operators are of particular importance; they are denoted by

$$\mathbb{J}_{xxx} \equiv \langle \partial_{xxx}, \partial_{xxy} \rangle_{LT} \quad \text{and} \quad \mathbb{J}_{xxy} \equiv \langle \partial_{xxy}, \partial_{xyy} \rangle_{LT}.$$

The subscripts of $\mathbb{J}$ denote the highest derivative occurring in the respective ideal. Both are generated by two third-order operators forming a Janet basis; their differential dimension is $(1, 2)$. For later use some of their properties are investigated next.

Lemma 2.2. *The ideal*

$$\mathbb{J}_{xxx} \equiv \langle L_1 \equiv \partial_{xxx} + p_1 \partial_{xyy} + p_2 \partial_{yyy} + p_3 \partial_{xx} + p_4 \partial_{xy} + p_5 \partial_{yy} + p_6 \partial_x$$
$$+ p_7 \partial_y + p_8,$$
$$L_2 \equiv \partial_{xxy} + q_1 \partial_{xyy} + q_2 \partial_{yyy} + q_3 \partial_{xx} + q_4 \partial_{xy} + q_5 \partial_{yy} + q_6 \partial_x$$
$$+ q_7 \partial_y + q_8 \rangle$$

is coherent if the coefficients of its generators obey the conditions

$$p_1 - q_2 + q_1^2 = 0, \quad p_2 + q_2 q_1 = 0,$$
$$q_{2,y} - q_{1,x} - q_{1,y} q_1 + p_4 - p_3 q_1 - q_5 + 2 q_4 q_1 + q_3 q_2 - 2 q_3 q_1^2 = 0,$$
$$q_{2,x} + q_{1,y} q_2 - p_5 + p_3 q_2 - q_5 q_1 - q_4 q_2 + 2 q_3 q_2 q_1 = 0,$$
$$p_{3,y} - q_{3,x} + q_{3,y} q_1 - q_6 + q_4 q_3 - q_3^2 q_1 = 0,$$
$$p_{4,y} - q_{4,x} + q_{4,y} q_1 + p_6 + p_4 q_3 - p_3 q_4 - q_7 + q_6 q_1 + q_4^2 - q_4 q_3 q_1 = 0,$$
$$p_{5,y} - q_{5,x} + q_{5,y} q_1 + p_7 + p_5 q_3 - p_3 q_5 + q_7 q_1 + q_5 q_4 - q_5 q_3 q_1 = 0,$$
$$p_{6,y} - q_{6,x} + q_{6,y} q_1 + p_6 q_3 - p_3 q_6 - q_8 + q_6 q_4 - q_6 q_3 q_1 = 0,$$
$$p_{7,y} - q_{7,x} + q_{7,y} q_1 + p_8 + p_7 q_3 - p_3 q_7 + q_8 q_1 + q_7 q_4 - q_7 q_3 q_1 = 0,$$
$$p_{8,y} - q_{8,x} + q_{8,y} q_1 + p_8 q_3 - p_3 q_8 + q_8 q_4 - q_8 q_3 q_1 = 0.$$

A grlex term order with $p_i \succ q_j$ for all i and j, $q_i \succ q_j$ and $p_i \succ p_j$ for $i < j$ is applied. There is a single syzygy between the generators.

$$L_{1,y} + q_3 L_1 - L_{2,x} + q_1 L_{2,y} - (p_3 + q_1 q_3 - q_4) L_2 = 0. \tag{2.28}$$

The proof of this lemma is considered in Exercise 2.9. An example of an ideal $\mathbb{J}_{xxx}$ is given next.

Example 2.15. The ideal

$$\langle\!\langle L_1 \equiv \partial_{xxx} - \partial_{xyy} + 2x\partial_{xy} - 3(3x^2 + 1)\partial_x + 4(2x^2 + 1)\partial_y - 8x^3 - 24x,$$
$$L_2 \equiv \partial_{xxy} + \partial_{xyy} + x\partial_{xx} - (x^2 + 1)\partial_x - 4(2x^2 + 1)\partial_y - 8x^3\rangle\!\rangle$$

is generated by a Janet basis. By the results of Chap. 6 it may be shown that both operators L_1 and L_2 do not have any first-order right factor. The single syzygy $L_{1,y} + xL_1 = L_{2,x} + xL_2$ is particularly simple in this case.   □

Lemma 2.3. *The ideal*

$$\mathbb{J}_{xxy} \equiv \langle K_1 \equiv \partial_{xxy} + p_1\partial_{yyy} + p_2\partial_{xx} + p_3\partial_{xy} + p_4\partial_{yy} + p_5\partial_x + p_6\partial_y + p_7,$$
$$K_2 \equiv \partial_{xyy} + q_1\partial_{yyy} + q_2\partial_{xx} + q_3\partial_{xy} + q_4\partial_{yy} + q_5\partial_x + q_6\partial_y + q_7\rangle$$

is coherent if the coefficients of its generators obey the conditions

$$q_2 = 0, \quad p_1 + q_1^2 = 0,$$
$$p_{7,y} - q_{7,x} + q_{7,y}q_1 - p_2p_7 - p_3q_7 - p_7q_1q_2 + p_7q_3 - q_1q_3q_7 + q_4q_7 = 0,$$
$$p_{6,y} - q_{6,x} + q_{6,y}q_1 - p_2p_6 - p_3q_6 - p_6q_1q_2$$
$$+ p_6q_3 + p_7 - q_1q_3q_6 + q_1q_7 + q_4q_6 = 0,$$
$$p_{5,y} - q_{5,x} + q_{5,y}q_1 - p_2p_5 - p_3q_5 - p_5q_1q_2$$
$$+ p_5q_3 - q_1q_3q_5 + q_4q_5 - q_7 = 0,$$
$$p_{4,y} - q_{4,x} + q_{4,y}q_1 - p_2p_4 - p_3q_4 - p_4q_1q_2$$
$$+ p_4q_3 + p_6 - q_1q_3q_4 + q_1q_6 + q_4^2 = 0,$$
$$p_{3,y} - q_{3,x} + q_{3,y}q_1 - p_2p_3 - p_3q_1q_2 + p_5 - q_1q_3^2 + q_1q_5 + q_3q_4 - q_6 = 0,$$
$$p_{2,y} - q_{2,x} + q_{2,y}q_1 - p_2^2 - p_2q_1q_2$$
$$+ p_2q_3 - p_3q_2 - q_1q_2q_3 + q_2q_4 - q_5 = 0,$$
$$p_{1,y} - q_{1,x} + q_{1,y}q_1 - p_1p_2 - p_1q_1q_2$$
$$+ p_1q_3 - p_3q_1 + p_4 - q_1^2q_3 + 2q_1q_4 = 0.$$

A grlex term order with $p_i \succ q_j$ for all i and j, $q_i \succ q_j$ and $p_i \succ p_j$ for $i < j$ is applied. There is a single syzygy between the generators.

$$K_{1,y} - (p_2 + q_1q_2 - q_3)K_1 - K_{2,x} + q_1K_{2,y} - (p_3 + q_1q_3 - q_4)K_2 = 0. \quad (2.29)$$

The proof is also considered in Exercise 2.9.

2.8 Lattice Structure of Ideals in $\mathbb{Q}(x, y)[\partial_x, \partial_y]$

In any ring, commutative or not, its ideals form a lattice if the join operation is defined as the sum of ideals, and the meet operation as its intersection. In order to understand the structure of this lattice, these two operations have to be studied in detail. The basics of lattice theory required for this purpose may be found in the books by Grätzer [19] or Davey and Priestley [15].

The first result deals with a special case that guarantees the existence of a principal intersection ideal of first order operators.

Proposition 2.6. *Let L be a partial differential operator in x and y with leading term ∂_{x^n}, and let $l_i \equiv \partial_x + a_i \partial_y + b_i$, $i = 1, \ldots, n$, $a_i \neq a_j$ for $i \neq j$, be n right divisors of L. Then the intersection ideal generated by the l_i is principal and is generated by L.*

Proof. Let $I_i = \langle l_i \rangle$ for $1 \leq i \leq n$ and $I = I_1 \cap \ldots \cap I_n$ be the intersection ideal. For any $P \in I$, $symb(P)$ is divided by $\prod_{1 \leq i \leq n}(\partial_x + a_i \partial_y)$, considered as algebraic polynomial in ∂_x and ∂_y; therefore $ord_x(P) \geq n$. On the other hand, according to Sit's relation (2.8) on page 29, for the typical differential dimension there follows $dim(I) \leq n$. Hence if $I = \langle L \rangle$ is principal then $ord_x(L) = dim(I) = n$. Conversely let $P \in I$ and divide P by L with remainder, i.e. $P = QL + R$. Then $ord_x(R) < n$, therefore $R = 0$. Thus $I = \langle L \rangle$. $\qquad\square$

The intersection ideals generated by two or three first-order operators in the plane are described in detail now.

Theorem 2.2. *Let the ideals $I_i = \langle \partial_x + a_i \partial_y + b_i \rangle$ for $i = 1, 2$ with $I_1 \neq I_2$ be given. Both ideals have differential dimension $(1, 1)$. There are three different cases for their intersection $I_1 \cap I_2$, all are of differential dimension $(1, 2)$.*

(i) If $a_1 \neq a_2$ and $\left(\dfrac{b_1 - b_2}{a_1 - a_2} \right)_x = \left(\dfrac{a_1 b_2 - a_2 b_1}{a_1 - a_2} \right)_y$ then

$$I_1 \cap I_2 = \langle \partial_{xx} \rangle_{LT} \text{ and } I_1 + I_2 = \left\langle \partial_x + \dfrac{a_1 b_2 - a_2 b_1}{a_1 - a_2}, \partial_y + \dfrac{b_1 - b_2}{a_1 - a_2} \right\rangle.$$

(ii) If $a_1 \neq a_2$ and $\left(\dfrac{b_1 - b_2}{a_1 - a_2} \right)_x \neq \left(\dfrac{a_1 b_2 - a_2 b_1}{a_1 - a_2} \right)_y$ then

$$I_1 \cap I_2 = \mathbb{J}_{xxx} \text{ and } I_1 + I_2 = \langle 1 \rangle.$$

(iii) If $a_1 = a_2 = a$ and $b_1 \neq b_2$ then

$$I_1 \cap I_2 = \langle \partial_{xx} \rangle_{LT} \text{ and } I_1 + I_2 = \langle 1 \rangle.$$

Case (ii) is the generic case for the intersection of two ideals I_1 and I_2.

Proof. The proof follows closely Grigoriev and Schwarz [22]. In accordance with Cox, Little and O'Shea [13], Theorem 11 on page 186, an auxiliary parameter u is introduced and the operators $u(\partial_x + a_1\partial_y + b_1)$ and $(1 - u)(\partial_x + a_2\partial_y + b_2)$ are considered. In order to compute generators for the intersection ideal, a Janet basis with u as the highest variable has to be generated. To this end, computationally it is more convenient to find the Janet basis with respect to the differential indeterminate z and a new indeterminate $w = uz$ with $w \succ z$ in a lexicographic term ordering. The intersection ideal is obtained from the expressions not involving w; the sum ideal is obtained by substituting $z = 0$. This yields the differential polynomials

$$w_x + a_1 w_y + b_1 w \quad \text{and} \quad w_x + a_2 w_y + b_2 w - z_x - a_2 z_y - b_2 z. \tag{2.30}$$

If $a_1 \neq a_2$ autoreduction leads to

$$w_x + \frac{a_1 b_2 - a_2 b_1}{a_1 - a_2} w - \frac{a_1}{a_1 - a_2}(z_x + a_2 z_y + b_2 z),$$
$$\tag{2.31}$$
$$w_y + \frac{b_1 - b_2}{a_1 - a_2} w + \frac{1}{a_1 - a_2}(z_x + a_2 z_y + b_2 z).$$

Defining $U \equiv z_x + a_2 z_y + b_2 z$, the integrability condition between these two elements has the form

$$\left[\left(\frac{a_1 b_2 - a_2 b_1}{a_1 - a_2}\right)_y - \left(\frac{b_1 - b_2}{a_1 - a_2}\right)_x\right]w - \frac{1}{a_1 - a_2}U_x - \frac{a_1}{a_1 - a_2}U_y$$
$$- \left[\left(\frac{1}{a_1 - a_2}\right)_x + \left(\frac{a_1}{a_1 - a_2}\right)_y + \frac{b_1}{a_1 - a_2}\right]U = 0.$$

If the coefficient of w vanishes, the remaining expression has the leading term z_{xx} and is the lowest element of a Janet basis. The sum ideal is obtained from (2.31). This is case (i).

If the coefficient of w does not vanish, this expression may be applied to eliminate w in (2.31). It yields two expressions with leading derivatives U_{xxx} and U_{xxy} respectively; they correspond to an intersection ideal $\mathbb{J}_{xxx}$. The sum ideal is trivial. This is case (ii).

Finally, if $a_1 = a_2 = a$, autoreduction of (2.30) yields two expressions of the type $w + o(z_x)$ and $o(z_{xx})$ respectively; they correspond to an intersection ideal $\langle \partial_{xx} \rangle_{LT}$ and a trivial sum ideal. This is case (iii).

Case (ii) is the generic case because it does not involve any constraints for the coefficients of the generators of I_1 and I_2. $\square$

The highest coefficients of the generators of the intersection ideal in case (ii) of the above theorem may be expressed explicitly in terms of the coefficients of the two first-order operators of the argument of the *Lclm* as shown in Exercise 2.10; the expressions for the lower coefficients are too voluminous to be given explicitly. However, if an ideal is a priori known to be principal its single generator may be given explicitly in terms of the coefficients of its factors or the arguments of the intersection as shown next.

Corollary 2.6. *Let the second-order operator*

$$L \equiv \partial_{xx} + A_1 \partial_{xy} + A_2 \partial_{yy} + A_3 \partial_x + A_4 \partial_y + A_5$$

be given.

(*i*) *Let L have the representation* $L = (\partial_x + a_2 \partial_y + b_2)(\partial_x + a_1 \partial_y + b_1)$. *Then*

$$L = \partial_{xx} + (a_1 + a_2)\partial_{xy} + a_1 a_2 \partial_{yy} + (b_1 + b_2)\partial_x$$
$$+ (a_{1,x} + a_{1,y}a_2 + a_1 b_2 + a_2 b_1)\partial_y + b_{1,x} + b_{1,y}a_2 + b_1 b_2.$$

$$(2.32)$$

(*ii*) *Let L have the representation* $L = Lclm(\partial_x + a_2\partial_y + b_2, \partial_x + a_1\partial_y + b_1)$.
If $a_1 \neq a_2$, *then*

$$L = \partial_{xx} + (a_1 + a_2)\partial_{xy} + a_1 a_2 \partial_{yy}$$
$$+ \left(b_1 + b_2 - \frac{1}{a_1 - a_2}(a_{1,x} - a_{2,x} + a_{1,y}a_2 - a_{2,y}a_1)\right)\partial_x$$
$$+ \left(a_1 b_2 + a_2 b_1 - \frac{1}{a_1 - a_2}(a_{1,x}a_2 - a_{2,x}a_1 + a_{1,y}a_2^2 - a_{2,y}a_1^2)\right)\partial_y$$
$$+ b_1 b_2 + b_{2,x} + a_1 b_{2,y} - \frac{b_2}{a_1 - a_2}(a_{1,x} - a_{2,x} + a_{1,y}a_2 - a_{2,y}a_1).$$

$$(2.33)$$

If $a_1 = a_2 = a$ *and* $b_1 \neq b_2$ *then*

$$L = \partial_{xx} + 2a\partial_{xy} + a^2\partial_{yy} + \left(b_1 + b_2 - \frac{(b_1 - b_2)_x}{b_1 - b_2} - a\frac{(b_1 - b_2)_y}{b_1 - b_2}\right)\partial_x$$
$$+ \left(a(b_1 + b_2) + a_x + aa_y - a\frac{(b_1 - b_2)_x}{b_1 - b_2} - a^2\frac{(b_1 - b_2)_y}{b_1 - b_2}\right)\partial_y$$
$$+ b_1 b_2 - \frac{1}{b_1 - b_2}(b_{1,x}b_2 - b_{2,x}b_1 - a(b_{1,y}b_2 - b_{2,y}b_1)).$$

$$(2.34)$$

Proof. In case (*i*), the representation of *L* is obtained by multiplication of the two first-order operators. In case (*ii*), reduction of *L* w.r.t. both first-order operators yields expressions in ∂_{yy}, ∂_y and a term without derivative. Its coefficients form a system of six linear algebraic equations for the coefficients of *L*. They may be solved for $A_1, \ldots, A_5$ with the given result if a constraint for a_1, b_1, a_2 and b_2 is satisfied. It is identical to the condition of case (*i*) in Theorem 2.2 assuring the existence of the principal intersection. □

This proof is simpler than that of the preceding theorem because the principality of the intersection ideal is part of the assumption.

There is a relation between the commutativity and the structure of the intersection ideal of two operators as shown next.

Lemma 2.4. *Two commuting operators* $l_1 \equiv \partial_x + a_1 \partial_y + b_1$ *and* $l_2 \equiv \partial_x + a_2 \partial_y + b_2$ *with* $a_1 \neq a_2$ *have a principal intersection ideal.*

Proof. The assertion is obvious from the representation

$$\left(\frac{b_1 - b_2}{a_1 - a_2} \right)_x - \left(\frac{a_1 b_2 - a_2 b_1}{a_1 - a_2} \right)_y = \left((a_1 - a_2)_x + a_{1,y} a_2 - a_{2,y} a_1 \right) \frac{b_1 - b_2}{(a_1 - a_2)^2}$$
$$+ \left((b_1 - b_2)_x + b_{1,y} a_2 - b_{2,y} a_1 \right) \frac{1}{a_1 - a_2}.$$

The left hand side is the condition of case (i) of Theorem 2.2. The coefficients of the fractions at the right hand side are just the conditions for commutativity given in Lemma 2.1.   □

Example 2.16. Consider the two ideals

$$I_1 = \langle \partial_x + 1 \rangle \text{ and } I_2 = \langle \partial_x + (y+1)\partial_y \rangle,$$

both of differential dimension $(1, 1)$. The condition for case (i) of Theorem 2.2 is satisfied. Hence

$$Lclm(I_1, I_2) = \langle\!\langle \partial_{xx} + (y+1)\partial_{xy} + \partial_x + (y+1)\partial_y \rangle\!\rangle,$$

$$Gcrd(I_1, I_2) = \langle\!\langle \partial_x + 1, \partial_y - \frac{1}{y+1} \rangle\!\rangle;$$

their differential dimension is $(1, 2)$ and $(0, 1)$ respectively.   □

Example 2.17. The two ideals $I_1 = \langle \partial_x + 1 \rangle$ and $I_2 = \langle \partial_x + x\partial_y \rangle$, both of differential dimension $(1, 1)$, do not satisfy the condition of case (i) of Theorem 2.2; furthermore $a_1 \neq a_2$. Therefore by case (ii) the intersection ideal is

$$Lclm(I_1, I_2) = \langle\!\langle \partial_{xxx} - x^2\partial_{xyy} + 3\partial_{xx} + (2x+3)\partial_{xy} - x^2\partial_{yy} + 2\partial_x + (2x+3)\partial_y,$$

$$\partial_{xxy} + x\partial_{xyy} - \tfrac{1}{x}\partial_{xy} + x\partial_{yy} - \tfrac{1}{x}\partial_x - \left(1 + \tfrac{1}{x}\right)\partial_y \rangle\!\rangle$$

of differential dimension $(1, 2)$; $Gcrd(I_1, I_2) = \langle 1 \rangle$.   □

The above Theorem 2.2 does not cover the case that the two operators generating I_1 and I_2 have different leading derivatives; it is considered next.

Theorem 2.3. *Let the ideals* $I_1 = \langle \partial_x + a_1 \partial_y + b_1 \rangle$ *and* $I_2 = \langle \partial_y + b_2 \rangle$ *be given,* $I_1 \neq I_2$. *There are two different cases for their intersection* $I_1 \cap I_2$.

(i) *If* $(b_1 - a_1 b_2)_y = b_{2,x}$ *then*

$$I_1 \cap I_2 = \langle \partial_{xy} \rangle_{LT} \text{ and } I_1 + I_2 = \langle \partial_x + b_1 - a_1 b_2, \partial_y + b_2 \rangle.$$

(ii) If the preceding case does not apply then

$$I_1 \cap I_2 = \mathbb{J}_{xxy} \text{ and } I_1 + I_2 = \langle 1 \rangle.$$

Proof. By a similar reasoning as in the preceding theorem, the differential polynomials $w_x + a_1 w_y + b_1 w$ and $w_y + b_2 w - z_y - b_2 z$ are obtained. Autoreduction yields

$$w_x + (b_1 - a_1 b_2)w + a_1 z_y + a_1 b_2 z, \quad w_y + b_2 w - z_y - b_2 z. \tag{2.35}$$

In order to make this into a Janet basis in a *lex* term order with $w \succ z$ and $x \succ y$, the single integrability condition has to be satisfied. Upon reduction w.r.t. (2.35) it assumes the form $[(b_1 - a_1 b_2)_y - b_{2,x}]w + o(z_{xy})$. If the coefficient of w vanishes, an expression with leading term z_{xy} remains; it corresponds to a principal intersection ideal. This is case (i).

If the coefficient of w does not vanish, this expression has to be applied to eliminate w from (2.35). The resulting polynomials with leading terms z_{xxy} and z_{xyy} correspond to the ideal in case (ii). □

The explicit expression for the generator of the principal intersection ideal of case (i) of Theorem 2.3 is determined in Exercise 2.11. In Exercise 2.12 the highest coefficients of the generators of the intersection ideal in case (ii) are determined.

All intersection ideals obtained in the above theorems are either principal, or they are generated by two operators in accordance with Stafford's theorem [67]. Furthermore, by construction the two generators form a Janet basis.

The above Theorem 2.2 and 2.3 are the key for understanding the decompositions of operators in the plane; in particular this is true for Blumberg's example that is considered later in Example 6.9 on page 128.

Intersecting three principal ideals generated by first order operators is much more involved as the next result shows.

Theorem 2.4. *Let the ideals $I_i = \langle \partial_x + a_i \partial_y + b_i \rangle$, $i = 1, 2, 3$ be given with $I_i \neq I_j$ for $i \neq j$. There are four different cases for their intersection $I_1 \cap I_2 \cap I_3$.*

(i) *Separable case $a_1 \neq a_2 \neq a_3$. If the constraints (2.41), (2.46) and (2.47) given below are satisfied the intersection is $I_1 \cap I_2 \cap I_3 = \langle \partial_{xxx} \rangle_{LT}$.*

    (a) *If in addition $\dfrac{b_1 - b_2}{a_1 - a_2} = \dfrac{b_2 - b_3}{a_2 - a_3}$, there is a nontrivial sum ideal*

$$I_1 + I_2 + I_3 = \langle \partial_x + \frac{a_2 b_3 - a_3 b_2}{a_2 - a_3}, \partial_y + \frac{b_2 - b_3}{a_2 - a_3} \rangle.$$

    (b) *If the condition of the preceding case does not apply the sum ideal is trivial, i.e. $I_1 + I_2 + I_3 = \langle 1 \rangle$.*

(ii)   *Double root* $a_1 = a_2 = a \neq a_3$, $b_1 \neq b_2$. *If the two constraints (2.41) and (2.50) given below are satisfied, it follows that* $I_1 \cap I_2 \cap I_3 = \langle \partial_{xxx} \rangle_{LT}$ *and* $I_1 + I_2 = \langle 1 \rangle$.

(iii) *Triple root* $a_1 = a_2 = a_3 = a$, $b_i \neq b_j$ *for* $i \neq j$. *If condition (2.52) is satisfied, the intersection ideal is* $\langle \partial_{xx} \rangle_{LT}$. *If the latter condition is not satisfied, the intersection ideal is* $\langle \partial_{xxx} \rangle_{LT}$. *In either case* $I_1 + I_2 + I_3 = \langle 1 \rangle$.

(iv)   *If the preceding three cases do not apply, the intersection ideal is generated by two operators. It may be* $\langle \partial_{xxxx}, \partial_{xxxy} \rangle_{LT}$, $\langle \partial_{xxxxy}, \partial_{xxxyy} \rangle_{LT}$ *or* $\langle \partial_{xxxxyy}, \partial_{xxxyyy} \rangle_{LT}$.

All intersection ideals have differential dimension $(1,3)$ *except the first alternative in case* (iii) *with differential dimension* $(1,2)$.

Proof. By a similar reasoning as in the preceding theorem the following differential polynomials are obtained.

$$w_{2,x} + a_3 w_{2,y} + b_3 w_2 + w_{1,x} + a_3 w_{1,y} + b_3 w_1 - z_x - a_3 z_y - b_3 z,$$
$$w_{2,x} + a_2 w_{2,y} + b_2 w_2, \quad w_{1,x} + a_1 w_{1,y} + b_1 w_1. \tag{2.36}$$

At first $a_1 \neq a_2 \neq a_3$ is assumed. Autoreduction yields the system

$$w_{2,x} + \frac{a_2 b_3 - a_3 b_2}{a_2 - a_3} w_2 - a_2 \frac{a_1 - a_3}{a_2 - a_3} w_{1,y} - a_2 \frac{b_1 - b_3}{a_2 - a_3} w_1 + o(z_x), \tag{2.37}$$

$$w_{2,y} + \frac{b_2 - b_3}{a_2 - a_3} w_2 + \frac{a_1 - a_3}{a_2 - a_3} w_{1,y} + \frac{b_1 - b_3}{a_2 - a_3} w_1 + o(z_x), \tag{2.38}$$

$$w_{1,x} + a_1 w_{1,y} + b_1 w_1. \tag{2.39}$$

It has to be transformed into a Janet basis in lex term ordering with $w_2 \succ w_1 \succ z$ and $x \succ y$. There is a single integrability condition $(2.37)_y - (2.38)_x$. Upon reduction w.r.t. to (2.37), (2.38) and (2.39) the following expression is obtained.

$$\left[\left(\frac{b_2 - b_3}{a_2 - a_3} \right)_x - \left(\frac{a_2 b_3 - a_3 b_2}{a_2 - a_3} \right)_y \right] w_2 - \frac{(a_1 - a_2)(a_1 - a_3)}{a_2 - a_3} w_{1,yy}$$

$$+ \left[\left(\frac{a_1 - a_3}{a_2 - a_3} \right)_x - (a_1 - a_2)_y \frac{a_1 - a_3}{a_2 - a_3} + a_2 \left(\frac{a_1 - a_3}{a_2 - a_3} \right)_y \right.$$

$$\left. - (a_1 - a_2) \frac{b_1 - b_3}{a_2 - a_3} - (b_1 - b_2) \frac{a_1 - a_3}{a_2 - a_3} \right] w_{1,y}$$

$$+ \left[\left(\frac{b_1 - b_3}{a_2 - a_3} \right)_x + \left(a_2 \frac{b_1 - b_3}{a_2 - a_3} \right)_y - \frac{(b_1 - b_2)(b_1 - b_3)}{a_2 - a_3} \right.$$

$$\left. - b_{1,y} \frac{a_1 - a_3}{a_2 - a_3} \right] w_1 + o(z_{xx}). \tag{2.40}$$

At first it is assumed that the coefficient of w_2 vanishes, i.e.

$$\left(\frac{b_2 - b_3}{a_2 - a_3} \right)_x - \left(\frac{a_2 b_3 - a_3 b_2}{a_2 - a_3} \right)_y = 0. \tag{2.41}$$

As a consequence the terms of (2.37) and (2.38) involving w_2 do not change any more. Defining

$$P = \frac{b_1 - b_2}{a_1 - a_2} + \frac{b_1 - b_3}{a_1 - a_3} + \frac{(a_1 - a_2)_y}{a_1 - a_2}$$
$$- \frac{1}{a_1 - a_2} \left(\frac{(a_1 - a_3)_x}{a_1 - a_3} + a_2 \frac{(a_1 - a_3)_y}{a_1 - a_3} - \frac{(a_2 - a_3)_x}{a_2 - a_3} - a_2 \frac{(a_2 - a_3)_y}{a_2 - a_3} \right) \tag{2.42}$$

and

$$Q = \frac{1}{a_1 - a_2} \frac{b_1 - b_3}{a_1 - a_3} \left(\frac{(a_2 - a_3)_x}{a_2 - a_3} + a_2 \frac{(a_2 - a_3)_y}{a_2 - a_3} + b_1 - b_2 - a_{2,y} \right)$$
$$- \frac{1}{a_1 - a_2} \left(\frac{(b_1 - b_3)_x}{a_1 - a_3} + a_2 \frac{(b_1 - b_3)_y}{a_1 - a_3} - b_{1,y} \right) \tag{2.43}$$

and dividing (2.40) by the coefficient of $w_{1,yy}$, the result may be written as

$$w_{1,yy} + P w_{1,y} + Q w_1 + o(z_{xx}). \tag{2.44}$$

Combined with (2.39) it must form a Janet basis for w_1 and z. To this end, the single integrability condition between (2.39) and (2.44) has to be satisfied. If all reductions are performed it assumes the form

$$[P_x + (a_1 P - a_{1,y} - 2b_1)_y] w_{1,y} + [Q_x + a_{1,y} Q - b_{1,y} P + (a_1 Q - b_{1,y})_y] w_1 + o(z_{xxx}). \tag{2.45}$$

The first alternative in order to make this expression into a Janet basis combined with (2.39) and (2.44) is to require that the coefficients of $w_{1,y}$ and w_1 vanish, i.e.

$$P_x + (a_1 P - a_{1,y} - 2b_1)_y = 0 \tag{2.46}$$

and

$$Q_x + a_{1,y} Q - b_{1,y} P + (a_1 Q - b_{1,y})_y = 0. \tag{2.47}$$

The resulting expression has the leading term z_{xxx}. It is the lowest element of a Janet basis of the full system and does not contain w_1 or w_2. Hence, the intersection ideal corresponding to this alternative is principal and has the form $\langle \partial_{xxx} \rangle_{LT}$.

If (2.46) or (2.47) is not satisfied, (2.39), (2.44) and (2.45) combined must be transformed into a Janet basis. In either case, autoreduction yields in several steps a basis the two lowest members of which have leading terms not higher than z_{xxxxy} or z_{xxxyy} respectively. If (2.41) is not satisfied, w_2 may be eliminated from (2.37) and (2.38). The resulting system has leading terms $w_{1,xyy}$, $w_{1,yyy}$ and $w_{1,x}$. Again autoreduction yields after several steps two members with leading terms not higher than z_{xxxxyy} and z_{xxxyyy}. This completes case (i).

If $a_1 = a_2 = a \neq a_3$ and $b_1 \neq b_2$ the coefficient of $w_{1,yy}$ in (2.40) vanishes. If condition (2.41) holds, the relation

$$w_{1,y} + Rw_1 + o(z_{xx}) \tag{2.48}$$

follows with

$$R = -\frac{1}{b_1 - b_2} \left[\left(\frac{b_1 - b_3}{a - a_3} \right)_x + \left(a\frac{b_1 - b_3}{a - a_3} \right)_y - \frac{(b_1 - b_2)(b_1 - b_3)}{a - a_3} \right]. \tag{2.49}$$

The single integrability condition between (2.39) and (2.48) yields after two reductions an expression of the form $[R_x - (b_1 - aR)_y]w_1 + o(z_{xxx})$. If the coefficient of w_1 vanishes, i.e. if

$$R_x - (b_1 - aR)_y = 0, \tag{2.50}$$

an expression with leading term proportional to z_{xxx} remains. It leads to a principal ideal with leading derivative ∂_{xxx}. If it does not vanish, it may be used for eliminating $w_{1,x}$ from (2.39) and $w_{1,y}$ from (2.48). The resulting expressions contain the leading derivatives z_{xxxx} and z_{xxxy} respectively corresponding to the ideal $\langle \partial_{xxxx}, \partial_{xxxy} \rangle_{LT}$.

If condition (2.41) is not valid, w_2 may be eliminated from (2.37) and (2.38) using (2.40). If the result is reduced w.r.t. (2.39) and then autoreduced, two expressions with leading derivatives z_{xxxxy} and z_{xxxyy} are obtained. They correspond to an ideal $\langle \partial_{xxxxy}, \partial_{xxxyy} \rangle_{LT}$. This completes case (ii).

Assume now $a_1 = a_2 = a_3 = a$ and $b_i \neq b_j$ for $i \neq j$ in (2.36). A single autoreduction step yields

$$w_{2,x} + aw_{2,y} + b_2 w_2 + o(z_x), \quad w_{1,x} + aw_{1,y} + b_1 w_1, \quad w_2 + \frac{b_3 - b_1}{b_3 - b_2}w_1 + o(z_x).$$

The last element may be applied to eliminate w_2 from the first one with the result

$$\left[\frac{(b_1 - b_2)(b_1 - b_3)}{b_2 - b_3} - \left(\frac{b_3 - b_1}{b_3 - b_2} \right)_x - a\left(\frac{b_3 - b_1}{b_3 - b_2} \right)_y \right] w_1 + o(z_{xx}). \tag{2.51}$$

If the coefficient of w_1 vanishes the condition

$$\left(\frac{b_3 - b_1}{b_3 - b_2} \right)_x + a\left(\frac{b_3 - b_1}{b_3 - b_2} \right)_y = \frac{(b_1 - b_2)(b_1 - b_3)}{b_2 - b_3} \tag{2.52}$$

follows; the expression (2.51) has a leading term proportional to z_{xx} corresponding to a principal ideal $\langle \partial_{xx} \rangle_{LT}$. If it does not vanish it may be used to eliminate w_1 from the second relation. The result is an expression with leading term z_{xxx} corresponding to a principal ideal $\langle \partial_{xxx} \rangle_{LT}$ without further constraint. This completes case (iii).

If $a_i \neq a_j$ for $i \neq j$, $i, j = 1, 2, 3$, autoreduction of the system $l_i = \partial_x + a_i \partial_y + b_i$, $i = 1, 2, 3$, leads to the system

$$\frac{b_1 - b_2}{a_1 - a_2} - \frac{b_2 - b_3}{a_2 - a_3}, \tag{2.53}$$

$$\partial_x + \frac{a_2 b_3 - a_3 b_2}{a_2 - a_3}, \quad \partial_y + \frac{b_2 - b_3}{a_2 - a_3}. \tag{2.54}$$

In order that $\{l_1, l_2, l_3\}$ form a nontrivial Janet basis, the expression (2.53) must vanish. In addition the integrability condition for the system (2.54) must be satisfied which is identical to (2.41). This yields the $Gcrd$ in case (i). In cases (i) and (iii) at least for one pair of coefficients there holds $a_i = a_j$, $b_i \neq b_j$ which leads to a trivial $Gcrd$. □

These result will be illustrated now by a few examples. The reader is encouraged to reproduce them by using the software provided on the website alltypes.de; as a useful exercise the effect of small variations of the input operators may be studied.

Example 2.18. Let three operators be given by

$$l_1 \equiv \partial_x + \frac{y}{x}\partial_y + \frac{1}{x}, \quad l_2 \equiv \partial_x + \frac{y}{x}\partial_y - \frac{xy + 1}{x^2 y - x}, \quad l_3 \equiv \partial_x + \frac{y}{x}\partial_y - \frac{x^2 + 1}{x^3 - x}.$$

Because $a_1 = a_2 = a_3$ case (iii) of Theorem 2.4 applies. Because the coefficients b_1, b_2 and b_3 satisfy (2.52), the intersection ideal is generated by

$$\partial_{xx} + \frac{2y}{x}\partial_{xy} + \frac{y^2}{x^2}\partial_{yy} + \frac{1}{x}\partial_x + \frac{y}{x^2}\partial_y - \frac{1}{x^2}. \qquad □$$

Example 2.19. Consider the three operators

$$l_1 = \partial_x + 2, \quad l_2 = \partial_x + \partial_y + 1, \quad l_3 = \partial_x + 2\partial_y$$

with $a_i \neq a_j$ for $i \neq j$ and $\frac{b_1 - b_2}{a_1 - a_2} \neq \frac{b_2 - b_3}{a_2 - a_3}$, i.e. subcase (a) of case (i) of Theorem 2.4 applies. Their intersection is

$$Lclm(l_1, l_2, l_3) = \langle \partial_{xxx} + 3\partial_{xxy} + 2\partial_{xyy} + 3\partial_{xx} + 8\partial_{xy} + 4\partial_{yy} + 2\partial_x + 4\partial_y \rangle$$

and their sum ideal $Gcrd(l_1, l_2, l_3) = \langle\!\langle \partial_x + 2, \partial_y - 1 \rangle\!\rangle$. □

Example 2.20. Consider the three operators

$$l_1 = \partial_x + 2, \quad l_2 = \partial_x + \partial_y + 2, \quad l_3 = \partial_x + 2\partial_y$$

with $a_1 \neq a_2 \neq a_3$ and $\dfrac{b_1 - b_2}{a_1 - a_2} = 0 \neq \dfrac{b_2 - b_3}{a_2 - a_3} = -2$, i.e. subcase (b) of case (i) of Theorem 2.4 applies. Their intersection is

$$Lclm(l_1, l_2, l_3) = \langle \partial_{xxx} + 3\partial_{xxy} + 2\partial_{xyy} + 4\partial_{xx} + 10\partial_{xy} + 4\partial_{yy} + 4\partial_x + 8\partial_y \rangle$$

and their sum ideal $Gcrd(l_1, l_2, l_3) = \langle 1 \rangle$.   □

The proof given above shows an additional feature of the intersection of three ideals. According to Theorem 2.2, condition (2.41) means that the ideals I_2 and I_3 have a principal intersection. If it is not satisfied, the subsequent Janet basis calculation leads into a branch which does not allow a principal intersection any more. Because this is true for all possible term orderings, the principality of the intersection of three ideals requires that the pairwise intersections of each pair be principal. However, the reverse is not true as may be seen from the following example.

Example 2.21. Let three principal ideals be given by

$$I_1 \equiv \langle \partial_x + \partial_y + x \rangle, \quad I_2 \equiv \langle \partial_x + x\partial_y + x \rangle \text{ and } I_3 \equiv \langle \partial_x + y\partial_y + x \rangle.$$

Any of its three pairwise intersections $I_i \cap I_j$, $i, j = 1, 2, 3$ is principal.

$$Lclm(I_1, I_2) = \left\langle\!\!\left\langle \partial_{xx} + (x+1)\partial_{xy} + x\partial_{yy} \right.\right.$$
$$\left.\left. +\Big(2x - \frac{1}{x-1}\Big)\partial_x + \Big(x^2 + x - \frac{1}{x-1}\Big)\partial_y + x^2 - \frac{1}{x-1} \right\rangle\!\!\right\rangle,$$

$$Lclm(I_1, I_3) = \left\langle\!\!\left\langle \partial_{xx} + (y+1)\partial_{xy} + y\partial_{yy} \right.\right.$$
$$\left.\left. +\Big(2x - \frac{1}{y-1}\Big)\partial_x + \Big(xy + x - \frac{1}{y-1}\Big)\partial_y + x^2 + 1 - \frac{1}{y-1} \right\rangle\!\!\right\rangle,$$

$$Lclm(I_2, I_3) = \left\langle\!\!\left\langle \partial_{xx} + (x+y)\partial_{xy} + xy\partial_{yy} \right.\right.$$
$$\left.\left. +\Big(2x + \frac{x-y}{y-1}\Big)\partial_x + \Big(x^2 - xy - \frac{x^2 - y}{x-y}\Big)\partial_y + x^2 + \frac{x^2 - y}{x-y} \right\rangle\!\!\right\rangle,$$

yet the intersection of all three ideals is a rather complicated non-principal ideal, it is $I_1 \cap I_2 \cap I_3 = \langle \partial_{xxxx}, \partial_{xxxy} \rangle_{LT}$.   □

The important property shown in this example is formulated as the next corollary.

Corollary 2.7. *In order that the intersection of the ideals $I_i = \langle \partial_x + a_i \partial_y + b_i \rangle$, $i = 1, 2, 3$ be principal it is necessary, but not sufficient, that their pairwise intersections are principal.*

Finally there remains the case of two operators with leading derivative ∂_y in the term ordering $grlex$ and $x \succ y$. The answer is given next.

Theorem 2.5. *Let the three ideals*

$$I_1 \equiv \langle \partial_y + b_1 \rangle, \quad I_2 \equiv \langle \partial_y + b_2 \rangle, \quad I_3 \equiv \langle \partial_x + a_3 \partial_y + b_3 \rangle$$

be given with $I_i \neq I_j$ for $i \neq j$ and $a_3 \neq 0$. There are three different cases for their intersection $I_1 \cap I_2 \cap I_3$. All intersection ideals have differential dimension $(1, 3)$.

(i)   *If $b_{1,x} - (b_3 - a_3 b_1)_y = 0$ and $b_{2,x} - (b_3 - a_3 b_2)_y = 0$ the intersection ideal is principal of the form $\langle \partial_{xyy} \rangle_{LT}$.*

(ii)   *If $b_{1,x} - (b_3 - a_3 b_1)_y = 0$ and $b_{2,x} - (b_3 - a_3 b_2)_y \neq 0$, or $b_{1,x} - (b_3 - a_3 b_1)_y \neq 0$ and $b_{2,x} - (b_3 - a_3 b_2)_y = 0$, the intersection ideal is not principal, it has the form $\langle \partial_{xyyy}, \partial_{xxyy} \rangle_{LT}$.*

(iii) *If $b_{1,x} - (b_3 - a_3 b_1)_y \neq 0$ and $b_{2,x} - (b_3 - a_3 b_2)_y \neq 0$ the intersection ideal is not principal, it has the form $\langle \partial_{xyyyy}, \partial_{xxyyy} \rangle_{LT}$ of differential dimension $(1, 3)$.*

Proof. By a similar reasoning as in the above theorems the following differential polynomials are obtained.

$$w_{1,x} + a_3 w_{1,y} + b_3 w_1 + w_{2,x} + a_3 w_{2,y} + b_3 w_2 - z_x - a_3 z_y - b_3 z,$$

$$w_{1,y} + b_1 w_1, \quad w_{2,y} + b_2 w_2.$$

By assumption the relation $b_1 \neq b_2$ is valid. The *lex* term order with $w_1 \succ w_2 \succ z$ and $x \succ y$ is always applied. Autoreduction yields the system

$$w_{1,x} + (b_3 - a_3 b_1) w_1 + w_{2,x} + (b_3 - a_3 b_2) w_2 - z_x - a_3 z_y - b_3 z,$$
$$w_{1,y} + b_1 w_1, \quad w_{2,y} + b_2 w_2. \tag{2.55}$$

The single integrability condition between the first two members yields after reduction

$$[(b_3 - a_3 b_1)_y - b_{1,x}] w_1 - (b_2 - b_1) w_{2,x}$$
$$+ [(b_3 - a_3 b_2)_y - b_{2,x} - (b_3 - a_3 b_2)(b_2 - b_1)] w_2 \tag{2.56}$$
$$- z_{xy} - b_1 z_x - a_3 z_{yy} - (a_{3,y} + a_3 b_1 + b_3) z_y - (b_{3,y} + b_1 b_3) z.$$

If the leading coefficient vanishes, i.e. if

$$b_{1,x} - (b_3 - a_3 b_1)_y = 0; \tag{2.57}$$

upon autoreduction the full system is

$$w_{1,x} + (b_3 - a_3 b_1) w_1 - \frac{1}{b_2 - b_1} [b_{2,x} - (b_3 - a_3 b_2)_y] w_2$$

$$- \frac{1}{b_2 - b_1} [z_{xy} + b_2 z_x + a_3 z_{yy} + (a_{3,y} + a_3 b_2 + b_3) z_y$$

$$+ (b_{3,y} + b_2 b_3) z], \quad w_{1,y} + b_1 w_1,$$

$$w_{2,x} + \frac{1}{b_2 - b_1} [b_{2,x} - (b_3 - a_3 b_2)_y + (b_3 - a_3 b_2)(b_2 - b_1)] w_2$$

$$+ \frac{1}{b_2 - b_1} [z_{xy} + b_1 z_x + a_3 z_{yy} + (a_{3,y} + a_3 b_1 + b_3) z_y$$

$$+ (b_{3,y} + b_1 b_3) z], \quad w_{2,y} + b_2 w_2.$$

The two integrability conditions for this system are satisfied if

$$b_{2,x} - (b_3 - a_3 b_2)_y = 0. \tag{2.58}$$

In addition a differential polynomial for z with leading derivative z_{xyy} has to be satisfied. This is case (i) yielding a principal intersection ideal. If (2.58) is not satisfied, the integrability condition for the last two equations of (2.8) is an expression of the form $z + o(z_{xyy})$. Reduction w.r.t. it yields a Janet basis the two lowest polynomials of which have leading derivatives z_{xyyy} and z_{xxyy}; this is case (ii). If neither (2.57) or (2.58) are satisfied, the two lowest polynomials of the resulting Janet basis have leading derivatives z_{xyyyy} and z_{xxyy}; this is case (iii). □

Subsequently three examples are given covering the three cases of the above theorem; they show that these alternatives do actually exist. Due to the Janet basis calculations involved the coefficients of the given operators I_1, I_2 and I_3 have to be fairly simple in order to obtain a manageable problem.

Example 2.22. The three operators $\partial_y + x$, $\partial_y + x + \frac{1}{x}$ and $\partial_x + \frac{y}{x} \partial_y + 2y$ have the principal intersection

$$\partial_{xyy} + \left(2x + \frac{1}{x}\right) \partial_{xy} + (x^2 + 1) \partial_x + \frac{y}{x} \partial_{yyy} + \left(4y + \frac{2}{x} + \frac{y}{y^2} \partial_{yy}\right)$$

$$+ \left(5xy + 6 + \frac{3y}{x} + \frac{1}{x^2}\right) \partial_y + 2xy + 4x + 2y + \frac{2}{x}$$

corresponding to case (i) of the above theorem.   □

Example 2.23. The non-vanishing coefficients of the operators $\partial_y + x$, $\partial_y + x + y$ and $\partial_x + \frac{y}{x} \partial_y + 2y$ are $b_1 = x$, $b_2 = x + y$, $b_3 = \frac{y}{x}$ and $a_3 = \frac{1}{x}$. They satisfy (2.57) but not (2.58). Therefore case (ii) of the above theorem applies. Their intersection ideal $\langle \partial_{xyyy}, \partial_{xxyy} \rangle_{LT}$ is not principal; its two generators are too voluminous to be given here.   □

Example 2.24. The coefficients of the operators $\partial_y + \frac{1}{x}$, $\partial_y + \frac{1}{y}$ and $\partial_x + \frac{y}{x}\partial_y + 2y$ satisfy neither (2.57) nor (2.58); thus by case (iii) their intersection ideal is not principal; it is generated by two huge operators with leading derivatives ∂_{xyyy} and ∂_{xxyy}. □

2.9 Exercises

Exercise 2.1. Prove Lemma 2.1 on page 22.

Exercise 2.2. What does the relation (2.8) on page 29 mean for ideals of differential dimension $(0, k)$?

Exercise 2.3. Determine the third-order terms of the generators of the intersection ideal in case (ii) of Theorem 2.2 on page 47 explicitly.

Exercise 2.4. The same problem for the generator of the principal ideal in case (i) of Theorem 2.3.

Exercise 2.5. Solve the equations $l_i z_i = 0$ with l_i, $i = 1, 2, 3$ from Example 2.19 and explain the solution of $Gcrd(l_1, l_2, l_3)z = 0$ in terms of the solutions of these three equations.

Exercise 2.6. Let three operators $l_i \equiv \partial_x + a_i \partial_y + b_i$ with constant coefficients be given, $a_i \neq a_j$ for $i \neq j$. Determine the general expressions for $Lclm(l_1, l_2, l_3)$ and $Gcrd(l_1, l_2, l_3)$ and distinguish subcases (a) and $b)$ of case (i) in Theorem 2.4.

Exercise 2.7. Determine the Hilbert-Kolchin polynomial H_I and the differential dimension d_I for the ideal $I = \langle \partial_{xxxxxx}, \partial_{xxxxyy} \rangle_{LT}$.

Exercise 2.8. Assume a module of differential dimension $(0, 1)$ in $\mathcal{D}^3$ is generated by the four elements $(\partial_x + a_1, a_2, a_3)$, (∂_y, b_2, b_3), $(c, 1, 0)$ and $(d, 0, 1)$. Determine the coherence conditions for its coefficients such that they form a Janet basis of a type $\mathbb{M}_3^{(0,1)}$ module defined on page 33.

Exercise 2.9. Prove Lemma 2.2 and Lemma 2.3 on page 45 and page 46 respectively.

Exercise 2.10. Determine explicit expressions for the coefficients p_1, p_2, q_1 and q_2 of the intersection ideal for case (ii) of Theorem 2.2; the notation is the same as in Lemma 2.2 on page 45.

Exercise 2.11. Determine the generator of the intersection ideal of $\partial_x + a_1 \partial_y + b_1$ and $\partial_y + b_2$ if case (i) of Theorem 2.3 applies.

Exercise 2.12. Determine explicit expressions for the coefficients p_1, p_2, q_1 and q_2 of the intersection ideal for case (ii) of Theorem 2.3; the notation is the same as in Lemma 2.3 on page 45.

Chapter 3
Equations with Finite-Dimensional Solution Space

Abstract In the preceding chapter on page 30 ideals of differential type zero have been introduced. The corresponding systems of pde's are considered now. They have the distinctive property that their general solution does not involve functions depending on one or more arguments, but only a finite number of constants. In other words, its general solution has the structure of a finite-dimensional vector space over constants like in the ordinary case. At first the Loewy decomposition of such systems in two independent variables containing derivatives of order not higher than three are discussed. Subsequently they are applied for finding its solutions.

3.1 Equations of Differential Type Zero

If the operators of a differential dimension $(0, k)$ ideal are applied to a differential indeterminate z, the differential polynomials obtained generate a system of linear pde's with a finite dimensional solution space. In this chapter systems with a solution space of dimension not higher than three are considered corresponding to ideals of differential dimension $(0, k)$ with $k \leq 3$; they have been introduced in Proposition 2.1.

At first the relations (1.4) are generalized. It does not seem to be possible to represent the coefficients of a Janet basis for any type $\mathbb{J}^{(0,k)}$ ideal in a single closed form expression. Rather, the representations for the individual cases have to be distinguished. The simplest one is the Janet basis type $\mathbb{J}^{(0,1)}$ with the corresponding system of pde's $z_x + az = 0$, $z_y + bz = 0$; if z_1 is a solution, there are the obvious representations $a = -\dfrac{z_{1,x}}{z_1}$ and $b = -\dfrac{z_{1,y}}{z_1}$. For the two Janet basis types of differential dimension $(0, 2)$ the answer is given next.

Lemma 3.1. *Let the coherent system $z_{yy} + a_1 z_y + a_2 z = 0$, $z_x + b_1 z_y + b_2 z = 0$ corresponding to Janet basis type $\mathbb{J}_1^{(0,2)}$ be given, and let $\{z_1, z_2\}$ be a basis for its*

F. Schwarz, *Loewy Decomposition of Linear Differential Equations*, Texts & Monographs in Symbolic Computation, DOI 10.1007/978-3-7091-1286-1_3,
© Springer-Verlag/Wien 2012

solution space; define $w \equiv \begin{vmatrix} z_1 & z_2 \\ z_{1,y} & z_{2,y} \end{vmatrix}$. *If* $w \neq 0$, *then*

$$a_1 = -\frac{1}{w}\begin{vmatrix} z_1 & z_2 \\ z_{1,yy} & z_{2,yy} \end{vmatrix}, \quad a_2 = \frac{1}{w}\begin{vmatrix} z_{1,y} & z_{2,y} \\ z_{1,yy} & z_{2,yy} \end{vmatrix},$$

$$b_1 = -\frac{1}{w}\begin{vmatrix} z_1 & z_2 \\ z_{1,x} & z_{2,x} \end{vmatrix}, \quad b_2 = -\frac{1}{w}\begin{vmatrix} z_{1,x} & z_{2,x} \\ z_{1,y} & z_{2,y} \end{vmatrix}.$$

Proof. Substituting the basis elements z_1 and z_2 into the two given equations yields the system $z_{i,yy} + a_1 z_{i,y} + a_2 z_i = 0$ and $z_{i,x} + b_1 z_{i,y} + b_2 z_i = 0, i = 1, 2$, comprising four equations altogether. The two subsystems for the a's and the b's respectively may be algebraically solved with the above result.   □

The same result for Janet basis type $\mathbb{J}_2^{(0,2)}$ is given without proof.

Lemma 3.2. *Let the coherent system* $z_{xx} + a_1 z_x + a_2 z = 0$ *and* $z_y + bz = 0$
corresponding to the Janet basis type $\mathbb{J}_2^{(0,2)}$ *be given, and let* $\{z_1, z_2\}$ *be a basis for
its solution space such that* $\left(\frac{z_1}{z_2}\right)_y = 0$; *define* $w \equiv \begin{vmatrix} z_1 & z_2 \\ z_{1,x} & z_{2,x} \end{vmatrix}$. *Then*

$$a_1 = -\frac{1}{w}\begin{vmatrix} z_1 & z_2 \\ z_{1,xx} & z_{2,xx} \end{vmatrix}, \quad a_2 = \frac{1}{w}\begin{vmatrix} z_{1,x} & z_{2,x} \\ z_{1,xx} & z_{2,xx} \end{vmatrix}, \quad b = -\frac{z_{1,y}}{z_1} = -\frac{z_{2,y}}{z_2}.$$

From the preceding two lemmata it becomes clear why in Proposition 2.1 the two Janet basis types $\mathbb{J}_1^{(0,2)}$ and $\mathbb{J}_2^{(0,2)}$ have been distinguished. The former corresponds to the generic case without constraints for a fundamental system; in the latter case the relation $z_1 = f(x)z_2$ must hold for any two basis elements, with $f(x)$ an undetermined function of the higher independent variable x. Similar constraints apply for Janet bases with higher dimensional solution spaces. The following two examples show this distinguishing feature.

Example 3.1. Let $z_1 = \frac{1}{xy}$ and $z_2 = \frac{1}{x+y}$ be given. Because $\left(\frac{z_1}{z_2}\right)_y = -\frac{1}{y^2} \neq 0$,
according to Lemma 3.1 the coefficients are

$$a_1 = \frac{2}{y}\frac{x+2y}{x+y}, \quad a_2 = \frac{2}{y(x+y)}, \quad b_1 = -\frac{y^2}{x^2}, \quad b_2 = \frac{x-y}{x^2}.$$

In *grlex*, $x \succ y$ term order they generate the Janet basis

$$z_{yy} + \frac{2}{y}\frac{x+2y}{x+y}z_y + \frac{2}{y(x+y)}z = 0, \quad z_x - \frac{y^2}{x^2}z_y + \frac{x-y}{x^2}z = 0.   \square$$

Example 3.2. Let $z_1 = \dfrac{x^2}{x+y}$ and $z_2 = \dfrac{x}{x+y}$ be given. Because now $\left(\dfrac{z_1}{z_2}\right)_y = 0$, Lemma 3.1 does not apply. Rather by Lemma 3.2 the coefficients

$$a_1 = -\frac{2}{x}\frac{y}{x+y}, \quad a_2 = \frac{2}{x^2}\frac{y}{x+y}, \quad b = -\frac{1}{x+y}$$

are obtained. In *grlex*, $x \succ y$ term order they generate the Janet basis

$$z_{xx} - \frac{2y}{x(x+y)}z_x + \frac{2y}{x^2(x+y)}z = 0, \quad z_y - \frac{1}{x+y}z = 0. \qquad \square$$

For higher order systems these relations become increasingly more complicated. In Exercise 3.1 the answer for ideals of type $\mathbb{J}_2^{(0,3)}$ is obtained.

3.2 Loewy Decomposition of Modules $\mathbb{M}^{(0,2)}$

The decompositions of modules in $\mathscr{D}^2$ are needed for the ideal decompositions in the subsequent section. There are altogether five modules of differential dimension $(0, 2)$; its decompositions into first-order components are discussed in detail.

Proposition 3.1. *Let the type* $\mathbb{M}_3^{(0,2)}$ *module*

$$M = \langle (\partial_x + A_1, A_2), (\partial_y + B_1, B_2), (C_1, \partial_x + C_2), (D_1, \partial_y + D_2) \rangle$$

be given where $A_i, B_i, C_i \in \mathbb{Q}(x, y)$ *for all i; it is assumed that the integrability conditions of Proposition 2.2 are satisfied. If it has a first-order divisor, the following cases are distinguished; $r(x, y)$, $r_i(x, y)$, $i = 1, 2$ and $r(x, y, C)$, C a constant, are rational functions of its arguments.*

(i) *The following divisors of type* $\mathbb{M}_1^{(0,1)}$ *may occur.*

   (a) *There is a single divisor*

$$\langle ((1, r(x, y)), (0, \partial_x + C_2 - C_1 r(x, y)), (0, \partial_y + D_2 - D_1 r(x, y))) \rangle;$$

   (b) *There are two divisors as above with $r(x, y)$ replaced by $r_1(x, y)$ and $r_2(x, y)$;*

   (c) *There are divisors as above with $r(x, y)$ replaced by $r(x, y, C)$.*

(ii) *If $C_1 = D_1 = 0$, there is the divisor $\langle (0, 1), (\partial_x + A_1, 0), (\partial_y + B_1, 0) \rangle$ of type* $\mathbb{M}_2^{(0,1)}$.

Proof. Dividing M by $\langle (1, a), (0, \partial_x + b), (0, \partial_y + c) \rangle$ of type $M_1^{(0,1)}$, the condition that this division be exact leads to the following constraints.

$$a_x + C_1 a^2 + (A_1 - C_2)a - A_2 = 0, \quad a_y + D_1 a^2 + (B_1 - D_2)a - B_2 = 0,$$

$$b = C_2 - C_1 a, \quad c = D_2 - D_1 a. \tag{3.1}$$

The rational solutions of the system for a determine the possible divisors, they have been described in Theorem B.1 and lead to the above cases in a straightforward way.

Proceeding similarly with a divisor $\langle (0, 1), (\partial_x + a, 0), (\partial_y + b, 0) \rangle$ of type $M_2^{(0,1)}$ yields the conditions $C_1 = D_1 = 0$, $a = A_1$ and $b = B_1$ of case (ii). $\qquad\square$

This proof shows that the first-order divisors of a type $M^{(0,2)}$ module may be obtained algorithmically. Applying this result its decomposition types are given next.

Theorem 3.1. *The possible Loewy decompositions of a module M of type $M^{(0,2)}$ into first-order components may be described as follows. M_1, M_2 and $M(C)$ are modules of differential type $M_1^{(0,1)}$ or $M_2^{(0,1)}$, C is a parameter; M_3 has type $M_3^{(0,1)}$.*

(i) *For type $M_3^{(0,2)}$ there are three decomposition types.*

$$\mathcal{L}^1_{x,y,x,y} : \quad M = M_3 M_1; \quad \mathcal{L}^2_{x,y,x,y} : \quad M = Lclm(M_2, M_1);$$

$$\mathcal{L}^3_{x,y,x,y} : \quad M = Lclm\,(M(C)).$$

(ii) *For types $M_1^{(0,2)}$ and $M_2^{(0,2)}$ there follows $\mathcal{L}^i_{1,yy,x} = \mathcal{L}^i_{x,y,x,y}$ for $i = 1, 2, 3$.*
(iii) *For types $M_4^{(0,2)}$ and $M_5^{(0,2)}$ there follows $\mathcal{L}^i_{1,xx,y} = \mathcal{L}^i_{x,y,x,y}$ for $i = 1, 2, 3$.*

Proof. The decompositions of a type $M_3^{(0,2)}$ module are a straightforward consequence of the preceding proposition. The decomposition for case (ii) and case (iii) follow from the decompositions of the corresponding ideals as given in Theorem 3.2. $\qquad\square$

Example 3.3. Consider the type $M_3^{(0,2)}$ module

$$M \equiv \left\langle\!\!\left\langle \left(\partial_x + \frac{1}{x}, -\frac{y}{x(x + y)} \right), \left(\partial_y, \frac{1}{x + y} \right), \left(0, \partial_x + \frac{1}{x + y} \right), \right.\right.$$

$$\left.\left. \left(0, \partial_y + \frac{1}{x + y} \right) \right\rangle\!\!\right\rangle. \tag{3.2}$$

Applying the notation of Proposition 3.1 the coefficients are

$$A_1 = \frac{1}{x}, \quad A_2 = -\frac{y}{x(x + y)}, \quad B_1 = C_1 = D_1 = 0, \quad B_2 = C_2 = D_2 = \frac{1}{x + y}.$$

By case (i) of this proposition the system for a is

$$a_x = \frac{y}{x(x+y)}a = -\frac{y}{x(x+y)}, \quad a_y - \frac{1}{x+y}a = \frac{1}{x+y}.$$

There is a single rational solution $a = 1 + \frac{2y}{x}$, it leads to the type $M_1^{(0,1)}$ divisor

$$M_1 \equiv \left\langle\!\left\langle \left(1, 1 + \frac{2y}{x}\right), \left(0, \partial_x + \frac{1}{x+y}\right), \left(0, \partial_y + \frac{1}{x+y}\right) \right\rangle\!\right\rangle.$$

By case (ii) of the same proposition the additional type $M_2^{(0,1)}$ divisor

$$M_2 = \left\langle\!\left\langle (0, 1), \left(\partial_x + \frac{1}{x}, 0\right), (\partial_y, 0) \right\rangle\!\right\rangle$$

follows. Consequently, M is completely reducible and may be represented as $M = Lclm(M_1, M_2)$, i.e. its decomposition type is $\mathscr{L}^2_{x,y,x,y}$. $\qquad\square$

Example 3.4. For the module M of type $M_3^{(0,2)}$ considered in Example 2.13 on page 33 the system (3.1) reads

$$a_x + \frac{4y(x^2+y)}{x^2(x^2-y)}a^2 + \frac{x^2+5y}{x(x^2-y)}a + \frac{1}{x^2-y} = 0, \quad a_y - \frac{2x}{x^2-y}a^2 - \frac{1}{x^2-y}a = 0.$$

Its only rational solution is $a = -\frac{x}{2y}$; there follows $b = 0$, $c = -\frac{1}{y}$. It yields the divisor $M_1 = \left\langle\!\left\langle \left(1, -\frac{x}{2y}\right), (0, \partial_x), \left(0, \partial_y - \frac{1}{y}\right) \right\rangle\!\right\rangle$ and the type $\mathscr{L}^1_{x,y,x,y}$ decomposition

$$M = \begin{pmatrix} \left(\partial_x - \frac{x^2-3y}{x(x^2-y)}, \frac{x}{2y}, 0\right) \\[2mm] \left(\partial_y, 0, \frac{x}{2y}\right) \\[2mm] \left(0, 1, \frac{4y(x^2+y)}{x^2(x^2-y)}\right) \\[2mm] \left(-\frac{2x}{x^2-y}, 0, 1\right) \end{pmatrix} \begin{pmatrix} w_1 \equiv z_1 - \frac{x}{2y}z_2 \\[2mm] w_2 \equiv z_{2,x} \\[2mm] w_3 \equiv z_{2,y} - \frac{1}{y}z_2 \end{pmatrix}; \quad (3.3)$$

the left factor represents a type $M_3^{(0,1)}$ module. Continued in Example 3.15. $\qquad\square$

3.3   Loewy Decomposition of Ideals $\mathbb{J}^{(0,2)}$ and $\mathbb{J}^{(0,3)}$

The first step in determining decompositions of any given ideal consists of listing its possible divisors. For ordinary operators a necessary condition for possible genuine divisors is that its order be lower than that of the given operator. Obviously this constraint involves only the leading derivatives. It results in a finite number of candidates for divisors.

It turns out that for ideals of partial differential operators, not necessarily principal, a similar reasoning leads to necessary conditions for its possible divisors as well. To this end, a partial order between left ideals is defined by the requirement that the leading derivatives ideal of a divisor candidate divides the ideal generated by the leading derivatives of the given ideal. The following shortened notation is introduced for describing these order relations.

$$a \prec \{b, c, \ldots\} \quad \text{means} \quad a \prec b, \ a \prec c, \ldots.$$

Lemma 3.3. *For left ideals of differential dimension $(0, k)$ with $k \leq 3$ the relations*

$$\langle \partial_{xx}, \partial_{xy}, \partial_{yy} \rangle_{LT} \prec \{\langle \partial_{xx}, \partial_y \rangle_{LT}, \ \langle \partial_x, \partial_y \rangle_{LT}\},$$

$$\langle \partial_{xx}, \partial_{xy}, \partial_{yy} \rangle_{LT} \prec \{\langle \partial_{xx}, \partial_y \rangle_{LT}, \ \langle \partial_{yy}, \partial_x \rangle_{LT}, \ \langle \partial_x, \partial_y \rangle_{LT}\},$$

$$\langle \partial_{yyy}, \partial_x \rangle_{LT} \prec \{\langle \partial_{yy}, \partial_x \rangle_{LT}, \ \langle \partial_x, \partial_y \rangle_{LT}\},$$

$$\langle \partial_{xx}, \partial_y \rangle_{LT} \prec \langle \partial_x, \partial_y \rangle_{LT}, \quad \langle \partial_{yy}, \partial_x \rangle_{LT} \prec \langle \partial_x, \partial_y \rangle_{LT}$$

generate a partial order indicating possible inclusion. It may be represented by the following diagram.

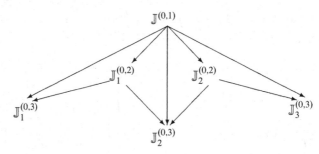

Proof. The generators of any ideal of differential dimension $(0, k)$ with $k > 1$ may be reduced w.r.t. the generators of $\langle \partial_x, \partial_y \rangle_{LT}$; therefore this latter ideal may be a divisor of the former under suitable constraints for the coefficients of the lower-order terms. The same is true for the pairs of ideals $\langle \partial_{yy}, \partial_x \rangle_{LT}$ and $\langle \partial_{yy}, \partial_x \rangle_{LT}$, $\langle \partial_{xx}, \partial_{xy}, \partial_{yy} \rangle_{LT}$ and $\langle \partial_{yy}, \partial_x \rangle_{LT}$, $\langle \partial_{xx}, \partial_{xy}, \partial_{yy} \rangle_{LT}$ and $\langle \partial_{xx}, \partial_y \rangle_{LT}$ and finally $\langle \partial_{xxx}, \partial_y \rangle_{LT}$ and $\langle \partial_{xx}, \partial_y \rangle_{LT}$.   □

If any ideal is given, the above diagram indicates possible divisors at its nodes. In order that such a divisor actually does exist it is necessary that the coefficients of the operators satisfy various constraints. In the first place, the operators generating the given ideal and the operators generating a divisor have rational coefficients and satisfy the coherence conditions of Sect. 2.4. In addition, certain relations between the coefficients of the operators must be satisfied. They are the subject of this section and are determined case by case. Ideals of type $\mathbb{J}^{(0,2)}$ are considered first.

Proposition 3.2. *The possible divisors of a type $\mathbb{J}_1^{(0,2)}$ ideal $\langle \partial_{yy} + A_1\partial_y + A_2, \partial_x + B_1\partial_y + B_2 \rangle$ are of type $\mathbb{J}^{(0,1)}$; A_i, $B_i \in \mathbb{Q}(x, y)$ for $i = 1, 2$; it is assumed that all integrability conditions of Proposition 2.1 are satisfied. The following cases are distinguished; $r(x, y)$, $r_1(x, y)$, $r_2(x, y)$ and $r(x, y, C)$ are rational functions of its arguments; C is a constant.*

  (i) There may be a single divisor $\langle \partial_x + B_2 - B_1 r(x, y), \partial_y + r(x, y) \rangle$;
 (ii) There may be two divisors $\langle \partial_x + B_2 - B_1 r_i(x, y), \partial_y + r_i(x, y) \rangle$, $i = 1, 2$;
(iii) There may be a divisor $\langle \partial_x + B_2 - B_1 r(x, y, C), \partial_y + r(x, y, C) \rangle$.

All divisors may be determined algorithmically.

Proof. The first statement follows from the above Lemma 3.3. Reduction of the given ideal w.r.t. $\langle \partial_x + a, \partial_y + b \rangle$ and autoreduction of the resulting constraints yields the system

$$b_x + B_1 b^2 + (B_{1,y} - A_1 B_1)b + A_2 B_1 - B_{2,y} = 0,$$
$$b_y - b^2 + A_1 b - A_2 = 0, \quad a + B_1 b - B_2 = 0. \tag{3.4}$$

The coherence of the system for b is shown in Exercise 3.2. Its solutions are described in Theorem B.1. The three cases given above are an immediate consequence. $\square$

Proposition 3.3. *With the same assumptions as in the preceding proposition the possible divisors of a type $\mathbb{J}_2^{(0,2)}$ ideal $\langle \partial_{xx} + A_1\partial_x + A_2, \partial_y + B \rangle$ may be described as follows.*

  (i) There may be a single divisor $\langle \partial_x + r(x, y), \partial_y + B \rangle$;
 (ii) There may be two divisors $\langle \partial_x + r_i(x, y), \partial_y + B \rangle$, $i = 1, 2$;
(iii) There may be a divisor $\langle \partial_x + r(x, y, C), \partial_y + B \rangle$ where C is a constant.

All divisors may be determined algorithmically.

Proof. The first statement follows from the above Lemma 3.3. Reduction of the given ideal w.r.t. $\langle \partial_x + a, \partial_y + b \rangle$ and autoreduction of the resulting constraints yields the system

$$a_x - a^2 + A_1 a - A_2 = 0, \quad a_y - B_x = 0, \quad b - B = 0. \tag{3.5}$$

The coherence of the system for a is also shown in Exercise 3.2. Its solutions are described in Theorem B.1. The three cases are an immediate consequence. □

Applying the preceding propositions, the Loewy decompositions of ideals of type $\mathbb{J}^{(0,2)}$ may be described as follows.

Theorem 3.2. *The Loewy decompositions of ideals of differential dimension $(0,2)$ may be described as follows.*

(i) *Let I be a type $\mathbb{J}_1^{(0,2)}$ ideal, I_1, I_2, and $I(C)$ be ideals of type $\mathbb{J}^{(0,1)}$, the latter depending on a parameter C; and M be a module of type $\mathbb{M}^{(0,1)}$. The following decompositions may occur.*

$$\mathscr{L}_{yy,x}^1: \quad I = MI_1; \quad \mathscr{L}_{yy,x}^2: \quad I = Lclm(I_1, I_2); \quad \mathscr{L}_{yy,x}^3: \quad I = Lclm(I(C)).$$

(ii) *The decompositions of a type $\mathbb{J}_2^{(0,2)}$ ideal are identical to those of an $\mathbb{J}_1^{(0,2)}$ ideal, i.e. $\mathscr{L}_{xx,y}^i = \mathscr{L}_{yy,x}^i$ for $i = 1, 2, 3$.*

The decomposition types $\mathscr{L}_{yy,x}^0$ and $\mathscr{L}_{xx,y}^0$ are assigned to irreducible ideals. For decomposition types $\mathscr{L}_{yy,x}^1$ and $\mathscr{L}_{xx,y}^1$ the Loewy divisors are M and I_1; the remaining decompositions are completely reducible.

Proof. The various factors obtained in Propositions 3.2 and 3.3 lead directly to the alternatives $\mathscr{L}_{yy,x}^i$ and $\mathscr{L}_{xx,y}^i$, $i = 1, 2, 3$. □

The remarks on page 7 apply here as well, i.e. if an ideal has decomposition type $\mathscr{L}^i$ for $i = 0, 1$ or 2 the type $\mathscr{L}^{i+1}$ is excluded.

The module M in decompositions $\mathscr{L}_{yy,x}^1$ and $\mathscr{L}_{xx,y}^1$ may have type $\mathbb{M}_1^{(0,1)}$ or $\mathbb{M}_2^{(0,1)}$, depending on the type of the right factor $Lclm$; details are discussed in Exercise 3.3.

Example 3.5. The type $\mathbb{J}_1^{(0,2)}$ ideal I of Example 2.7 on page 27 is considered again. According to Proposition 3.2 the system

$$b_x - \frac{y^2}{x^2}b^2 + \frac{2y^2}{x^2(x+y)}b + \frac{x-y}{x^2(x+y)} = 0,$$

$$b_y - b^2 + \frac{2}{y}\frac{x+2y}{x+y}b - \frac{2}{y(x+y)} = 0$$

for b is obtained. According to Theorem B.1 of Appendix B the general solution of this system is rational, containing a constant C. The result may be written in the form

$$b = \frac{1}{y} - \left(\frac{x}{x+y}\right)^2 \frac{1}{\frac{xy}{x+y}+C}.$$

It yields the decomposition

$$I = Lclm\left(\left\langle\!\!\left\langle \partial_x + \frac{1}{x} + \left(\frac{y}{x+y}\right)^2 \frac{1}{\frac{xy}{x+y}+C}, \partial_y + \frac{1}{y}\right.\right.\right.$$
$$\left.\left.\left. -\left(\frac{x}{x+y}\right)^2 \frac{1}{\frac{xy}{x+y}+C}\right\rangle\!\!\right\rangle\right)$$

of type $\mathcal{L}_{yy,x}^3$. In order to generate the originally given ideal I from it, the $Lclm$ of this divisor for two special values of the constant C has to be taken. Choosing $C = 0$ and $C \to \infty$ the values $a = \frac{1}{x}, b = \frac{1}{y}$ and $a = b = \frac{1}{x+y}$ are obtained. There follows

$$I = Lclm\left(I_1 \equiv \left\langle\!\!\left\langle \partial_x + \frac{1}{x}, \partial_y + \frac{1}{y}\right\rangle\!\!\right\rangle, \ I_2 \equiv \left\langle\!\!\left\langle \partial_x + \frac{1}{x+y}, \partial_y + \frac{1}{x+y}\right\rangle\!\!\right\rangle\right).$$

Continued in Example 3.10. □

Example 3.6 (Li et al. [42, 43]). Consider the ideal in *grlex*, $x \succ y$ term order $I = \left\langle\!\!\left\langle \partial_{yy} - \frac{xy-1}{y}\partial_y - \frac{x}{y}, \partial_x - \frac{y}{x}\partial_y \right\rangle\!\!\right\rangle$. According to case *(i)* of the preceding proposition the system

$$b_x - \frac{y}{x}b^2 - yb + 1 = 0, \quad b_y - b^2 - \left(x - \frac{1}{y}\right)b + \frac{x}{y} = 0$$

for b follows. There is a single rational solution $b = -x$; it yields $a = -y$ and the single divisor $J \equiv \langle \partial_x - y, \partial_y - x \rangle$, hence the type $\mathcal{L}_{yy,x}^1$ decomposition

$$I = \left\langle\!\!\left\langle \begin{pmatrix} 0 & \partial_x \\ 0 & \partial_y + \frac{1}{y} \\ 1 & -\frac{y}{x} \end{pmatrix} \begin{pmatrix} \partial_x - y \\ \partial_y - x \end{pmatrix} \right\rangle\!\!\right\rangle$$

is obtained. Continued in Example 3.12. □

Ideals of type $\mathbb{J}_2^{(0,3)}$ at the center bottom of the above diagram are considered next. The corresponding equations have been discussed in detail by Liouville [45].

Proposition 3.4. *Let the type $\mathbb{J}_2^{(0,3)}$ ideal*

$$I \equiv \langle \partial_{xx} + A_1\partial_x + A_2\partial_y + A_3,$$
$$\partial_{xy} + B_1\partial_x + B_2\partial_y + B_3, \partial_{yy} + C_1\partial_x + C_2\partial_y + C_3 \rangle$$

be given by a Janet basis in grlex term order with $x \succ y$; $A_i, B_i, C_i \in \mathbb{Q}(x, y)$; it is assumed that its coefficients satisfy the coherence conditions of Proposition 2.1.

If it has any first-order divisor $J \equiv \langle \partial_x + a, \partial_y + b \rangle$, *the following alternatives may occur.*

   (*i*) *There may be a divisor depending on two parameters.*
  (*ii*) *There may be a divisor depending on a single parameter.*
 (*iii*) *There is a single divisor, there are two or there are three divisors.*
 (*iv*) *There is a divisor depending on a parameter and in addition a single divisor.*

All divisors may be determined algorithmically.

Proof. Reduction of the ideal I w.r.t. to J and taking into account the coherence conditions of the former, the following system of equations for the coefficients of the divisor J is obtained.

$$a_x - a^2 + A_1 a + A_2 b - A_3 = 0, \quad a_y - ab + B_1 a + B_2 b - B_3 = 0,$$
$$b_x - ab + B_1 a + B_2 b - B_3 = 0, \quad b_y - b^2 + C_1 a + C_2 b - C_3 = 0. \tag{3.6}$$

In Theorem B.2 of Appendix B the solutions of this partial Riccati-like system are described; the above cases (*i*)–(*iv*) are an immediate consequence. $\qquad\square$

   The preceding proposition is applied now to determine the various decompositions of an ideal of type $\mathbb{J}_2^{(0,3)}$.

Theorem 3.3. *The possible Loewy decompositions of an ideal I of type $\mathbb{J}_2^{(0,3)}$ into first-order components may be described as follows. Let I_1, I_2, I_3, $I(C)$, and $I(C_1, C_2)$ be ideals of type $\mathbb{J}^{(0,1)}$; the latter ideals depend on parameters C or C_1 and C_2 respectively. Furthermore, let M_1, M_2, M and $M(C)$ be modules of type $\mathbb{M}^{(0,1)}$, the latter depending on a parameter C, and M_3 a module of $\mathbb{M}_3^{(0,1)}$. The following alternatives are distinguished.*

$$\mathscr{L}_{xx,xy,yy}^1 : \quad I = M_3 M_1 I_1; \quad \mathscr{L}_{xx,xy,yy}^2 : \quad I = Lclm(M_2, M_1) I_1;$$

$$\mathscr{L}_{xx,xy,yy}^3 : \quad I = Lclm(M(C)) I_1; \quad \mathscr{L}_{xx,xy,yy}^4 : \quad I = MLclm(I_2, I_1);$$

$$\mathscr{L}_{xx,xy,yy}^5 : \quad I = Lclm(I_1, I_2, I_3) \quad \mathscr{L}_{xx,xy,yy}^6 : \quad I = MLclm(I(C));$$

$$\mathscr{L}_{xx,xy,yy}^7 : \quad I = Lclm(I(C), I_1); \quad \mathscr{L}_{xx,xy,yy}^8 : \quad I = Lclm(I(C_1, C_2)).$$

Proof. The single right divisor of decomposition types $\mathscr{L}_{xx,xy,yy}^i$ for $i = 1, 2, 3$ follow from case (*iii*) of Proposition 3.4; the three decompositions of the exact quotient have been described in Theorem 3.1. Decomposition types $\mathscr{L}_{xx,xy,yy}^4$ and $\mathscr{L}_{xx,xy,yy}^5$ follow also from case (*iii*) of Proposition 3.4. Case (*i*), case (*ii*) and case (*iv*) yield the types $\mathscr{L}_{xx,xy,yy}^8$, $\mathscr{L}_{xx,xy,yy}^6$ and $\mathscr{L}_{xx,xy,yy}^7$ respectively. The types $\mathscr{L}_{xx,xy,yy}^k$, $k = 5, 7, 8$ are completely reducible. $\qquad\square$

Example 3.7. For the ideal

$$I \equiv \left\langle\!\!\left\langle \partial_{xx} + \frac{1}{x}\partial_x - \frac{y}{x(x+y)}\partial_y, \ \partial_{xy} + \frac{1}{x+y}\partial_y, \ \partial_{yy} + \frac{1}{x+y}\partial_y \right\rangle\!\!\right\rangle$$

of type $\mathbb{J}_2^{(0,3)}$ system (3.6) has the form

$$a_x - a^2 + \tfrac{1}{x}a - \frac{y}{x(x+y)}b = 0, \quad a_y - ab + \frac{1}{x+y}b = 0,$$

$$b_x - ab + \frac{1}{x+y}b = 0, \quad b_y - b^2 + \frac{1}{x+y}b = 0$$

with the single rational solution $a = b = 0$, i.e. there is the divisor $I_1 = \langle\partial_x, \partial_y\rangle$.
Dividing it out yields the type $\mathbb{M}_3^{(0,2)}$ exact quotient module

$$M \equiv \left\langle\!\!\left\langle \left(\partial_x + \frac{1}{x}, -\frac{y}{x(x+y)}\right), \left(\partial_y, \frac{1}{x+y}\right), \left(0, \partial_x + \frac{1}{x+y}\right), \right.\right.$$

$$\left.\left. \left(0, \partial_y + \frac{1}{x+y}\right) \right\rangle\!\!\right\rangle. \tag{3.7}$$

In Example 3.3 it has been shown that this module is completely reducible and
may be represented as $Lclm$ of the two first-order modules M_1 and M_2 given there.
Hence I has the type $\mathscr{L}_{xx,xy,yy}^2$ decomposition $I = Lclm(M_2, M_1)I_1$. Continued in
Example 3.11 □

Example 3.8. Consider the ideal

$$I \equiv \left\langle\!\!\left\langle \partial_{xx} + \frac{2}{x}\partial_x + \frac{y}{x^2}\partial_y - \frac{1}{x^2}, \ \partial_{xy} - \frac{1}{y}\partial_x, \ \partial_{yy} - \frac{1}{y}\partial_y + \frac{1}{y^2} \right\rangle\!\!\right\rangle$$

of type $\mathbb{J}_2^{(0,3)}$, generated by a Janet basis in *grlex*, $x \succ y$ term order. System (3.6) is

$$a_x - a^2 + \tfrac{2}{x}a + \frac{y}{x^2}b + \frac{1}{x^2} = 0, \quad a_y - ab - \tfrac{1}{y}a = 0,$$

$$b_x - ab - \tfrac{1}{y}a = 0, \quad b_y - b^2 - \tfrac{1}{y}b - \frac{1}{y^2} = 0.$$

The last equation is an ode for b with general solution $b = -\tfrac{1}{y}\dfrac{\log y + C}{\log y + C - 1}$.
The only rational solution $b = -\tfrac{1}{y}$ is obtained for $C \to \infty$. Substitution into
the remaining part of the system yields $a_y = 0$ and $a_x - a^2 + \tfrac{2}{x}a = 0$ with
the general solution $a = \dfrac{1}{Cx^2 + x}$; C is a new parameter now. The Loewy

divisor $Lclm\left(\left\langle\!\left\langle\partial_x + \dfrac{1}{Cx^2 + x}, \partial_y - \dfrac{1}{y}\right\rangle\!\right\rangle\right)$ is obtained, it generates a type $\mathscr{L}^6_{xx,xy,yy}$ decomposition. Choosing $C = 0$ and $C \to \infty$ the two divisors

$$I_1 \equiv \left\langle\!\left\langle\partial_x + \tfrac{1}{x}, \partial_y - \tfrac{1}{y}\right\rangle\!\right\rangle \text{ and } I_2 \equiv \left\langle\!\left\langle\partial_x, \partial_y - \tfrac{1}{y}\right\rangle\!\right\rangle$$

are obtained with $Lclm(I_1, I_2) = \left\langle\!\left\langle\partial_{xx} + \tfrac{2}{x}\partial_x, \partial_y - \tfrac{1}{y}\right\rangle\!\right\rangle$. It yields the decomposition

$$I = \left\langle\!\!\left\langle \begin{pmatrix} (0, \ \partial_x) \\[4pt] (0, \ \partial_y) \\[4pt] (1, \ \tfrac{y}{x^2}) \end{pmatrix} \begin{pmatrix} \partial_{xx} + \tfrac{2}{x}\partial_x \\[8pt] \partial_y - \tfrac{1}{y} \end{pmatrix} \right\rangle\!\!\right\rangle.$$

The solutions of the system $Iz = 0$ will be considered later in this section in Example 3.14.   □

There remains to be discussed how the remaining ideals of differential type $(0, 3)$ decompose into first-order components. The answer is given next.

Corollary 3.1. *Ideals of type* $\mathbb{J}_1^{(0,3)}$ *and* $\mathbb{J}_3^{(0,3)}$ *decompose into first order components exactly as ideals of type* $\mathbb{J}_2^{(0,3)}$. *The two alternatives may be described as follows.*

(i) *For type* $\mathbb{J}_1^{(0,3)}$ *ideals there follows* $\mathscr{L}^i_{yyy,x} = \mathscr{L}^i_{xx,xy,yy}$ *for* $i = 1, \ldots, 8$.
(ii) *For type* $\mathbb{J}_3^{(0,3)}$ *ideals there follows* $\mathscr{L}^i_{xxx,y} = \mathscr{L}^i_{xx,xy,yy}$ *for* $i = 1, \ldots, 8$.

Proof. In either case, the right divisors are obtained from a second-order Riccati equation. According to Lemma B.2 the structure of its solutions is the same as for system (3.6). Hence the discussion of the various cases is analogous to Theorem 3.3.   □

Example 3.9. Consider the ideal $I \equiv \left\langle\!\left\langle\partial_{yyy} + \tfrac{3}{x}\partial_{yy}, \partial_x + \tfrac{y}{x}\partial_y\right\rangle\!\right\rangle$. The equation for b in case (i) of the above lemma is $b_{yy} - 3bb_y + b^3 - \tfrac{3}{x}(b_y - b^2) = 0$ with the two rational solutions $b = 0$ and $b = -\tfrac{1}{y}$. They yield the two divisors $I_1 \equiv \left\langle\!\left\langle\partial_x, \partial_y\right\rangle\!\right\rangle$ and $I_2 \equiv \left\langle\!\left\langle\partial_x + \tfrac{1}{x}, \partial_y - \tfrac{1}{y}\right\rangle\!\right\rangle$ with $Lclm(I_1, I_2) = \left\langle\!\left\langle\partial_{yy}, \partial_x + \tfrac{y}{x}\partial_y\right\rangle\!\right\rangle$. It yields the decomposition

$$L = \left\langle\!\!\left\langle \begin{pmatrix} \left(\partial_x + \tfrac{2}{x} - \tfrac{3y}{x^2}, \ 0\right) \\[6pt] \left(\partial_y + \tfrac{3}{x}, \ 0\right) \\[6pt] (0, \ 1) \end{pmatrix} \begin{pmatrix} \partial_{yy} \\[8pt] \partial_x + \tfrac{y}{x}\partial_y \end{pmatrix} \right\rangle\!\!\right\rangle.$$

The solutions of the system $Iz = 0$ will be considered later in Exercise 3.6.   □

3.4 Solving Homogeneous Equations

Similar as for ordinary differential equations there remains to be discussed how the existence of divisors of an ideal or a module of partial differential operators simplifies the solution procedure; compare the discussion on page 11. The following result is due to Grigoriev and Schwarz [23] where also the proof may be found.

Theorem 3.4. *Let* $I \equiv \langle f_1, \ldots, f_p \rangle$ *and* $J \equiv \langle g_1, \ldots, g_q \rangle$ *be two submodules of* $\mathscr{D}^m$ *such that* $I \subseteq J$, *with solution space* V_I *and* V_J *of finite dimension* d_I *and* d_J *respectively. Let the exact quotient module* $Exquo(I, J) \equiv K \subset \mathscr{D}^q$ *be generated by* $\langle h_1, \ldots, h_r \rangle$, *and* $z = (z_1, \ldots, z_m)$ *and* $\bar{z} = (\bar{z}_1, \ldots, \bar{z}_q)$ *be differential indeterminates. A basis for the solution space* V_I *may be constructed as follows.*

(i)   *Determine a basis of* V_J *from* $g_1 z = 0, \ldots, g_q z = 0$, *let its elements be*
      $v_k = (v_{k,1}, \ldots, v_{k,m})$, $k = 1, \ldots, d_J$.
(ii) *Determine a basis of* V_K *from* $h_1 \bar{z} = 0, \ldots, h_r \bar{z} = 0$, *let its elements be*
      $w_k = (w_{k,1}, \ldots, w_{k,q})$, $k = 1, \ldots, d_K = d_I - d_J$.
(iii) *For each inhomogeneous system* $g_1 \bar{z} = w_{k,1}, \ldots, g_q \bar{z} = w_{k,q}$, $k = 1, \ldots, d_K$
      determine a special solution $\bar{v}_k = (\bar{v}_{k,1}, \ldots, \bar{v}_{k,m})$.

A basis of the solution space V_I *is* $\{v_1, \ldots, v_{d_J}, \bar{v}_1, \ldots, \bar{v}_{d_K}\}$.

Whenever a system is not completely reducible the exact quotient module is indispensable for finding a basis of the solution space. However, it may also be applied if a single divisor is already known in order to simplify the remaining calculations.

Example 3.10. The system

$$z_{yy} + \frac{2}{y}\frac{x+2y}{x+y}z_y + \frac{2}{y(x+y)}z = 0, \ z_x - \frac{y^2}{x^2}z_y + \frac{x-y}{x^2}z = 0$$

has been considered in Example 2.7. It allows two first-order divisors I_1 and I_2 as shown in Example 3.5; they yield the fundamental system $\left\{\dfrac{1}{xy}, \dfrac{1}{x+y}\right\}$.   □

Example 3.11. Consider the system

$$z_{xx} + \frac{1}{x}z_x - \frac{y}{x(x+y)}z_y = 0, \ z_{xy} + \frac{1}{x+y}z_y = 0, \ z_{yy} + \frac{1}{x+y}z_y = 0;$$

in Example 3.7 it has been shown that the corresponding ideal of operators allows the single divisor $\langle\!\langle \partial_x, \partial_y \rangle\!\rangle$, it yields the solution $z = 1$. The quotient module (3.7) has been shown in Example 3.3 to be completely reducible, it is the *Lclm* of the modules M_1 and M_2. The latter yields the system $z_2 = 0$, $z_1 + \frac{1}{x}z_1 = 0$, $z_{1,y} = 0$, with the special solution $z_1 = \frac{1}{x}$. Substitution into the right divisor equations $z_x = \frac{1}{x}$, $z_y = 0$ yields $z = \log x$. The module M_1 leads to the system

$$z_1 + \left(1 + \frac{2y}{x}\right)z_2 = 0, \ z_{2,x} + \frac{1}{x+y}z_2 = 0, \ z_{2,y} + \frac{1}{x+y}z_2 = 0,$$

with the solution $z_1 = -\dfrac{x + 2y}{x(x + y)}$, $z_2 = \dfrac{1}{x + y}$. With these inhomogeneities the right divisor equations $z_x = -\dfrac{x + 2y}{x(x + y)}$, $z_y = \dfrac{1}{x + y}$ yield $z = \log(x + y) - 2\log x$. Thus a fundamental system for the original system is $\{1, \log x, \log(x + y)\}$.

□

The systems occurring in the next two examples are not completely reducible, i.e. step (ii) and (iii) of Theorem 3.4 are necessary.

Example 3.12. In Example 3.6 the system of pde's corresponding to the only divisor J is $z_x - yz = 0$, $z_y - xz = 0$, it yields the basis $\{e^{xy}\}$ for V_J. The exact quotient module leads to the system $z_2 - \frac{x}{y}z_1 = 0$, $z_{1,x} + \frac{1}{x}z_1 = 0$ and $z_{1,y} = 0$; its one-dimensional solution space is generated by $z_1 = \frac{1}{x}$, $z_2 = \frac{1}{y}$. There is the inhomogeneous system $z_x - yz = \frac{1}{x}$, $z_y - xz = \frac{1}{y}$ with the special solution $\bar{v}_1 = e^{xy} Ei(-xy)$. The exponential integral is defined by $Ei(ax) = \int e^{ax} \frac{dx}{x}$. A basis for the solution space of the original system is $\{e^{xy}, e^{xy} Ei(-xy)\}$.

□

Example 3.13. In Example 2.9 on page 28, the first-order divisor corresponding to the system $z_x = 0$, $z_y = 0$, yields the basis $\{1\}$ for V_J. The two arguments of the *Lclm* lead to the systems $\bar{z}_1 = 0$, $\bar{z}_{2,x} = 0$, $\bar{z}_{2,y} + \frac{1}{y}\bar{z}_2 = 0$, with the solution $\bar{z}_{1,1} = 0$, $\bar{z}_{1,2} = 0$, and $\bar{z}_2 + \frac{x}{y}\bar{z}_1 = 0$, $\bar{z}_{1,x} + \frac{1}{x}\bar{z}_1 = 0$, $\bar{z}_{1,y} = 0$, with the solution $\bar{z}_{2,1} = \frac{1}{x}$, $\bar{z}_{2,2} = -\frac{1}{y}$. Substituting them into the inhomogeneous system yields the two special solutions $\log y$ and $\log \frac{x}{y}$; a basis for the given system is $\{1, \log x, \log y\}$.

□

Example 3.14. In Example 3.8 the system of equations $Iz = 0$ is

$$z_{xx} + \frac{2}{x}z_x + \frac{y}{x^2}z_y - \frac{1}{x^2}z = 0, \quad z_{xy} - \frac{1}{y}z_x = 0, \quad z_{yy} - \frac{1}{y}z_y + \frac{1}{y^2}z = 0. \quad (3.8)$$

The two equations $I_1z = 0$ and $I_2z = 0$ yield the basis elements $\frac{y}{x}$ and y. According to $Lclm(I_1, I_2)z = 0$ there follows $z_1 \equiv z_{xx} + \frac{2}{x}z_x = 0$, $z_2 \equiv z_y - \frac{1}{y}z = 0$. The exact quotient leads to $z_1 + \frac{y}{x^2}z_2 = 0$, $z_{2,x} = z_{2,y} = 0$ with the solution $z_2 = C$, $z_1 = -C\frac{y}{x^2}$. It yields the inhomogeneous system $z_{xx} + \frac{2}{x}z_x = -\frac{y}{x^2}$ and $z_y - \frac{1}{y}z = 1$ from which the third basis element $y \log \frac{x}{y}$ follows.

□

Example 3.15. The module considered in Example 2.13 on page 33 corresponds to the system of pde's

$$z_{1,x} - \frac{x^2 - 3y}{x(x^2 - y)}z_1 - \frac{1}{x^2 - y}z_2 = 0, \quad z_{1,y} = 0,$$

$$z_{2,x} + \frac{4y(x^2 + y)}{x^2(x^2 - y)}z_1 - \frac{2(x^2 + y)}{x(x^2 - y)}z_2 = 0, \quad z_{2,y} - \frac{2x}{x^2 - y}z_1 + \frac{1}{x^2 - y}z_2 = 0$$

$$(3.9)$$

for the two unknown functions $z_1(x, y)$ and $z_2(x, y)$. The first-order right divisor M_1 determined in Example 3.4 leads to the system

$$w_1 \equiv z_1 - \frac{x}{2y}z_2 = 0, \quad w_2 \equiv z_{2,x} = 0, \quad w_3 \equiv z_{2,y} - \frac{1}{y}z_2 = 0$$

with the solution $z_{1,1} = x$, $z_{2,1} = 2y$. From the left factor module in (3.3) the system

$$w_{1,x} - \frac{x^2 - 3y}{x(x^2 - y)}w_1 + \frac{x}{2y}w_2 = 0, \quad w_{1,y} + \frac{x}{2y}w_3 = 0,$$

$$w_2 + \frac{4y(x^2 + y)}{x^2(x^2 - y)}w_1 = 0, \quad w_3 - \frac{2x}{x^2 - y}w_1 = 0$$

is obtained. A solution is $w_1 = \frac{x(x^2 - y)}{y}$, $w_2 = -\frac{4(x^2 + y)}{x}$, $w_3 = \frac{2x^2}{y}$. Taking it as inhomogeneity in the right-factor equations leads to

$$z_1 - \frac{x}{2y}z_2 = \frac{x(x^2 - y)}{y}, \quad z_{2,x} = -\frac{4(x^2 + y)}{x}, \quad z_{2,y} - \frac{1}{y}z_2 = \frac{4x^2}{y}.$$

The special solution $z_{1,2} = x \log x$, $z_{2,2} = 2y \log x + x^2 - y$ is the second element of a fundamental system for the original system (3.9).   □

3.5 Solving Inhomogeneous Equations

The coherence conditions for any system of linear homogeneous pde's have to be supplemented by certain constraints if there are non-vanishing right hand sides; if they are not satisfied, the equations are inconsistent. The complete answer for type $\mathbb{J}_1^{(0,2)}$ ideals is given next in detail. The following auxiliary results are needed.

Lemma 3.4. *Let*

$$I = \langle \partial_{yy} + A_1 \partial_y + A_2, \partial_x + B_1 \partial_y + B_2 \rangle$$

$$= Lclm(\langle \partial_x + a_1, \partial_y + b_1 \rangle, \langle \partial_x + a_2, \partial_y + b_2 \rangle)$$

be a completely reducible ideal with a type $\mathcal{L}^2_{yy,x}$ *decomposition. Then*

$$A_1 = b_1 + b_2 - \frac{b_{1,y} - b_{2,y}}{b_1 - b_2}, \quad A_2 = b_1 b_2 + \frac{b_1 b_{2,y} - b_{1,y} b_2}{b_1 - b_2},$$

$$B_1 = -\frac{a_1 - a_2}{b_1 - b_2}, \quad B_2 = -\frac{a_1 b_2 - a_2 b_1}{b_1 - b_2}.$$

$$(3.10)$$

Proof. Reduction of I w.r.t. the two arguments of the *Lclm* yields a linear algebraic system for A_1, A_2, B_1, and B_2 with the above solution.   □

Lemma 3.5. *An inhomogeneous system corresponding to a type* $\mathbb{J}_1^{(0,2)}$ *ideal*

$$z_{yy} + A_1 z_y + A_2 z = R, \quad z_x + B_1 z_y + B_2 z = S \tag{3.11}$$

is consistent if the inhomogeneities satisfy

$$R_x + B_1 R_y + (B_2 + 2B_{1,y})R = S_{yy} + A_1 S_y + A_2 S. \tag{3.12}$$

Proof. The calculation is the same as for Proposition 2.1 with the inhomogeneities carried along.   □

After the consistency of an inhomogeneous system has been established, its solutions may be determined as follows.

Proposition 3.5. *With the same notation as in Lemma 3.4, the solution of the system (3.11) may be obtained as the sum* $z = \alpha z_1 + \beta z_2$ *where* α *and* β *are constants;* z_1 *and* z_2 *are solutions of*

$$z_{1,x} + a_1 z_1 = r_1, \quad z_{1,y} + b_1 z_1 = s_1, \quad \text{and} \quad z_{2,x} + a_2 z_2 = r_2, \quad z_{2,y} + b_2 z_2 = s_2.$$

The inhomogeneities r_1 *and* s_1 *are determined by*

$$s_{1,x} + \left(a_2 - \frac{a_{1,y} - a_{2,y}}{b_1 - b_2}\right) s_1 = S_y + b_1 S + \frac{a_1 - a_2}{b_1 - b_2} R,$$

$$s_{1,y} + \left(b_2 - \frac{b_{1,y} - b_{2,y}}{b_1 - b_2}\right) s_1 = R, \tag{3.13}$$

$$r_1 - \frac{a_1 - a_2}{b_1 - b_2} s_1 = S,$$

whereas the inhomogeneities r_2 *and* s_2 *are determined by*

$$s_{2,x} + \left(a_1 - \frac{a_{1,y} - a_{2,y}}{b_1 - b_2}\right) s_2 = S_y + b_2 S + \frac{a_1 - a_2}{b_1 - b_2} R,$$

$$s_{2,y} + \left(b_1 - \frac{b_{1,y} - b_{2,y}}{b_1 - b_2}\right) s_2 = R, \tag{3.14}$$

$$r_2 - \frac{a_1 - a_2}{b_1 - b_2} s_2 = S.$$

Proof. Reducing the first equation (3.11) w.r.t. the system for z_1 and z_2, and using (3.10) yields the two equations with leading terms $s_{1,x}$ and $s_{2,x}$. Similarly, reducing the second equation of (3.11) yields the expressions for r_1 and r_2.   □

This result implies that s_i and r_i, $i = 1, 2$, are Liouvillian over the extended base field of (3.11) because they are determined by first-order linear ode's and algebraic relations.

Example 3.16. Consider the system

$$z_{yy} + \frac{2x + 2y}{y} \frac{x + y}{x + y} z_y + \frac{2}{y(x+y)} z = \frac{1}{xy^5}, \quad z_x - \frac{y^2}{x^2} z_y + \frac{x-y}{x^2} z = \frac{1}{xy}. \quad (3.15)$$

The fundamental system $z_1(x, y) = \frac{1}{xy}$ and $z_2(x, y) = \frac{1}{x+y}$ for the corresponding homogeneous system may be obtained from the type $\mathcal{L}^4_{yy,x}$ Loewy decomposition considered in Example 3.5. The coefficients of (3.15) are $a_1 = \frac{1}{x}$, $b_1 = \frac{1}{y}$ and $a_2 = b_2 = \frac{1}{x+y}$. The inhomogeneities $R = \frac{1}{xy^5}$ and $S = \frac{1}{xy}$ satisfy the constraint (3.12). The system (3.13) is

$$s_{1,x} + \frac{1}{x(x+y)} s_1 = \frac{x^2 y^4 - xy - 1}{y^2}, \quad s_{1,y} + \frac{xy+2}{y(x+y)} s_1 = x^3 y$$

with the solution

$$s_1 = C_1 \frac{xy+1}{xy^2} - \frac{xy+1}{xy^2} \log(xy+1)$$

$$+ \frac{1}{6y^2} \left(2x^3 y^4 - x^2 y^3 - 3x^2 y + 3xy^2 - 3x + 6y\right).$$

According to Proposition 3.5 the system

$$z_{1,x} + \frac{1}{x} z_1 = \frac{1}{x^4 y^2 (x+y)^2} \left(x^5 y + 2x^4 y^2 + \frac{4}{3} x^3 y^3 - \frac{1}{3} x^3 - x^2 y - xy^2 - \frac{1}{3} y^3\right),$$

$$z_{1,y} + \frac{1}{y} z_1 = \frac{1}{x^2 y^4 (x+y)^2} \left(\frac{1}{3} x^3 y^3 - \frac{1}{3} x^3 - x^2 y - xy^2 - \frac{1}{3} y^3\right)$$

is obtained; a special solution is

$$z_{1,0} = \frac{1}{x^3 y^3 (x+y)} \left(x^4 y^2 + \frac{4}{3} x^3 y^3 - \frac{1}{2} x^3 y^2 + \frac{1}{6} x^3 - \frac{1}{2} x^2 y^3\right.$$

$$\left. + \frac{1}{2} x^2 y + \frac{1}{2} xy^2 + \frac{1}{6} y^3\right).$$

The system (3.14) is

$$s_{2,x} + \frac{1}{x+y} s_2 = -\frac{x^3 y - x - y}{x^3 y^3 (x+y)}, \quad s_{2,y} + \frac{2x+3y}{y(x+y)} s_2 = \frac{1}{xy^5}$$

with the solution

$$s_2 = C_2 \frac{1}{y^2(x+y)} - \frac{1}{2x^2 y^4(x+y)} \left(2x^3 y^2 - x^2 y^2 + x^2 + 2xy - y^2\right).$$

It yields

$$z_{2,x} + \frac{1}{x+y} z_2 = \frac{1}{x^4 y^2} x + y \left(x^2 y + \tfrac{1}{2} x^2 y^2 - \tfrac{1}{2} x^2 - xy - \tfrac{1}{2} y^2\right),$$

$$z_{2,y} + \frac{1}{x+y} z_2 = -\frac{1}{x^2 y^4(x+y)} \left(x^3 y^2 - \tfrac{1}{2} x^2 y^2 + \tfrac{1}{2} x^2 - xy - \tfrac{1}{2} y^2\right),$$

with the special solution

$$z_{2,0} = \frac{1}{x^3 y^3(x+y)} \left(x^4 y^2 + \tfrac{1}{3} x^3 y^3 - \tfrac{1}{2} x^3 y^2 + \tfrac{1}{6} x^3 - \tfrac{1}{2} x^2 y^3\right.$$
$$\left. + \tfrac{1}{2} x^2 y + \tfrac{1}{2} xy^2 + \tfrac{1}{6} y^3\right).$$

A simple calculation yields $\alpha = \beta = \tfrac{1}{2}$; thus the special solution $z_0 = \tfrac{1}{2}(z_{1,0} + z_{2,0})$ of (3.15) is obtained as

$$z_0 = \frac{1}{x^3 y^3(x+y)} \left(2x^4 y^2 + \tfrac{5}{3} x^3 y^3 - x^3 y^2 + \tfrac{1}{3} x^3 - x^2 y^3 + x^2 y + xy^2 + \tfrac{1}{3} y^3\right).$$

The second term in the bracket may be removed if $\tfrac{5}{3} z_2$ is added; the third and the fifth term by subtraction of z_1; this comes down to changing the basis in the solution space of the homogeneous system. The simplified expression is

$$z_0 = \frac{1}{2x^3 y^3(x+y)} \left(2x^4 y^2 + \tfrac{1}{3} x^3 + x^2 y + xy^2 + \tfrac{1}{3} y^3\right). \qquad \square$$

3.6 Exercises

Exercise 3.1. Let the system

$$z_{xx} + a_1 z_x + a_2 z_y + a_3 z = 0,$$
$$z_{xy} + b_1 z_x + b_2 z_y + b_3 z = 0,$$
$$z_{yy} + c_1 z_x + c_2 z_y + c_3 z = 0$$

be given; assume that the coherence conditions of Proposition 2.1 for Janet basis type $\mathbb{J}_2^{(0,3)}$ are satisfied. Express the coefficients a_i, b_i and c_i, $i = 1, 2, 3$ in terms of a fundamental system $\{z_1, z_2, z_3\}$.

Exercise 3.2. Show that systems (3.4) and (3.5) are coherent.

Exercise 3.3. Discuss the relation between the type of I_1 and the type of the left factor M in decomposition types $\mathscr{L}^1_{yyx}$ and $\mathscr{L}^1_{xx,y}$ of Theorem 3.2 on page 68.

Exercise 3.4. Assume an ideal $I \equiv \langle \partial_{yy} + A_1\partial_y + A_2, \partial_x + B_1\partial_y + B_2 \rangle$ allows a decomposition of type $\mathscr{L}^1_{yy,x}$ according to Theorem 3.2. Determine the corresponding exact quotient module explicitly.

The same problem for an ideal $I \equiv \langle \partial_{xx} + A_1\partial_x + A_2, \partial_y + B \rangle$ with a decomposition of type $\mathscr{L}^1_{xx,y}$.

Exercise 3.5. Determine the divisors of the ideal

$$I \equiv \left\langle\!\!\left\langle \partial_{xx} + \frac{4}{x}\partial_x + \frac{2}{x^2}, \; \partial_{xy} + \frac{1}{x}\partial_y, \; \partial_{yy} - \frac{x}{y^2}\partial_x + \frac{1}{y}\partial_y - \frac{2}{y^2} \right\rangle\!\!\right\rangle$$

and generate a fundamental system from them.

Exercise 3.6. Determine a fundamental system for $Iz = 0$ where I is the same as in Example 3.9 on page 72.

Chapter 4
Decomposition of Second-Order Operators

Abstract This chapter deals with a genuine extension of Loewy's theory. The ideals under consideration have differential type greater than zero. This means that the corresponding differential equations have a general solution involving not only constants but undetermined functions of varying numbers of arguments. Loewy's results are applied to individual linear pde's of second order in the plane with coordinates x and y, and the principal ideals generated by the corresponding operators. These equations have been considered extensively in the literature of the nineteenth century [14, 18, 28, 40, 44]. Like in the classical theory, equations with leading derivatives ∂_{xx} or ∂_{xy} are distinguished.

4.1 Operators with Leading Derivative ∂_{xx}

At first the generic second-order operator with leading derivative ∂_{xx} is considered. It is not assumed that any coefficient of a lower derivative vanishes. The reason for this assumption and the relation to operators with leading derivatives ∂_{xy} will become clear later in the subsection on transformation theory. As usual, factorizations in the base field $\mathbb{Q}(x, y)$ are considered.

Proposition 4.1. *Let the second-order partial differential operator*

$$L \equiv \partial_{xx} + A_1\partial_{xy} + A_2\partial_{yy} + A_3\partial_x + A_4\partial_y + A_5 \qquad (4.1)$$

be given with $A_i \in \mathbb{Q}(x, y)$ for all i. Its first order factors $\partial_x + a\partial_y + b$ with $a, b \in \mathbb{Q}(x, y)$ are determined by the roots a_1 and a_2 of $a^2 - A_1a + A_2 = 0$. The following alternatives may occur.

(i) *If $a_1 \neq a_2$ are two different rational solutions, and b_1 and b_2 are determined by (4.5), a factor $l_i = \partial_x + a_i\partial_y + b_i$ exists if the pair a_i, b_i satisfies (4.6). If there are two factors l_1 and l_2, the operator (4.1) is completely reducible, and $L = Lclm(l_1, l_2)$.*

F. Schwarz, *Loewy Decomposition of Linear Differential Equations*, Texts & Monographs in Symbolic Computation, DOI 10.1007/978-3-7091-1286-1_4, © Springer-Verlag/Wien 2012

(ii) If $a_1 = a_2 = a$ is a double root and

$$A_{1,x} + \tfrac{1}{2}A_1 A_{1,y} + A_1 A_3 = 2A_4, \tag{4.2}$$

the factorization depends on the rational solutions of the partial Riccati equation

$$b_x + \tfrac{1}{2}A_1 b_y - b^2 + A_3 b = A_5. \tag{4.3}$$

(a) *A right factor $l(\Phi) \equiv \partial_x + \tfrac{1}{2}A_1 \partial_y + R(x, y, \Phi(\varphi))$ exists if (4.3) has a rational general solution $R(x, y, \Phi(\varphi))$; $\varphi(x, y)$ is a rational first integral of $\dfrac{dy}{dx} = \tfrac{1}{2}A_1(x, y)$; Φ is an undetermined function. L is completely reducible, there holds $L = Lclm(l(\Phi_1), l(\Phi_2))$ for any two choices Φ_1 and Φ_2 such that $\Phi_1 \neq \Phi_2$.*

(b) *A right factor $l \equiv \partial_x + \tfrac{1}{2}A_1 \partial_y + r(x, y)$ exists if (4.3) has the single rational solution $r(x, y)$.*

(c) *Two right factors $l_i \equiv \partial_x + \tfrac{1}{2}A_1 \partial_y + r_i(x, y)$ exist if (4.3) has the rational solutions $r_1(x, y)$ and $r_2(x, y)$. The operator (4.1) is completely reducible, there holds $L = Lclm(l_2, l_1)$.*

Proof. Dividing the operator (4.1) by $\partial_x + a\partial_y + b$, the condition that this division be exact leads to the following set of equations between the coefficients.

$$a^2 - A_1 a + A_2 = 0, \tag{4.4}$$

$$a_x + (A_1 - a)a_y + A_3 a + (A_1 - 2a)b = A_4, \tag{4.5}$$

$$b_x + (A_1 - a)b_y - b^2 + A_3 b = A_5. \tag{4.6}$$

The first equation determines the coefficient a.

Case (i). Assume that (4.4) has two simple rational roots a_1 and a_2. Then $a_1 \neq a_2$ and $a_i \neq \tfrac{1}{2}A_1$ for $i = 1, 2$; the second equation (4.5) determines rational values of b_1 and b_2. The third equation (4.6) is a constraint. Those pairs a_i, b_i which satisfy it lead to a factor. There may be none, a single one l_1, or two factors l_1 and l_2. In the latter case, by Proposition 2.6 there follows $L = Lclm(l_2, l_1)$.

Case (ii) If $a_1 = a_2 = \tfrac{1}{2}A_1$ is a twofold root then $A_2 = \tfrac{1}{4}A_1^2$. The coefficient of b in (4.5) vanishes, it becomes the constraint (4.2). If it is not satisfied, factors cannot exist in any field extension. If it is satisfied, b is determined by (4.3) which is obtained from (4.6) by simplification. Depending on the type of its rational solutions, see Appendix B, the three subcases (a), (b), or (c) occur. If there are two factors, by case (iii) of Theorem 2.2 their intersection equals L. □

In order to apply this result for solving any given differential equation involving the operator (4.1) the question arises whether its first-order factors may be determined algorithmically. The subsequent corollary provides the answer for factors with coefficients either in the base field $\mathbb{Q}(x, y)$ or a universal extension.

Corollary 4.1. *In general, first-order right factors of (4.1) in the base field $\mathbb{Q}(x, y)$ cannot be determined algorithmically; however, absolute irreducibility may always be decided. In more detail the answer is as follows.*

(i) Separable symbol polynomial. Any factor may be determined.
(ii) Double root of symbol polynomial. In general it is not possible to determine the right factors over the base field. The existence of factors in a universal field may always be decided.

Proof. In the separable case (i), solving equation (4.4) and testing condition (4.6) requires only differentiations and arithmetic in the base field or in a quadratic function field; this can always be performed. In the non-separable case (ii), testing condition (4.2) requires only arithmetic and differentiations in the base field. If it is not satisfied, factors cannot exist in any field extension. If it is satisfied, factors are determined by the solutions of the partial Riccati equation (4.3). However, in general no algorithm is known at present for determining the rational solutions of (4.3). The same is true for solutions in a field extension although they always do exists if (4.2) is satisfied. □

Applying Proposition 4.1, Loewy's Theorem 1.1 may be generalized to operators of the form (4.1) as follows.

Theorem 4.1. *Let the differential operator L be defined by*

$$L \equiv \partial_{xx} + A_1 \partial_{xy} + A_2 \partial_{yy} + A_3 \partial_x + A_4 \partial_y + A_5 \tag{4.7}$$

such that $A_i \in \mathbb{Q}(x, y)$ for all i. Let $l_i \equiv \partial_x + a_i \partial_y + b_i$ for $i = 1$ and $i = 2$, and $l(\Phi) \equiv \partial_x + a \partial_y + b(\Phi)$ be first-order operators with $a_i, b_i, a \in \mathbb{Q}(x, y)$; Φ is an undetermined function of a single argument. Then L may be decomposed according to one of the following types.

$$\mathscr{L}_{xx}^1 : L = l_2 l_1; \quad \mathscr{L}_{xx}^2 : L = Lclm(l_2, l_1); \quad \mathscr{L}_{xx}^3 : L = Lclm(l(\Phi)). \tag{4.8}$$

If L does not have any first-order factor in the base field, its decomposition type is defined to be $\mathscr{L}_{xx}^0$. Decompositions $\mathscr{L}_{xx}^0$, $\mathscr{L}_{xx}^2$ and $\mathscr{L}_{xx}^3$ are completely reducible. For decomposition $\mathscr{L}_{xx}^1$ the first-order right factor is a Loewy divisor.

Proof. It is based on Proposition 4.1. In the separable case (i) there are either two first order factors with a principal intersection corresponding to decomposition type $\mathscr{L}_{xx}^2$; or a single first-order factor corresponding to type $\mathscr{L}_{xx}^1$; or no factor at all corresponding to type $\mathscr{L}_{xx}^0$. In case (ii), depending on the rational solutions of the Riccati equation (4.6), a similar distinction as in case (i) occurs. In addition there may be a factor containing an undetermined function yielding decomposition type $\mathscr{L}_{xx}^3$. □

The subsequent examples show that each decomposition type does actually occur. They will be applied in the next chapter for determining the solutions of the respective equations as indicated.

Example 4.1. The operator $L \equiv \partial_{xx} + \frac{2}{x}\partial_x + \frac{y}{x^2}\partial_y - \frac{1}{x^2}$ has been considered before in Example 2.5 on page 26. Equation 4.4 reads $a^2 = 0$, i.e. case (ii) of the above theorem applies. Because $A_1 = 0$, $A_4 \neq 0$, condition (4.2) is violated; hence there does not exist a first-order factor in any field extension.   □

The operator of the preceding example has two divisors of differential type zero as has been shown previously, yet it is irreducible according to the definition applied in this monograph.

Although the next three examples are separable, there is only a single first order factor.

Example 4.2. Consider the operator

$$L \equiv \partial_{xx} + (x+1)\partial_{xy} + x\partial_{yy} + (y+1)\partial_x + (xy+1)\partial_y + x + y.$$

The rational solutions of $a^2 - (x+1)a + x = 0$ are $a_1 = 1$ and $a_2 = x$, i.e. case (i) of Proposition 4.1 applies. It leads to $b_1 = y$ and $b_2 = \frac{x}{x-1}$. Only the first alternative satisfies condition (4.6). Hence there is the single right factor $l_1 = \partial_x + \partial_y + y$; dividing it out yields the type $\mathcal{L}^1_{xx}$ decomposition

$$L = (\partial_x + x\partial_y + 1)(\partial_x + \partial_y + y);$$

continued in Example 5.1.   □

Example 4.3. For the operator

$$L \equiv \partial_{xx} + (y^2 - 2x)\partial_{xy} + x(x - y^2)\partial_{yy} - (y^2 + 1)\partial_y - 1$$

the equation $a^2 - (y^2 - 2x)a + x(x - y^2) = 0$ has the two rational roots $a_1 = -x$ and $a_2 = y^2 - x$, i.e. case (i) of Proposition 4.1 applies. They yield $b_1 = -1$ and $b_2 = 1 + \frac{2}{y}(y^2 - 2x)$. Only a_1, b_1 satisfy condition (4.6), i.e. there is a single factor yielding the type $\mathcal{L}^1_{xx}$ decomposition

$$L = (\partial_x + (y^2 - x)\partial_y + 1)(\partial_x - x\partial_y - 1);$$

continued in Example 5.2.   □

Example 4.4 (Forsyth [17]). Forsyth [17], vol. VI, page 16, considered the differential equation $Lz = 0$ where

$$L \equiv \partial_{xx} - \partial_{yy} + \frac{4}{x+y}\partial_x.$$

The rational solutions of $a^2 - 1 = 0$ are $a_1 = 1$, $a_2 = -1$, i.e. case (*i*) of Proposition 4.1 applies. From (4.5) there follows $b_{1,2} = \dfrac{2}{x+y}$. Only $a_2 = -1$, $b_2 = \dfrac{2}{x+y}$ satisfy (4.6). There is a single right factor leading to the type $\mathscr{L}^1_{xx}$ decomposition

$$L = \left(\partial_x + \partial_y + \frac{2}{x+y} \right) \left(\partial_x - \partial_y + \frac{2}{x+y} \right);$$

continued in Example 5.3. □

The next two examples show that complete reducibility may occur for separable and non-separable operators.

Example 4.5. Consider the operator

$$L \equiv \partial_{xx} - \frac{2y}{x} \partial_{xy} + \frac{y^2}{x^2} \left(1 - x^4 y^2 \right) \partial_{yy} + \frac{2y}{x^2} \partial_y.$$

The rational solutions of $a^2 + \dfrac{2y}{x} a + \dfrac{y^2}{x^2} \left(1 - x^2 y^2 \right) = 0$ are $a_1 = xy^2 - \dfrac{y}{x}$ and $a_2 = -xy^2 - \dfrac{y}{x}$, i.e. case (*i*) of Proposition 4.1 applies. It leads to $b_1 = -xy$ and $b_2 = xy$. Both alternatives satisfy condition (4.6). Hence there are two factors $l_{1,2} = \partial_x - \left(\dfrac{y}{x} \pm xy^2 \right) \partial_y \pm xy$; they yield the representation

$$Lclm(l_2, l_1) = L \quad \text{and} \quad Gcrd(l_2, l_1) = \left\langle\!\!\left\langle \partial_x - \frac{1}{x}, \partial_y - \frac{1}{y} \right\rangle\!\!\right\rangle,$$

i.e. L has decomposition type $\mathscr{L}^2_{xx}$; completed in Example 5.4. □

Example 4.6 (Miller [49]). Let the operator

$$L \equiv \partial_{xx} + \frac{2y}{x} \partial_{xy} + \frac{y^2}{x^2} \partial_{yy} + \frac{1}{x} \partial_x + \frac{y}{x^2} \partial_y - \frac{1}{x^2}$$

be given. Because $\frac{1}{4} A_1^2 - A_2 = 0$, case (*ii*) of Proposition 4.1 applies. It yields $a = \dfrac{y}{x}$ and leads to equation $b_x + \dfrac{y}{x} b_y - b^2 + \dfrac{1}{x} b + \dfrac{1}{x^2} = 0$ for b with general solution $b = \dfrac{1}{x} \dfrac{1 + x^2 \Phi(\varphi)}{1 - x^2 \Phi(\varphi)}$ where $\varphi = \dfrac{y}{x}$; Φ is an undetermined function of its argument. Therefore the given second-order operator has an infinite number of first-order right factors of the form $l(\Phi) \equiv \partial_x + \dfrac{y}{x} \partial_y + \dfrac{1}{x} \dfrac{1 + x^2 \Phi(\varphi)}{1 - x^2 \Phi(\varphi)}$ which are parameterized by Φ; the decomposition type is $\mathscr{L}^3_{xx}$; completed in Example 5.5. □

4.2 Operators with Leading Derivative ∂_{xy}

If an operator does not contain a derivative ∂_{xx} but ∂_{yy} does occur, permuting the variables x and y leads to an operator of the form (4.1) such that the above theorem may be applied. If there is neither a derivative ∂_{xx} or ∂_{yy}, possible factors must obviously be of the form $\partial_x + a$ or $\partial_y + b$; the same is true for the arguments of a representation as an intersection due to Theorem 2.3. Hence the possible factorizations may be described as follows.

Proposition 4.2. *Let the second-order operator*

$$L \equiv \partial_{xy} + A_1 \partial_x + A_2 \partial_y + A_3 \tag{4.9}$$

be given with $A_i \in \mathbb{Q}(x, y)$ for all i. The following factorizations may occur.

(i) *If $A_3 - A_1 A_2 = A_{2,y}$ then $L = (\partial_y + A_1)(\partial_x + A_2)$.*
(ii) *If $A_3 - A_1 A_2 = A_{1,x}$ then $L = (\partial_x + A_2)(\partial_y + A_1)$.*
(iii) *If $A_3 - A_1 A_2 = A_{1,x}$ and $A_{1,x} = A_{2,y}$ there are two right factors; then $L = Lclm(\partial_x + A_2, \partial_y + A_1)$.*
(iv) *There may exist a Laplace divisor $\mathbb{L}_{y^n}(L)$ for $n \geq 2$.*
(v) *There may exist a Laplace divisor $\mathbb{L}_{x^m}(L)$ for $m \geq 2$.*
(vi) *There may exist both Laplace divisors $\mathbb{L}_{x^m}(L)$ and $\mathbb{L}_{y^n}(L)$. In this case L is completely reducible; L is the left intersection of two Laplace divisors.*

Proof. Dividing the operator (4.9) by $\partial_x + a\partial_y + b$, the condition that this division be exact leads to the following set of equations between the coefficients

$$a = 0, \quad A_2 - A_1 a - a_y - b = 0, \quad A_3 - A_1 b - b_y = 0$$

with the solution $a = 0$, $b = A_2$ and the constraint $A_3 - A_1 A_2 = A_{2,y}$. Dividing out the right factor $\partial_x + A_2$ yields the left factor $\partial_y + A_1$. This is case (i). Dividing (4.9) by $\partial_y + c$, the condition that this division be exact leads to $c = A_1$ and the constraint $A_3 - A_1 A_2 = A_{1,x}$. This is case (ii). Finally if the conditions for cases (i) and (ii) are satisfied simultaneously, a simple calculation shows that L is the left intersection of its right factors. This is case (iii).

The possible existence of the Laplace divisors in cases (iv) to (vi) is a consequence of Proposition 2.3 and the constructive proof given there. $\qquad\square$

Case (iv), $n = 1$ and case (v), $m = 1$ are covered by case (i), (ii) and (iii). The corresponding ideals are maximal and principal because they are generated by $\partial_y + a_1$ and $\partial_x + b_1$ respectively. The term *factorization* applies in these cases in the proper sense because the obvious analogy to ordinary differential operators where all ideals are principal.

Laplace divisors obey $\mathbb{L}_{x^{m_2}}(L) \subset \mathbb{L}_{x^{m_1}}(L)$ for $m_1 < m_2$; and similarly $\mathbb{L}_{y^{n_2}}(L) \subset \mathbb{L}_{y^{n_1}}(L)$ for $n_1 < n_2$. These relations become clear from the graph shown in Fig. 4.1. The heavy dot at $(1, 1)$ represents the leading derivative ∂_{xy} of the

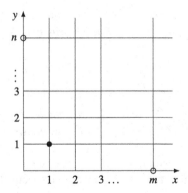

Fig. 4.1 The dots represent leading derivatives

given operator. If a second operator with leading derivative ∂_{x^m} represented by the circle at $(m, 0)$ exists, the ideal is enlarged by the corresponding operator. For $m = 1$ this ideal contains the original operator with leading derivative ∂_{xy}, i.e. this operator disappears by autoreduction. This shows clearly how the conventional factorization corresponding to a first-order operator is obtained as special case for $m = 1$. A similar discussion applies for an additional operator with leading derivative ∂_{x^n}.

The following corollary describes to what extent the factorizations described above may be determined algorithmically.

Corollary 4.2. *The coefficients of any first-order factor or Laplace divisor of fixed order are in the base field; they may be determined algorithmically. However, a bound for the order of a Laplace divisor is not known.*

Proof. For the first-order factors in cases (i), (ii) and (iii) this is obvious. For any Laplace divisor of fixed order this follows from the constructive proof of Proposition 2.3 and Corollary 2.1.   □

It should be emphasized that according to this corollary finding a Laplace divisor in general is not algorithmic. To this end, an upper bound for the order of a possible divisor would be required; at present such a bound is not known.

Goursat [18], Sect. 110, describes a method for constructing a linear ode in involution with a given second order equation $z_{xy} + az_x + bz_y + cz = 0$. The advantage of the method given above is that it may be applied to more general operators; the same reasoning works for third-order operators discussed in Chap. 6. It is not obvious how to generalize Goursat's scheme to any other case beyond the special second-order equation considered by him.

Applying the preceding results, Loewy decompositions of (4.9) involving first-order principal factors may be described as follows.

Theorem 4.2. *Let the differential operator L be defined by*

$$L \equiv \partial_{xy} + A_1 \partial_x + A_2 \partial_y + A_3 \tag{4.10}$$

with $A_i \in \mathbb{Q}(x, y)$ for all i; $l \equiv \partial_x + A_2$ and $k \equiv \partial_y + A_1$ are first-order operators. L may decompose into first-order principal divisors according to one of the following types.

$$\mathcal{L}^1_{xy} : L = kl; \quad \mathcal{L}^2_{xy} : L = lk; \quad \mathcal{L}^3_{xy} : L = Lclm(k, l).$$

The decompositions of type $\mathcal{L}^0_{xy}$ and $\mathcal{L}^3_{xy}$ are completely reducible; the first-order factors in decompositions $\mathcal{L}^1_{xy}$ and $\mathcal{L}^2_{xy}$ are Loewy divisors.

Proof. It is based on Proposition 4.2. If the conditions for case (i) or (ii) or are satisfied, the decomposition type is $\mathcal{L}^1_{xy}$ or $\mathcal{L}^2_{xy}$ respectively; if both are satisfied the decomposition type $\mathcal{L}^3_{xy}$ is obtained.   □

The following examples provide applications of the above results. In the next chapter it will be shown how a particular decomposition may lead to explicit solutions.

Example 4.7 (Goursat 1898). The equation $Lz \equiv (\partial_{xy} - y\partial_y)z = 0$ has been considered in [18], vol II, page 212. There is the obvious right factor ∂_y, it yields the decomposition $L = (\partial_x - y)\partial_y$ of type $\mathcal{L}^2_{xy}$. Continued in Example 5.7.   □

Example 4.8. The operator

$$L \equiv \partial_{xy} + (x + y)\partial_x + \left(y + \frac{1}{x}\right)\partial_y + xy + y^2 + 2 + \frac{y}{x}$$

obeys the conditions of case (iii) of Proposition 4.2. Therefore the type $\mathcal{L}^3_{xy}$ decomposition $L = Lclm\left(\partial_x + y + \frac{1}{x}, \partial_y + x + y\right)$ is obtained. Continued in Example 5.8.   □

Loewy decompositions of (4.9) involving non-principal divisors, possibly in addition to principal ones, are considered next.

Theorem 4.3. *Let the differential operator L be defined by*

$$L \equiv \partial_{xy} + A_1\partial_x + A_2\partial_y + A_3 \tag{4.11}$$

with $A_i \in \mathbb{Q}(x, y)$ for all i. $\mathbb{L}_{x^m}(L)$ and $\mathbb{L}_{y^n}(L)$ as well as $\mathfrak{l}_m$ and $\mathfrak{k}_n$ are defined in Definition 2.3; furthermore $l \equiv \partial_x + a$, $k \equiv \partial_y + b$, $a, b \in \mathbb{Q}(x, y)$. L may decompose into Laplace divisors according to one of the following types where $m, n \geq 2$.

$$\mathcal{L}^4_{xy} : L = Lclm\left(\mathbb{L}_{x^m}(L), \mathbb{L}_{y^n}(L)\right);$$

$$\mathcal{L}^5_{xy} : L = Exquo\left(L, \mathbb{L}_{x^m}(L)\right)\mathbb{L}_{x^m}(L) = \begin{pmatrix} 1 & 0 \\ 0 & \partial_y + A_1 \end{pmatrix}\begin{pmatrix} L \\ \mathfrak{l}_m \end{pmatrix};$$

$$\mathscr{L}_{xy}^6 : L = Exquo\left(L, \mathbb{L}_{y^n}(L)\right) \mathbb{L}_{y^n}(L) = \begin{pmatrix} 1 & 0 \\ 0 & \partial_x + A_2 \end{pmatrix} \begin{pmatrix} L \\ \mathfrak{k}_n \end{pmatrix};$$

$$\mathscr{L}_{xy}^7 : L = Lclm\left(k, \mathbb{L}_{x^m}(L)\right); \qquad \mathscr{L}_{xy}^8 : L = Lclm\left(l, \mathbb{L}_{y^n}(L)\right).$$

If L neither has a first-order principal factor according to Theorem 4.2 nor a Laplace divisor of any order its decomposition type is $\mathscr{L}_{xy}^0$. The decompositions $\mathscr{L}_{xy}^0$, $\mathscr{L}_{xy}^4$, $\mathscr{L}_{xy}^7$ and $\mathscr{L}_{xy}^8$ are completely reducible. The Laplace divisors and the exact quotients in decompositions $\mathscr{L}_{xy}^5$ and $\mathscr{L}_{xy}^6$ are Loewy divisors.

Proof. Decomposition types $\mathscr{L}_{xy}^4$, $\mathscr{L}_{xy}^7$ and $\mathscr{L}_{xy}^8$ are completely reducible with the obvious representation given above. For decomposition type $\mathscr{L}_{xy}^5$, dividing L by $\mathbb{L}_{x^m}(L)$ yields the exact quotient $(1, 0)$. The single syzygy of $\mathbb{L}_{x^m}(L)$ has been determined in Corollary 2.3. Autoreduction w.r.t. $(1, 0)$ yields the generators $(1, 0)$ and $(0, \partial_y + A_1)$. The calculation for decomposition type $\mathscr{L}_{xy}^6$ is similar. $\qquad\square$

The following example taken from Forsyth shows how complete reducibility has its straightforward generalization although if there are Laplace divisors involved.

Example 4.9 (Forsyth [17]). Define

$$L \equiv \partial_{xy} + \frac{2}{x - y}\partial_x - \frac{2}{x - y}\partial_y - \frac{4}{(x - y)^2}$$

generating the principal ideal $\langle L \rangle$ of differential dimension $(1, 2)$. The equation $Lz = 0$ has been considered in [17], vol. VI, page 80, Ex. 5 (iii). By Proposition 4.2, a first-order factor does not exist. However, by Proposition 2.3 or Corollary 2.1 there exist Laplace divisors

$$\mathbb{L}_{x^2}(L) \equiv \left\langle\!\!\left\langle \mathfrak{l}_2 \equiv \partial_{xx} - \frac{2}{x - y}\partial_x + \frac{2}{(x - y)^2}, L \right\rangle\!\!\right\rangle$$

and

$$\mathbb{L}_{y^2}(L) \equiv \left\langle\!\!\left\langle L, \mathfrak{k}_2 \equiv \partial_{yy} + \frac{2}{x - y}\partial_y + \frac{2}{(x - y)^2} \right\rangle\!\!\right\rangle,$$

each of differential dimension $(1, 1)$. There follows $\langle L \rangle = Lclm\left(\mathbb{L}_{x^2}(L), \mathbb{L}_{y^2}(L)\right)$, i.e. L is completely reducible; its decomposition type is $\mathscr{L}_{xy}^4$. Continued in Example 5.9. $\qquad\square$

The next example due to Imschenetzky is not completely reducible because it has been shown before to allow a single Laplace divisor.

Example 4.10. Imschenetzky's operator $L = \partial_{xy} + xy\partial_x - 2y$ has been considered already in Example 2.14 on page 37; see also Exercise C.2. Using these results the decomposition

$$L = \begin{pmatrix} 1 & 0 \\ 0 & \partial_y + xy \end{pmatrix} \begin{pmatrix} \partial_{xy} + xy\partial_x - 2y \\ \partial_{xxx} \end{pmatrix}$$

of type $\mathcal{L}_{xy}^5$ is obtained. Continued in Example 5.10.   □

Example 4.11. The operator $L \equiv \partial_{xy} - \frac{1}{x+y}\partial_y$ has the obvious factor ∂_y; in addition there is the Laplace divisor $\mathbb{L}_{x^2}(L) \equiv \langle \partial_{xx}, L \rangle$, i.e. its decomposition type is $\mathcal{L}_{xy}^7$. Continued in Example 5.11.   □

For linear ode's of low order it has been shown in Exercise 1.1 how the coefficients of the equation may be expressed in terms of a fundamental system. This is generalized in Exercise 4.2 to pde's allowing a low order Laplace divisor.

4.3   Exercises

Exercise 4.1. Determine the coefficients $A_1, \ldots, A_5$ in (4.1) in terms of the coefficients a_i and b_i, $i = 1, 2$, of the right factors.

Exercise 4.2. Assume an operator (4.9) allows two Laplace divisors of second order according to Proposition 4.2. The solution of the equation $Lz = 0$ has the structure $z = f_0 F(y) + f_1 F'(y) + g_0 G(x) + g_1 G'(x)$ with F and G undetermined functions. Express the coefficients A_1, A_2 and A_3 of (4.9) in terms of f_0, f_1, g_0 and g_1.

Exercise 4.3. For a given operator $L \equiv \partial_{xy} + A_1\partial_x + A_2\partial_y + A_3$ determine the coefficients of $l_2 \equiv \partial_{xx} + a_1\partial_x + a_2$ such that the two operators combined form a Janet basis and find the constraints for A_1, A_2 and A_3 such that this Janet basis may exist. Determine the exact quotient module $Exquo(\langle L \rangle, \langle l_2 \rangle)$.

Exercise 4.4. Discuss the various Loewy decompositions of an operator $L \equiv \partial_{xx} + A_1\partial_{xy} + A_2\partial_{yy} + A_3\partial_x + A_4\partial_y + A_5$ if all coefficients $A_1, \ldots, A_5$ are constant.

Exercise 4.5. Determine the Loewy decomposition of

$$L \equiv \partial_{xy} + \left(x + \frac{1}{y} \right) \partial_x + y\partial_y + xy + 2.$$

Chapter 5
Solving Second-Order Equations

Abstract The results of the preceding chapters are applied now for solving differential equations of second order for an unknown function $z(x, y)$. At first some general properties of the solutions are discussed.

For linear ode's, or systems of linear pde's with a finite dimensional solution space considered in Chap. 3, the undetermined elements in the general solution are constants. By contrast, for general pde's the undetermined elements are described by Theorem 2.1, due to Kolchin. The left-hand sides of the equations considered in this chapter allow a decomposition into first-order principal divisors, or certain non-principal divisors. It will be shown that the solution of such equations with differential dimension $(1, n)$ has the form

$$z(x, y) = z_1\big(x, y, F_1(\varphi_1(x, y))\big) + \ldots + z_n\big(x, y, F_n(\varphi_n(x, y))\big); \qquad (5.1)$$

each z_i is a sum of terms containing an undetermined function F_i depending on an argument $\varphi_i(x, y)$. Collectively the $\{z_1, \ldots, z_n\}$ are called a *differential fundamental system* or simply *fundamental system*. For the decomposition types considered in this chapter the following properties of a fundamental system are characteristic.

- Each $F_i\big(\varphi_i(x, y)\big)$, or derivatives or integrals thereof, occurs linearly in the corresponding z_i.
- The arguments $\varphi_i(x, y)$ of the undetermined functions are determined by the coefficients of the given equation.

It will turn out that the detailed structure of the solutions is essentially determined by the decomposition type of the equation.

5.1 Solving Homogeneous Equations

Reducible equations with leading derivative ∂_{xx} are considered first. Because there are only principal divisors the answer is similar to ordinary second-order equations.

F. Schwarz, *Loewy Decomposition of Linear Differential Equations*, Texts & Monographs in Symbolic Computation, DOI 10.1007/978-3-7091-1286-1_5,
© Springer-Verlag/Wien 2012

Proposition 5.1. *Let a reducible second-order equation*

$$Lz \equiv (\partial_{xx} + A_1\partial_{xy} + A_2\partial_{yy} + A_3\partial_x + A_4\partial_y + A_5)z = 0$$

be given with $A_1, \ldots, A_5 \in \mathbb{Q}(x, y)$. *Define* $l_i \equiv \partial_x + a_i\partial_y + b_i$, $a_i, b_i \in \mathbb{Q}(x, y)$
for $i = 1, 2$; $\varphi_i(x, y) = const$ *is a rational first integral of* $\frac{dy}{dx} = a_i(x, y)$;
$\bar{y} \equiv \varphi_i(x, y)$ *and the inverse* $y = \psi_i(x, \bar{y})$; *both* φ_i *and* ψ_i *are assumed to exist.*
Furthermore define

$$\mathscr{E}_i(x, y) \equiv \exp\left(-\int b_i(x, y)\big|_{y=\psi_i(x,\bar{y})} dx\right)\Big|_{\bar{y}=\varphi_i(x,y)} \tag{5.2}$$

for $i = 1, 2$. *A differential fundamental system has the following structure for the
various decompositions into first-order components.*

$$\mathscr{L}_{xx}^1 : \begin{cases} z_1(x, y) = \mathscr{E}_1(x, y) F_1(\varphi_1), \\[2mm] z_2(x, y) = \mathscr{E}_1(x, y) \displaystyle\int \frac{\mathscr{E}_2(x, y)}{\mathscr{E}_1(x, y)} F_2\big(\varphi_2(x, y)\big)\big|_{y=\psi_1(x,\bar{y})} dx\Big|_{\bar{y}=\varphi_1(x,y)}; \end{cases}$$

$$\mathscr{L}_{xx}^2 : z_i(x, y) = \mathscr{E}_i(x, y) F_i\big(\varphi_i(x, y)\big), \quad i = 1, 2;$$

$$\mathscr{L}_{xx}^3 : z_i(x, y) = \mathscr{E}_i(x, y) F_i\big(\varphi(x, y)\big), \quad i = 1, 2.$$

The F_i *are undetermined functions of a single argument;* φ, φ_1 *and* φ_2 *are rational
in all arguments; they are determined by the coefficients* A_1, A_2 *and* A_3 *of the given
equation.*

Proof. It is based on Theorem 4.1 and Lemma B.3. For a decomposition
$L = l_2 l_1$ of type $\mathscr{L}_{xx}^1$, (B.5) applied to the factor l_1 yields the above given
$z_1(x, y)$. The left factor equation $l_2 w = 0$ yields $w = \mathscr{E}_2(x, y) F_2(\varphi_2)$. Taking
it as inhomogeneity for the right factor equation, by (B.4) the given expression for
$z_2(x, y)$ is obtained.

For a decomposition $L = Lclm(l_2, l_1)$ of type $\mathscr{L}_{xx}^2$, similar arguments as for the
right factor l_1 in the preceding decomposition lead to the above solutions $z_1(x, y)$
and $z_2(x, y)$. It may occur that $\varphi_1 = \varphi_2$; this is always true if the decomposition
originates from case (ii) of Proposition 4.1 instead of case (i).

Finally in a decomposition $L = Lclm\big(l(\Phi)\big)$ of type $\mathscr{L}_{xx}^3$, two special independent functions Φ_1 and Φ_2 in the operator $l(\Phi)$ may be chosen. Both first-order
operators obtained in this way have the same coefficient of ∂_y; as a consequence the
arguments of the undetermined functions are the same, i.e. $\varphi_1 = \varphi_2 = \varphi$. $\square$

The following taxonomy may be seen from this result. Whenever an operator
is not completely reducible, undetermined functions occur under an integral sign,
in general with shifted arguments. If an undetermined function occurs in the
decomposition of the operator, the undetermined functions occurring in the solutions

have the same arguments in both members of a differential fundamental system. The following examples show that all decomposition types actually do occur.

Example 5.1. The type $\mathcal{L}_{xx}^1$ decomposition of the operator considered in Example 4.2 leads to the equation

$$Lz = l_2 l_1 z = (\partial_x + x\partial_y + 1)(\partial_x + \partial_y + y)z = 0.$$

There follows

$$\varphi_1(x, y) = x - y, \quad \psi_1(x, \bar{y}) = x - \bar{y}, \quad \mathscr{E}_1(x, y) = \exp\left(\tfrac{1}{2}x^2 - xy\right),$$

$$\varphi_2(x, y) = x^2 - 2y, \quad \psi_2(x, \bar{y}) = \tfrac{1}{2}(x^2 - \bar{y}), \quad \mathscr{E}_2(x, y) = e^{-x}.$$

Hence

$$z_1(x, y) = \exp\left(\tfrac{1}{2}x^2 - xy\right) F_1(x - y),$$

$$z_2(x, y) = \exp\left(\tfrac{1}{2}x^2 - xy\right) \int \exp\left(\tfrac{1}{2}x^2 - x\bar{y} - x\right) F_2(2\bar{y} + x^2 - 2x) dx \Big|_{\bar{y}=y-x};$$

F_1 and F_2 are undetermined functions. □

Example 5.2. The operator considered in Example 4.3 has a type $\mathcal{L}_{xx}^1$ decomposition; it leads to the equation

$$Lz = (\partial_x + (y^2 - x)\partial_y + 1)(\partial_x - x\partial_y - 1)z = 0.$$

There follows

$$\varphi_1(x, y) = y + \tfrac{1}{2}x^2, \quad \psi_1(x, \bar{y}) = \bar{y} - \tfrac{1}{2}x^2, \quad \text{and} \quad \mathscr{E}_1(x, y) = e^x.$$

Thus $z_1(x, y) = F(y + \tfrac{1}{2}x^2)e^x$; F is an undetermined function. The left factor yields

$$\varphi_2 = \frac{Ai'(x) + yAi(x)}{Bi'(x) + yBi(x)} \quad \text{and} \quad \mathscr{E}_2(x, y) = e^{-x}.$$

$Ai(x)$ and $Bi(x)$ are Airy functions defined in Abramowitz and Stegun [1], Sect. 10.4. The inverse function $\psi_2(x, \bar{y})$ does not exist in closed form. There follows

$$z_2(x, y) = e^x \left[\int e^{-2x} G\left(\frac{Ai'(x) - \tfrac{1}{2}x^2 Ai(x) + \bar{y}Ai(x)}{Bi'(x) - \tfrac{1}{2}x^2 Bi(x) + \bar{y}Bi(x)} \right) dx \right]_{\bar{y}=y+\tfrac{1}{2}x^2}. \quad \square$$

In Exercise 5.1 an equation closely related to the one of the preceding example is considered.

Example 5.3. Forsyth's operator considered in Example 4.4 has also a type $\mathcal{L}_{xx}^1$ decomposition leading to the equation

$$Lz \equiv l_2 l_1 z = \left(\partial_x + \partial_y + \frac{2}{x+y}\right)\left(\partial_x - \partial_y + \frac{2}{x+y}\right)z = 0.$$

There follows

$$\varphi_1(x, y) = x + y, \quad \psi_1(x, y) = \bar{y} - x, \quad \mathscr{E}_1(x, y) = \exp\left(\frac{2y}{x+y}\right),$$

$$\varphi_2(x, y) = x - y, \quad \psi_2(x, y) = x - \bar{y}, \quad \mathscr{E}_2(x, y) = -\frac{1}{x+y}.$$

Thus

$$z_1(x, y) = \exp\left(\frac{2y}{x+y}\right)F(x + y),$$

$$z_2(x, y) = \frac{1}{x+y}\exp\left(\frac{2y}{x+y}\right)\int \exp\left(\frac{2x - \bar{y}}{\bar{y}}\right)G(2x - \bar{y})dx\bigg|_{\bar{y}=x+y}.$$

F and G are undetermined functions.   □

Example 5.4. The type $\mathcal{L}_{xx}^2$ decomposition of the operator $L = Lclm(l_2, l_1)$ considered in Example 4.5 leads to the equations

$$l_2 z = \left(\partial_x - \left(\frac{y}{x} + xy^2\right)\partial_y + xy\right)z = 0, l_1 z = \left(\partial_x - \left(\frac{y}{x} - xy^2\right)\partial_y - xy\right)z = 0.$$

There follows

$$\varphi_{1,2}(x, y) = \frac{1}{xy} \pm x, \quad \psi_{1,2}(x, y) = \frac{1}{x(\bar{y} \mp x)}, \quad \mathscr{E}_{1,2}(x, y) = xy$$

with the result

$$z_1(x, y) = xyF_1\left(\frac{1}{xy} + x\right) \quad \text{and} \quad z_2(x, y) = xyF_2\left(\frac{1}{xy} - x\right);$$

F_1 and F_2 are undetermined functions. The special solution of $Gcrd(l_2, l_1)z = 0$ is $z = xy$, it is obtained by choosing $F = G = 1$.   □

Example 5.5. Miller's operator has been considered in Example 4.6. Its type $\mathcal{L}_{xx}^3$ decomposition yields the factor

$$l(\Phi) = \partial_x + \frac{y}{x}\partial_y + \frac{1}{x}\frac{1 + x^2\Phi(\varphi)}{1 - x^2\Phi(\varphi)}$$

where $\varphi = \frac{y}{x}$ and Φ is an undetermined function. There follows

$$\varphi(x, y) = \frac{y}{x}, \quad \psi(x, \bar{y}) = x\bar{y}, \quad \mathscr{E}(x, y) = \Phi(\varphi)x - \frac{1}{x}.$$

Choosing $\Phi = 0$ and $\Phi \to \infty$ the solutions

$$z_1(x, y) = xF_1(\varphi) \quad \text{and} \quad z_2(x, y) = \frac{1}{x}F_2(\varphi)$$

are obtained; F_1 and F_2 are undetermined functions. □

Solutions of reducible equations with mixed leading derivative ∂_{xy} and principal divisors are considered next.

Proposition 5.2. *Let a reducible second-order equation*

$$Lz \equiv (\partial_{xy} + A_1\partial_x + A_2\partial_y + A_3)z = 0$$

be given with $A_i \in \mathbb{Q}(x, y)$, $i = 1, 2, 3$. Define $l \equiv \partial_x + b$, $k \equiv \partial_y + c$; $b, c \in \mathbb{Q}(x, y)$,

$$\varepsilon_l(x, y) \equiv \exp\left(-\int b(x, y)dx\right) \quad and \quad \varepsilon_k(x, y) \equiv \exp\left(-\int c(x, y)dy\right).$$

A differential fundamental system has the following structure for decompositions into principal divisors.

$$\mathscr{L}^1_{xy} : z_1(x, y) = F(y)\varepsilon_l(x, y), \quad z_2(x, y) = \varepsilon_l \int \frac{\varepsilon_k(x, y)}{\varepsilon_l(x, y)}G(x)dx;$$

$$\mathscr{L}^2_{xy} : z_1(x, y) = F(x)\varepsilon_k(x, y), \quad z_2(x, y) = \varepsilon_k \int \frac{\varepsilon_l(x, y)}{\varepsilon_k(x, y)}G(y)dy;$$

$$\mathscr{L}^3_{xy} : z_1(x, y) = F(y)\varepsilon_l(x, y), \quad z_2(x, y) = G(x)\varepsilon_k(x, y).$$

F and G are undetermined functions of a single argument.

Proof. It is based on Theorem 4.2 and Lemma B.1; the notation is the same as in this theorem. For decomposition type $\mathscr{L}^1_{xy}$, (B.5) applied to the factor l yields the above solution $z_1(x, y)$. The left factor equation $w_y + cw = 0$ has the solution $w = G(x)\varepsilon_k(x, y)$. The second solution $z_2(x, y)$ then follows from $z_x + bz = G(x)\varepsilon_k(x, y)$. Interchanging k and l yields the result for decomposition type $\mathscr{L}^2_{xy}$. The two first-order right factors of decomposition type $\mathscr{L}^3_{xy}$ yield the given expressions for both $z_1(x, y)$ and $z_2(x, y)$ as it is true for $z_1(x, y)$ in the previous case. □

The subsequent examples show solutions for all three decomposition types as described above.

Example 5.6. Consider the equation

$$z_{xy} - \frac{x}{y}z_x + yz_y - (x-1)z = \left(\partial_y - \frac{x}{y}\right)(\partial_x + y)z = 0$$

with a type $\mathcal{L}_{xy}^1$ decomposition. The right factor yields $z_1(x,y) = F(y)\exp(-xy)$. The homogeneous left-factor equation $w_x - \frac{x}{y}w = 0$ has the solution $w = G(x)y^x$. Taking it as inhomogeneity of the right-factor equation yields

$$z_2(x,y) = \exp(-xy)\int G(x)\exp(xy)y^x dx.$$

Permuting the two factors leads to the closely related equation

$$z_{xy} - \frac{x}{y}z_x + yz_y - \left(x - \frac{1}{y}\right)z = (\partial_x + y)\left(\partial_y - \frac{x}{y}\right)z = 0,$$

with a type $\mathcal{L}_{xy}^2$ decomposition. By similar arguments its solutions

$$z_1(x,y) = F(x)y^x \quad \text{and} \quad z_2(x,y) = y^x \int G(y)\exp(-xy)\frac{dy}{y^x}$$

follow; F and G are undetermined functions.   □

Example 5.7. The $\mathcal{L}_{xy}^2$ decomposition of Goursat's equation in Example 4.7 yields the equation $(\partial_x - y)\partial_y z = 0$. The right factor equation $\partial_y z = 0$ has the solution $z_1(x,y) = G(x)$; G is an undetermined function. The homogeneous equation $w_x - yw = 0$ corresponding to the left factor yields $w = F(y)\exp(xy)$; F is an undetermined function. Taking it as inhomogeneity of the right factor equation $z_y = F(y)\exp(xy)$, the solution $z_2(x,y) = \int F(y)\exp(xy)dy$ follows.   □

Example 5.8. The two arguments of the type $\mathcal{L}_{xy}^3$ decomposition of Example 4.8 yield the two solutions

$$z_1(x,y) = \exp\left(-xy - \tfrac{1}{2}y^2\right)F(x) \quad \text{and} \quad z_2(x,y) = \exp(-xy)\frac{1}{x}G(y);$$

F and G are undetermined functions.   □

According to Proposition 4.2 there are five more decompositions with non-principal divisors. They are considered next.

Proposition 5.3. *With the same notation as in Proposition 5.2 a differential fundamental system for the various decomposition types involving non-principal divisors may be described as follows.*

$$\mathcal{L}_{xy}^4 : z_1(x, y) = \sum_{i=0}^{m-1} f_i(x, y) F^{(i)}(y), z_2(x, y) = \sum_{i=0}^{n-1} g_i(x, y) G^{(i)}(x);$$

$$\mathcal{L}_{xy}^5 : z_1(x, y) = \sum_{i=0}^{m-1} f_i(x, y) F^{(i)}(y), z_2(x, y) = \sum_{i=0}^{m-1} g_i(x, y) \int h_i(x, y) G(x) dx;$$

$$\mathcal{L}_{xy}^6 : z_1(x, y) = \sum_{i=0}^{m-1} f_i(x, y) F^{(i)}(x), z_2(x, y) = \sum_{i=0}^{m} g_i(x, y) \int h_i(x, y) G(y) dx.$$

$$\mathcal{L}_{xy}^7 : z_1(x, y) = F(x)\varepsilon_k(x, y), \quad z_2(x, y) = \sum_{i=0}^{m-1} f_i(x, y) F^{(i)}(y),$$

$$\mathcal{L}_{xy}^8 : z_1(x, y) = F(y)\varepsilon_l(x, y), \quad z_2(x, y) = \sum_{i=0}^{m-1} g_i(x, y) F^{(i)}(x),$$

F and G are undetermined functions of a single argument; f, g and h are Liouvillian over the base field; they are determined by the coefficients A_1, A_2 and A_3 of the given equation.

Proof. Let $\mathbb{L}_{x^m}(L)$ be a Laplace divisor as defined in Proposition 4.2. The linear ode $\mathfrak{l}_m z = 0$ has the general solution $z = C_1 f_1(x, y) + \ldots + C_m f_m(x, y)$. The C_i are constants w.r.t. x; they are undetermined functions of y. This expression for z must also satisfy the equation $Lz = 0$. Because the operators L and $\mathfrak{l}_m$ combined are a Janet basis generating an ideal of differential dimension $(1, 1)$, by Kolchin's Theorem 2.1 it must be possible to express $C_1, \ldots, C_m$ in terms of a single function $F(y)$ and its derivatives $F', F'', \ldots, F^{(m-1)}$. This yields the first sum of the solution for decomposition type $\mathcal{L}_{xy}^4$. For the second Laplace divisor $\mathbb{L}_{y^n}(L)$ the same steps with x and y interchanged yield the second sum.

If there is a single Laplace divisor $\mathbb{L}_{x^m}(L)$ as in decomposition type $\mathcal{L}_{xy}^5$, the solution $z_1(x, y)$ is the same as above. In order to obtain the second solution, according to Theorem 4.3 define $w_1 \equiv Lz$ and $w_2 = \mathfrak{l}_m z$; then the quotient equations are $w_1 = 0$, $w_{2,y} + A_1 w_2 = 0$. The solution of the latter leads to $\mathfrak{l}_m = G(x) \exp\left(-\int A_1 dy\right)$, G an undetermined function. This is an inhomogeneous linear ode; according to the discussion in Chap. 1 on page 17 a special solution yields $z_2(x, y)$ as given above; the $g_i(x, y)$ and $h_i(x, y)$ are determined by the coefficients of $\mathfrak{l}_m$. The discussion for decomposition type $\mathcal{L}_{xy}^6$ is similar. $\square$

The preceding propositions subsume all results that are known for a linear pde with leading derivative ∂_{xy} from the classical literature back in the nineteenth century under the general principle of determining divisors of various types; there is no heuristics involved whatsoever, and the selection of possible divisors is complete. These results will be illustrated now by several examples. The first example taken from Forsyth shows how complete reducibility has its straightforward generalization if there are no principal divisors.

Example 5.9. For Forsyth's equation with type $\mathcal{L}_{xy}^4$ decomposition considered in Example 4.9 the sum ideal of the two divisors is

$$J = \left\langle\!\!\left\langle \mathfrak{l}_2 \equiv \partial_{xx} - \frac{2}{x-y}\partial_x + \frac{2}{(x-y)^2}, L, \mathfrak{k}_2 \equiv \partial_{yy} + \frac{2}{x-y}\partial_y + \frac{2}{(x-y)^2} \right\rangle\!\!\right\rangle;$$

its differential dimension is $(0,3)$. Thus $Jz = 0$ has a three-dimensional solution space with basis $\{x-y, (x-y)^2, xy(x-y)\}$.

The general solution of $\mathfrak{l}_2 z = 0$ is $z = C_1(x-y) + C_2 x(x-y)$ where C_1 and C_2 are undetermined functions of y. Substitution into $Lz = 0$ leads to the constraint $C_{1,y} + yC_{2,y} - C_2 = 0$ with the solution $C_1 = 2F(y) - yF'(y)$ and $C_2 = F'(y)$. Thus

$$z_1(x, y) = 2(x-y)F(y) + (x-y)^2 F'(y)$$

follows. The equation $\mathfrak{k}_2 z = 0$ has the solution $C_1(y-x) + C_2 y(y-x)$ where $C_{1,2}$ are undetermined functions of x. By a similar procedure as above there follows

$$z_2(x, y) = 2(y-x)G(x) + (y-x)^2 G'(x). \qquad\qquad \square$$

Example 5.10. A Laplace divisor has been determined for Imschenetzky's equation $Lz = (\partial_{xy} + xy\partial_x - 2y)z = 0$ in Example 2.14 on page 37; it yields a $\mathcal{L}_{xy}^5$ type decomposition. The equation $\partial_{xxx}z = 0$ has the general solution $C_1 + C_2 x + C_3 x^2$ where the C_i, $i = 1,2,3$ are constants w.r.t. x. Substituting it into $Lz = 0$ and equating the coefficients of x to zero leads to the system $C_{2,y} - 2yC_1 = 0$, $C_{3,y} - \frac{1}{2}yC_2 = 0$. The C_i may be represented as

$$C_1 = \frac{1}{y^2}F'' - \frac{1}{y^3}F', \quad C_2 = \frac{2}{y}F', \quad C_3 = F;$$

F is an undetermined function of y, $F' \equiv \dfrac{dF}{dy}$. It yields the solution

$$z_1(x, y) = x^2 F(y) + \frac{2xy^2 - 1}{y^3}F'(y) + \frac{1}{y^2}F''(y). \qquad (5.3)$$

From the decomposition

$$Lz = \begin{pmatrix} 1 & 0 \\ 0 & \partial_y + xy \end{pmatrix}\begin{pmatrix} w_1 \equiv z_{xy} + xyz_x - 2yz \\ w_2 \equiv z_{xxx} \end{pmatrix}$$

the equations $w_1 = 0$, $w_{2,y} + xyw_2 = 0$ are obtained with the solution $w_1 = 0$, $w_2 = G(x)\exp\left(-\frac{1}{2}xy^2\right)$. The resulting equation $z_{xxx} = G(x)\exp\left(-\frac{1}{2}xy^2\right)$ yields the second member

$$z_2(x, y) = \tfrac{1}{2} \int G(x) \exp\left(-\tfrac{1}{2}xy^2\right)x^2 dx$$

$$-x \int G(x) \exp\left(-\tfrac{1}{2}xy^2\right)x \, dx + \tfrac{1}{2}x^2 \int G(x) \exp\left(-\tfrac{1}{2}xy^2\right)dx \quad (5.4)$$

of a fundamental system. □

Example 5.11. The factor $\partial_y z = 0$ of the operator L in Example 4.11 yields the solution $z_1(x, y) = F(x)$. The Laplace divisor equation $\partial_{xx} z = 0$ leads to $z = C_1(y) + C_2(y)x$; from $Lz = 0$ the relation $y C_2' - C_1' = 0$ follows with the solution $C_1(y) = G(y) - y G'(y)$ and $C_2(y) = G(y)$, i.e. $z_2(xy) = (x+1)G(y) - y G'(y)$. □

Whenever an equation is solved by factorization the following problem occurs: How does the solution procedure proceed if not all irreducible right factors are known? This is particularly acute if there is a Laplace divisor involved because in general there is no guarantee that it may be found. This problem is studied in Exercise 5.9.

5.2 Solving Inhomogeneous Second Order Equations

Up to now homogeneous pde's have been solved. However, in various applications there occur inhomogeneous equations $Lz = R$ with $R \neq 0$, e.g. when third-order homogeneous equations are considered that are reducible but not completely reducible. The coefficients of L at the left hand side of an inhomogeneous equation determine the base field. Like in the case of ordinary equations, the smallest field containing also R is called the *extended base field*. Fundamental systems for homogeneous equations $Lz = 0$ have been determined in the preceding section. For inhomogeneous equations in addition a special solution z_0 has to be found which is characterized by the relation $Lz_0 = R$. This is the subject of the present section. At first reducible equations with leading derivative ∂_{xx} are considered. A method similar to that in Proposition 1.3 is applied.

Proposition 5.4. *The notation is the same as in Proposition 5.1. Let a reducible equation*

$$Lz \equiv (\partial_{xx} + A_1\partial_{xy} + A_2\partial_{yy} + A_3\partial_x + A_4\partial_y + A_5)z = R \quad (5.5)$$

be given with $A_1, \ldots, A_5 \in \mathbb{Q}(x, y)$. A special solution $z_0(x, y)$ satisfying $Lz_0 = R$ may be obtained by solving first-order equations; $\mathscr{E}_i(x, y)$ for $i = 1, 2$ are given by (B.3). Three cases are distinguished.

(i) Type $\mathscr{L}_{xx}^1$ decomposition, $L = l_2 l_1$.

$$r(x, y) = \mathscr{E}_2(x, y) \int \frac{R(x, y)}{\mathscr{E}_2(x, y)}\Big|_{y=\psi_2(x,\bar{y})} dx \Big|_{\bar{y}=\varphi_2(x,y)}, \quad (5.6)$$

$$z_0(x, y) = \mathscr{E}_1(x, y) \int \frac{r(x, y)}{\mathscr{E}_1(x, y)}\Big|_{y=\psi_1(x,\bar{y})} dx\Big|_{\bar{y}=\varphi_1(x,y)}. \qquad (5.7)$$

(ii) *Type $\mathscr{L}_{xx}^2$ decomposition, $L = Lclm(l_2, l_1)$. If $a_1 \neq a_2$ then*

$$r = r_0 \int \frac{R}{a_2 - a_1} \frac{dy}{r_0} \quad and \quad r_0 = \exp\left(-\int \frac{b_1 - b_2}{a_1 - a_2} dy\right), \qquad (5.8)$$

$$z_0(x, y) = \mathscr{E}_1(x, y) \int \frac{r(x, y)}{\mathscr{E}_1(x, y)}\Big|_{y=\psi_1(x,\bar{y})} dx\Big|_{\bar{y}=\varphi_1(x,y)}$$
$$-\mathscr{E}_2(x, y) \int \frac{r(x, y)}{\mathscr{E}_2(x, y)}\Big|_{y=\psi_2(x,\bar{y})} dx\Big|_{\bar{y}=\varphi_2(x,y)}. \qquad (5.9)$$

(iii) *Type $\mathscr{L}_{xx}^3$ decomposition, $L = Lclm(l(\Phi))$ and $r(x, y) = \dfrac{R}{b_2 - b_1}$; z_0 is again given by (5.9); now $\varphi_1 = \varphi_2 = \varphi$ and $\psi_1 = \psi_2 = \psi$.*

Proof. In case (*i*), $Lz = l_2 l_1 z = R$. Defining $r(x, y)$ by $l_1 z = r$, the first-order equation $l_2 r = R$ for r follows. Its solution (5.6) has to be substituted into $l_1 z = r$. By construction, the special solution (5.7) of this equation is the desired solution of (5.5).

In case (*ii*), $L = Lclm(l_2, l_1)$; $r_i(x, y)$ are defined by $l_i z = r_i$, $i = 1, 2$. Reduction of $Lz = R$ w.r.t. these equations yield if $b_1 \neq b_2$

$$r_{1,x} + a_2 r_{1,y} + r_{2,x} + a_1 r_{2,y} + b_2 r_1 + b_1 r_2$$

$$-\frac{1}{b_1 - b_2}(b_{1,x} - b_{2,x} + a_2 b_{1,y} - a_1 b_{2,y})(r_1 + r_2) = R. \qquad (5.10)$$

If $a_1 \neq a_2$, choosing $r_1 = r$ and $r_2 = -r$ leads to

$$r_y + \frac{b_2 - b_1}{a_2 - a_1} r = \frac{R}{a_2 - a_1}$$

with solution (5.8). The sum of the special solutions of the equations $l_1 z = r$ and $l_2 z = -r$ is the desired solution $z_0(x, y)$ given in (5.9).

In case (*iii*), $a_1 = a_2 = a$; the transformation functions φ_i and ψ_i for $i = 1$ and $i = 2$ are pairwise identical to each other. Choosing again $r_1 = -r_2 = r$ the expression $r = \dfrac{R}{b_2 - b_1}$ is obtained from (5.10) for $b_1 \neq b_2$. $\square$

The values $r_1 = r$, $r_2 = -r$ chosen in case (*ii*) of the preceding proposition are not the only possible choices. In some applications the alternatives given in the subsequent corollary may be more favourable.

Corollary 5.1. *Two more choices for r_1 and r_2 in (5.10) are considered for the decomposition types $\mathscr{L}_{xx}^1$ and $\mathscr{L}_{xx}^2$ of the preceding proposition.*

(*i*) $r_1 = 0$; *then r_2 follows from*

$$r_{2,x} + a_1 r_{2,y} + \left(b_1 - \frac{1}{b_1 - b_2}(b_{1,x} - b_{2,x} + a_2 b_{1,y} - a_1 b_{2,y})\right) r_2 = R.$$

(*ii*) $r_2 = 0$; *then r_1 follows from*

$$r_{1,x} + a_2 r_{1,y} + \left(b_2 - \frac{1}{b_1 - b_2}(b_{1,x} - b_{2,x} + a_2 b_{1,y} - a_1 b_{2,y})\right) r_1 = R.$$

In case (i) the inhomogeneity is completely shifted to the equation for z_2; in case (ii) the same is true for z_1.

The similarity between the expressions (1.18) and (1.19) on the one hand, and (5.7) and (5.9) on the other should be noted. In either case, the special solution contains the inhomogeneity linearly under an integral sign; there are as many terms as there are factors. These results are illustrated now by some examples.

Example 5.12. Consider the equation

$$z_{xx} + (x + y - 1)z_{xy} - (x + y)z_{yy} + \tfrac{1}{x}z_x - \tfrac{1}{x}z_y$$

$$= \left(\partial_x + (x + y)\partial_y + \tfrac{1}{x}\right)(\partial_x - \partial_y)z = xy$$

with $a_1 = 1$, $b_1 = 0$, $a_2 = x + y$ and $b_2 = \tfrac{1}{x}$. The $\mathscr{L}_{xx}^1$ type decomposition of its left-hand side yields the differential fundamental system

$$z_1(x, y) = F(x + y), \quad z_2(x, y) = \int G\big((\bar{y} + 1)e^{-x}\big)\frac{dx}{x}\Big|_{\bar{y}=x+y}$$

for the homogeneous equation. Furthermore

$$\varphi_1(x, y) = y + x, \quad \psi_1(x, \bar{y}) = \bar{y} - x, \quad \mathscr{E}_1(x, y) = 1,$$

$$\varphi_2(x, y) = (x + y + 1)e^{-x}, \quad \psi_2(x, \bar{y}) = \bar{y}e^x - x - 1, \quad \mathscr{E}_2(x, y) = \tfrac{1}{x}.$$

Equation 5.6 leads to

$$r(x, y) = -\tfrac{1}{4}x^3 + \tfrac{2}{3}x^2 + xy - x - 2y + \tfrac{2}{x}(y + 1).$$

Using this expression (5.7) yields

$$z_0(x, y) = 2(x + y + 1)\log(x) - \tfrac{1}{16}x^4 + \tfrac{7}{18}x^3 + \tfrac{1}{2}x^2 y - \tfrac{3}{2}x^2 - 2xy - 2x. \quad \square$$

Example 5.13. Consider the equation

$$z_{xx} + (y+1)z_{xy} + z_x + (y+1)z_y = Lclm(\partial_x + (y+1)\partial_y, \partial_x + 1)z = xy \quad (5.11)$$

with type $\mathcal{L}_{xx}^2$ decomposition; $a_1 = b_2 = 0$, $a_2 = y + 1$, $b_1 = 1$ and $R = xy$. A differential fundamental system for the homogeneous equation is

$$z_1(x, y) = F(y)e^{-x}, \quad z_2(x, y) = G\big((y + 1)e^{-x}\big).$$

Furthermore

$$\varphi_1(x, y) = y, \quad \psi_1(x, \bar{y}) = \bar{y}, \quad \mathscr{E}_1(x, y) = e^{-x},$$
$$\varphi_2(x, y) = (y + 1)e^{-x}, \quad \psi_2(x, \bar{y}) = \bar{y}e^x - 1, \quad \mathscr{E}_2(x, y) = 1.$$

Equation 5.8 yields

$$r = (y + 1)\int \frac{xy}{y+1}\frac{dy}{y+1} = x(y + 1)\log(y + 1) - xy.$$

Finally substitution of these values into (5.9) leads to

$$z_0(x, y) = xy - \tfrac{1}{2}x^2 + 2(x - y); \qquad\qquad (5.12)$$

F and G are undetermined functions.   □

The next example is a first application of Corollary 5.1.

Example 5.14. The same equation as in the preceding example is considered. Now the two alternatives of Corollary 5.1 are applied. By case (i), $r_1 = 0$ and $r_{2,x} + r_2 = xy$; the latter equation yields $r_2 = (x - 1)y$.

$$z_{1,x} + z_1 = 0 \quad \text{and} \quad z_{2,x} + (y + 1)z_{2,y} = (x - 1)y$$

with the solutions

$$z_1(x, y) = F(y)e^{-x} \quad \text{and} \quad z_2(x, y) = G\big((y + 1)e^{-x}\big) + xy - \tfrac{1}{2}x^2 + 2(x - y); \qquad (5.13)$$

F and G are undetermined functions.

By case (ii), $r_2 = 0$ and $r_{1,x} + (y + 1)r_{1,y} = xy$; the latter equation yields $r_1 = xy - \tfrac{1}{2}x^2 + x - y$. Consequently,

$$z_{1,x} + z_1 = xy - \tfrac{1}{2}x^2 + x - y \quad \text{and} \quad z_{2,x} + (y + 1)z_{2,y} = 0$$

with the solutions

$$z_1(x, y) = F(y)e^{-x} + xy - \tfrac{1}{2}x^2 + 2(x - y) \quad \text{and} \quad z_2(x, y) = G\big((y + 1)e^{-x}\big);$$

their sum $z_1 + z_2$ is the same as in the previous case.   □

Second-order inhomogeneous equations with leading derivative ∂_{xy} are considered next. It is assumed that the corresponding homogeneous equation allows a nontrivial decomposition according to Theorem 4.2; at first operators with principal divisors are considered.

For an operator with a type $\mathcal{L}_{xy}^3$ decomposition the following explicit representation will be useful.

Lemma 5.1. *Let an operator L be completely reducible with type $\mathcal{L}_{xy}^3$ decomposition*

$$L \equiv \partial_{xy} + A_1\partial_x + A_2\partial_y + A_3 = Lclm(\partial_x + b, \partial_y + c);$$

then $b_y = c_x$ and

$$L = \partial_{xy} + c\partial_x + b\partial_y + \tfrac{1}{2}(b_y + c_x + 2bc). \tag{5.14}$$

The proof is similar to that of Lemma 1.2 and is omitted. The condition on b and c guarantees the principality of the intersection, see Theorem 2.3. It will be applied next for solving the corresponding inhomogeneous equations.

Proposition 5.5. *Let a reducible equation*

$$Lz \equiv (\partial_{xy} + A_1\partial_x + A_2\partial_y + A_3)z = R \tag{5.15}$$

be given with $A_1, A_2, A_3 \in \mathbb{Q}(x, y)$. Define $l \equiv \partial_x + b$ and $k \equiv \partial_y + c$ whereupon $b, c \in \mathbb{Q}(x, y)$; define

$$\varepsilon_l(x, y) \equiv \exp\Big(-\int b(x, y)dx\Big) \quad \text{and} \quad \varepsilon_k(x, y) \equiv \exp\Big(-\int c(x, y)dy\Big).$$

A special solution $z_0(x, y)$ satisfying $Lz_0 = R$ may be obtained as follows.

(i) Type $\mathcal{L}_{xy}^1$ decomposition, $L = kl$.

$$z_0(x, y) = \varepsilon_l(x, y) \int \frac{\varepsilon_k(x, y)}{\varepsilon_l(x, y)} \int \frac{R(x, y)}{\varepsilon_k(x, y)} dy dx. \tag{5.16}$$

(ii) Type $\mathcal{L}_{xy}^2$ decomposition, $L = lk$.

$$z_0(x, y) = \varepsilon_k(x, y) \int \frac{\varepsilon_l(x, y)}{\varepsilon_k(x, y)} \int \frac{R(x, y)}{\varepsilon_l(x, y)} dx dy. \tag{5.17}$$

(iii) *Type $\mathcal{L}^3_{xy}$ decomposition $L = Lclm(k,l)$. Special solutions are given by (5.16) and (5.17).*

Proof. In case (i) $Lz = klz = R$; $r(x, y)$ is defined by $lz = r$; it has to satisfy $kr = R$ with the solution $r(x, y) = \varepsilon_k(x, y) \int \dfrac{R(x, y)}{\varepsilon_k(x, y)} dx$. Taking it as inhomogeneity in the first-order equation $lz = r$ yields (5.16).

Case (ii) is essentially obtained by exchanging x and y.

In case (iii) $Lz = Lclm(k,l)z = R$; let $z \equiv z_k + z_l$. Define $r_k(x, y)$ by $kz_k = r_k$ and $r_l(x, y)$ by $lz_l = r_l$. Reduction of $Lz = R$ w.r.t. to these first-order equations yields

$$r_{k,x} + r_{l,y} + br_k + cr_l = R. \tag{5.18}$$

Choosing either $r_k(x, y) = 0$ or $r_l(x, y) = 0$, the above solutions (5.16) or (5.17) are obtained.   □

Relation (5.18) shows clearly how the inhomogeneity R may be generated from r_1 and r_2; depending on the particular problem a different choice than the one given above may be more favourable.

Although the expressions (5.16) and (5.17) have a similar structure as (5.7), they are in fact much simpler; from the definition of the $\varepsilon_i(x, y)$ it is obvious that they are Liouvillian over the extended base field. This is not true for (5.7) due to the occurrence of the $\mathscr{E}_i(x, y)$; in general the latter require the first integrals of a first-order ode.

The exponential integral $Ei(x)$ that occurs in the subsequent examples is defined in Abramowitz and Stegun [1], page 227.

Example 5.15. The equation

$$\left(\partial_{xy} - \frac{1}{x}\partial_x + y\partial_y - \frac{1}{x^2}(xy - 1)\right)z = (\partial_x + y)\left(\partial_y - \frac{1}{x}\right)z = x + y + 1 \tag{5.19}$$

has a type $\mathcal{L}^2_{xy}$ decomposition. With the notation of the above proposition there follows $b = y$, $c = -\frac{1}{x}$ and $R = x + y + 1$. Furthermore $\varepsilon_k(x, y) = \exp\left(\frac{y}{x}\right)$ and $\varepsilon_l(x, y) = \exp(-xy)$. Substitution into (5.16) yields

$$z_0(x, y) = \frac{1}{x}(x^2 + x + 1)\exp\left(\frac{y}{x}\right)Ei\left(-\frac{y}{x}\right) + \frac{1}{y} - x. \qquad\qquad □$$

Example 5.16. Consider the equation

$$z_{xy} + xz_x + yz_y + (xy + 1)z = Lclm(\partial_y + x, \partial_x + y)z = \frac{x}{y} + 1$$

with a type $\mathcal{L}^3_{xy}$ decomposition. Here $b = y$ and $c = x$. A differential fundamental system for the corresponding homogeneous equation is

$$z_1(x, y) = F(x)\exp(-xy), \quad z_2(x, y) = G(y)\exp(-xy).$$

The integrals yield $\varepsilon_k(x, y) = \varepsilon_l(x, y) = \exp(-xy)$. Substitution into (5.16) or (5.17) yields the special solution

$$z_0(x, y) = \tfrac{1}{2}(x^2 + 2)\exp(-xy)Ei(xy) - \frac{1}{2y^2}(xy - 1).$$ □

Example 5.17. The equation considered now differs slightly from the preceding example.

$$z_{xy} + (x + 1)z_x + yz_y + (xy + y + 1)z$$
$$= Lclm(\partial_y + x + 1, \partial_x + y)z = \tfrac{x}{y} + 1.$$ (5.20)

Here $b = y$ and $c = x + 1$. A differential fundamental system is

$$z_1(x, y) = F(x)\exp(-xy - y), \quad z_2(x, y) = G(y)\exp(-xy).$$

The integrals yield $\varepsilon_k = \exp(-xy - y)$ and $\varepsilon_l = \exp(-xy)$. The special solutions $z_{0,1}$ and $z_{0,2}$ obtained from (5.16) and (5.17) appear different now.

$$z_{0,1} = \frac{1}{2y^2}(y + 1 - xy) + \tfrac{1}{2}(x^2 + 1)\exp(-xy - y)Ei(xy + y),$$

$$z_{0,2} = \exp(-xy - y)\left(Ei(xy + y) + \int Ei(xy + y)x\,dx\right).$$

Yet, they are essentially the same; their difference $z_{0,1} - z_{0,2}$ satisfies the homogeneous equation corresponding to (5.20), i.e. they are different by the choice of F and G.   □

Operators with a type $\mathscr{L}_{xy}^4$, $\mathscr{L}_{xy}^5$ or $\mathscr{L}_{xy}^6$ decomposition involving a Laplace divisor of order 2 or 3 are considered next.

Proposition 5.6. *Let the equation*

$$(\partial_{xy} + A_1\partial_x + A_2\partial_y + A_3)z = R$$ (5.21)

be given. If the corresponding homogeneous equation has a Laplace divisor $\mathbb{L}_{x^m}(L)$, $m = 2$ or $m = 3$, the following linear inhomogeneous ode's exist.

$$m = 2 : z_{xx} + a_1 z_x + a_0 z = r \quad \text{where} \quad r_y + A_1 r = R_x + (a_1 - A_2)R,$$

$$m = 3 : \begin{cases} z_{xxx} + a_2 z_{xx} + a_1 z_x + a_0 z = r \quad \text{where} \\ r_y + A_1 r = R_{xx} + (a_2 - A_2)R_x + (a_1 - a_1 A_2 + A_2^2 - 2A_{2,x})R. \end{cases}$$

If there is a Laplace divisor $\mathbb{L}_{y^n}(L)$, $n = 2$ or $n = 3$, the following linear inhomogeneous ode's exist.

$$n = 2 : z_{yy} + b_1 z_y + b_0 z = r \quad \text{where} \quad r_x + A_2 r = R_y + (b_1 - A_1)R,$$

$$n = 3 : \begin{cases} z_{yyy} + b_2 z_{yy} + b_1 z_y + b_0 z = r \quad \text{where} \\ r_x + A_2 r = R_{yy} + (b_2 - A_1)R_y + (b_1 - b_2 A_1 + A_1^2 - 2A_{1.y})R. \end{cases}$$

Proof. For $m = 2$ a second-order linear ode $z_{xx} + a_1 z_x + a_0 z = r$ must exist such that it forms a coherent system combined with (5.21). There are three constraints not involving R and r; they are satisfied due to the assumption that a Laplace divisor for the homogeneous problem does exist. The single condition involving the inhomogeneities is the first-order ode for r given above for $m = 2$. Similar arguments apply for any $m = 3$ and for Laplace divisors $\mathbb{L}_{y^n}(L)$.   □

The first-order equations for the inhomogeneities r occurring in the above proposition guarantee that they are Liouvillian over the extended base field of (5.21). A special solution of this equation is the desired inhomogeneity. If the first-order equation for r is homogeneous, there follows $r = 0$.

A special solution of (5.21) is obtained by a modification of the proceeding described in the proof of Proposition 5.3; essentially the non-vanishing inhomogeneities have to be taken into account. This is explained in the subsequent examples.

Example 5.18. The solution of the inhomogeneous equation

$$z_{xy} + \frac{2}{x - y} z_x - \frac{2}{x - y} z_y - \frac{4}{(x - y)^2} z = 1 \tag{5.22}$$

has to be determined. A differential fundamental system for the corresponding homogeneous equation with type $\mathscr{L}_{xy}^4$ decomposition has already been obtained in Example 5.9.

$$z_1(x, y) = 2(x - y)F(y) + (x - y)^2 F'(y),$$
$$z_2(x, y) = 2(x - y)G(x) + (x - y)^2 G'(x).$$

The Laplace divisor $\mathfrak{l}_2$ given in Example 4.9 is chosen. By Proposition 5.6, the inhomogeneity r has to satisfy $r_y + \frac{2}{x - y} r = 0$. With the special solution $r = 0$ the inhomogeneous Laplace divisor equation becomes

$$z_{xx} - \frac{2}{x - y} z_x + \frac{2}{(x - y)^2} z = 0;$$

its general solution is $z = C_1(x - y) + C_2 x(x - y)$ with C_1 and C_2 undetermined functions of y. Substituting this expression into the given equation leads to the constraint $C_{1.y} + y C_{2.y} - C_2 + 1 = 0$. Choosing $C_1 = 0, C_2 = 1$ yields the special solution $z_0 = x(x - y)$ of (5.22).   □

A modification of (5.22) is considered in Exercise 5.2. In the next example there occurs a non-vanishing inhomogeneity r in the Laplace divisor equation.

Example 5.19. Imschenetzky's equation has been considered before in Example 5.10. An inhomogeneous version is

$$z_{xy} + xyz_x - 2yz = \frac{y}{x}.$$

According to Proposition 5.6 the inhomogeneity r is determined by $r_y + xyr = \frac{2y}{x^3}$ with the special solution $r = \frac{2}{x^4}$. It leads to $z_{xxx} = \frac{2}{x^4}$ with the general solution

$$z = C_1 + C_2 x + C_3 x^2 - \frac{1}{3x}.$$

The three undetermined functions C_i of y must satisfy $C_{3,y} - \frac{1}{2}yC_2 = 0$ and $C_{2,y} - 2yC_1 = 0$. Three sets of solutions and the corresponding special solution $z_0(x, y)$ are listed in the table below.

C_1	C_2	C_3	$z_0(x, y)$
0	0	0	$-\frac{1}{3x}$
0	0	1	$\frac{3x^3 - 1}{3x}$
$-\frac{1}{y^3}$	$\frac{2}{y}$	y	$x^2 y + \frac{2x}{y} - \frac{1}{y^3} - \frac{1}{3x}$

The relation between the various special solutions is discussed in Exercise 5.3.   □

Special solutions for inhomogeneous versions of Imschenetzky's equation may be rather involved, some additional examples are discussed in Exercise 5.4.

5.3   Solving Equations Corresponding to the Ideals $\mathbb{J}_{xxx}$ and $\mathbb{J}_{xxy}$

Inhomogeneous systems corresponding to the ideals $\mathbb{J}_{xxx}$ and $\mathbb{J}_{xxy}$ (see page 45, Lemmas 2.2 and 2.3) occur when solving reducible third-order equations with non-principal left intersection ideal. The following results will be applied in this chapter on decompositions of such equations; ideals $\mathbb{J}_{xxx}$ are considered first.

Lemma 5.2. *Consider the system*

$$L_1 z \equiv (\partial_{xxx} + p_1 \partial_{xyy} + p_2 \partial_{yyy} + p_3 \partial_{xx} + p_4 \partial_{xy} + p_5 \partial_{yy}$$
$$+ p_6 \partial_x + p_7 \partial_y + p_8)z = R_1, \tag{5.23}$$
$$L_2 z \equiv \partial_{xxy} + q_1 \partial_{xyy} + q_2 \partial_{yyy} + q_3 \partial_{xx} + q_4 \partial_{xy} + q_5 \partial_{yy}$$
$$+ q_6 \partial_x + q_7 \partial_y + q_8)z = R_2.$$

It is assumed that the coefficients p_i and q_j satisfy the coherence conditions of Lemma 2.2 such that L_1 and L_2 generate an ideal $\mathbb{J}_{xxx}$. Consistency requires that the right-hand sides R_1 and R_2 obey

$$R_{1,y} - R_{2,x} + q_1 R_{2,y} + q_3 R_1 - (p_3 + q_1 q_3 - q_4) R_2 = 0. \qquad (5.24)$$

Proof. Due to the coherence of the left-hand sides, the single integrability condition reduces to constraint (5.34). □

It is assumed now that the ideal $\langle L_1, L_2 \rangle$ generated by the operators defined in the above lemma is the intersection of two first-order operators with leading derivative ∂_x as described in Theorem 2.2 on page 47. As a consequence, a differential fundamental system for the corresponding homogeneous equations is known. The following theorem provides the means for obtaining a special solution of the inhomogeneous system (5.23).

Theorem 5.1. *Let the system (5.23) be given as in the above lemma. Furthermore, let both L_1 and L_2 allow first-order right factors $l_1 \equiv \partial_x + a_1 \partial_y + b_1$ and $l_2 \equiv \partial_x + a_2 \partial_y + b_2$; the $\mathscr{E}_i(x, y)$, $i = 1, 2$, are given by (B.3). Then a special solution $z_0(x, y)$ is*

$$z_0(x, y) = \mathscr{E}_1(x, y) \int \left. \frac{r(x, y)}{\mathscr{E}_1(x, y)} \right|_{y = \psi_1(x, \bar{y})} dx \Big|_{\bar{y} = \varphi_1(x, y)}$$

$$- \mathscr{E}_2(x, y) \int \left. \frac{r(x, y)}{\mathscr{E}_2(x, y)} \right|_{y = \psi_2(x, \bar{y})} dx \Big|_{\bar{y} = \varphi_2(x, y)}. \qquad (5.25)$$

The inhomogeneity $r(x, y)$ is determined by the system

$$r_{xy} + \frac{b_1 - b_2}{a_1 - a_2} r_x + \left(p_3 + q_3(a_1 + a_2) - b_1 - b_2 + 2 \frac{(a_1 - a_2)_x}{a_1 - a_2} + \frac{a_{1,y} a_2 - a_{2,y} a_1}{a_1 - a_2} \right) r_y$$

$$+ \frac{1}{a_1 - a_2} \left((p_3 + q_3(a_1 + a_2) - b_1 - b_2)(b_1 - b_2) + 2(b_1 - b_2)_x + b_{1,y} a_2 - b_{2,y} a_1 \right) r$$

$$= -\frac{1}{a_1 - a_2} (R_1 + (a_1 + a_2) R_2),$$

$$r_{yy} + \left(q_3 + \frac{(a_1 - a_2)_y}{a_1 - a_2} + \frac{b_1 - b_2}{a_1 - a_2} \right) r_y + \left(\frac{(b_1 - b_2)_y}{a_1 - a_2} + q_3 \frac{b_1 - b_2}{a_1 - a_2} \right) r = -\frac{R_2}{a_1 - a_2}.$$

$$(5.26)$$

By construction, the left-hand sides form a Janet basis and Proposition 5.6 applies.

Proof. Define the functions $r_i(x, y)$ and $z_i(x, y)$ by $l_i z_i = r_i$ for $i = 1, 2$ and the requirement that l_1 and l_2 are right factors of both L_1 and L_2. Reducing the system (5.23) for $z = z_1 + z_2$ w.r.t. these first-order equations, and expressing p_1, q_1, p_2 and q_2 in terms of a_1, b_1, a_2 and b_2 by means of (A.2) leads to a system that is linear and homogeneous in r_1, r_2, R_1, and R_2; transforming it into a Janet basis in *grlex* term order with $r_1 \succ r_2 \succ R_1 \succ R_2$ and $x \succ y$ yields

$$r_{1,xx} - a_1 r_{1,xy} - (a_1 a_2 + a_2^2) r_{1,yy} + r_{2,xx} - a_2 r_{2,xy} - (a_1^2 + a_1 a_2) r_{2,yy}$$

$$+(p_3 - b_1) r_{1,x} - (2a_{1,x} - a_{1,y} a_1 + p_3 a_1 - p_4 - 2a_1 b_1) r_{1,y}$$

$$+(p_3 - b_2) r_{2,x} - (2a_{2,x} - a_{2,y} a_2 + p_3 a_2 - p_4 - 2a_2 b_2) r_{2,y} \qquad (5.27)$$

$$-(2b_{1,x} - b_{1,y} a_1 + p_3 b_1 - p_6 - b_1^2) r_1$$

$$-(2b_{2,x} - b_{2,y} a_2 + p_3 b_2 - p_6 - b_2^2) r_2 = R_1,$$

$$r_{1,xy} + a_2 r_{1,yy} + r_{2,xy} + a_1 r_{2,yy}$$

$$+q_3 r_{1,x} - (a_{1,y} + q_3 a_1 - q_4 + b_1) r_{1,y}$$

$$+q_3 r_{2,x} - (a_{2,y} + q_3 a_2 - q_4 + b_2) r_{2,y} - (b_{1,y} + q_3 b_1 - q_6) r_1$$

$$-(b_{2,y} + q_3 b_2 - q_6) r_2 = R_2. \qquad (5.28)$$

Choosing $r_1 = -r_2 = r$, the system (5.26) for $r(x, y)$ follows. Due to this choice, a special solution $z_0(x, y)$ of (5.23) is obtained as sum of two special solutions of $l_1 z_1 = r$ and $l_2 z_2 = -r$ as given by (5.25). □

The above theorem guarantees that a special solution of the system (5.23) is Liouvillian over the base field of a generic system of this type; this follows from Proposition 5.6, i.e. the existence of two first-order factors L_1 and L_2, and using both factors for the solution procedure. If only a single factor is known, this is not true in general.

Example 5.20. Let the system

$$z_{xxx} - x^2 z_{xyy} + 3z_{xx} + (2x + 3)z_{xy} - x^2 z_{yy} + 2z_x + (2x + 3)z_y$$
$$= 2x^2 + 5x + 2y + 3,$$

$$z_{xxy} + x z_{xyy} - \frac{1}{x} z_{xx} - \frac{1}{x} z_{xy} + x z_{yy} - \frac{1}{x} z_x - \left(1 + \frac{1}{x}\right) z_y$$

$$= -\frac{1}{x}(x^2 + x + y + 1) \qquad (5.29)$$

be given. The right-hand sides obey the constraint (5.24). Both L_1 and L_2 have the right divisors $l_1 \equiv \partial_x + 1$ and $l_2 \equiv \partial_x + x\partial_y$, i.e. Theorem 5.1 applies. There follows $\mathcal{E}_1(x, y) = e^{-x}$ and $\mathcal{E}_2(x, y) = 1$. A differential fundamental system is

$$z_1(x, y) = F(y)e^{-x} \quad \text{and} \quad z_2(x, y) = G\left(y - \tfrac{1}{2}x^2\right).$$

System (5.26) leads to

$$r_{xy} - \frac{1}{x}r_x + \frac{x+2}{x}r_y - \frac{1}{x}r = \frac{1}{x}(x^2 + 4x + y + 2),$$

$$r_{yy} - \frac{2}{x}r_y + \frac{1}{x^2}r = -\frac{1}{x^2}(x^2 + x + y + 1). \tag{5.30}$$

Applying Proposition 5.6 the special solution $r(x, y) = -x^2 - 3x - y - 1$ is obtained. Substitution into (5.25) and some simplification yields $z_0 = xy$. □

It may occur that for a particular system (5.23) case (i) or (ii) of the following corollary leads to a simpler system for the desired right-hand side of the chosen first order equation.

Corollary 5.2. *Two additional choices for $r_1(x, y)$ and $r_2(x, y)$ in (5.27) and (5.28) are as follows.*

(i) $r_1 = 0$, *then r_2 obeys the system*

$$r_{2,xx} - a_1^2 r_{2,yy} + (p_3 + q_3 a_2 - b_2)r_{2,x}$$
$$-(2a_{2,x} + p_3 a_2 - p_4 + q_3 a_2^2 - q_4 a_2 - a_2 b_2)r_{2,y}$$
$$-(2b_{2,x} + p_3 b_2 - p_6 + q_3 a_2 b_2 - q_6 a_2 - b_2^2)r_2 = R_1 + a_2 R_2,$$
$$r_{2,xy} + a_1 r_{2,yy} + q_3 r_{2,x} - (a_{2,y} + q_3 a_2 - q_4 + b_2)r_{2,y}$$
$$-(b_{2,y} + q_3 b_2 - q_6)r_2 = R_2. \tag{5.31}$$

(ii) $r_2 = 0$, *then r_1 obeys the system*

$$r_{1,xx} - a_2^2 r_{1,yy} + (p_3 + q_3 a_1 - b_1)r_{1,x}$$
$$-(2a_{1,x} + p_3 a_1 - p_4 + q_3 a_1^2 - q_4 a_1 - a_1 b_1)r_{1,y}$$
$$-(2b_{1,x} + p_3 b_1 - p_6 + q_3 a_1 b_1 - q_6 a_1 - b_1^2)r_1 = R_1 + a_1 R_2,$$
$$r_{1,xy} + a_2 r_{1,yy} + q_3 r_{1,x} - (a_{1,y} + q_3 a_1 - q_4 + b_1)r_{1,y}$$
$$-(b_{1,y} + q_3 b_1 - q_6)r_1 = R_2. \tag{5.32}$$

Example 5.21. The system (5.29) of the preceding example is considered again. Choosing $r_1 = 0$ according to case (i) of Corollary 5.2 leads to the system

$$r_{2,xx} + 2r_{2,x} + r_2 = x^2 + 4x + y + 2,$$

$$r_{2,xy} - \frac{1}{x}r_{2,x} + \left(1 - \frac{1}{x}\right)r_{2,y} - \frac{1}{x}r_2 = -\frac{1}{x}(x^2 + x + y + 1).$$

The general solution of the first equation is $r_2 = F(y)e^{-x} + G(y)xe^{-x} + x^2 + y$ where F and G are undetermined functions of y. Substitution into the latter leads to the constraint $F' + G = 0$; it yields the general solution of the full system $r_2 = F(y)e^{-x} - F'(y)xe^{-x} + x^2 + y$. The special solution corresponding to $F = 0$ leads to the equation $z_x + xz_y = x^2 + y$. By Corollary B.1 a special solution is $z_0 = xy$. $\square$

In the preceding example the choice of alternative (i) was particularly favourable because the equation with leading derivative $r_{2,xx}$ is an ordinary equation with constant coefficients; this may not be true in general.

Systems of linear pde's corresponding to the ideal $\mathbb{J}_{xxy}$ are the next subject to be considered. Due to the divisor with leading derivative ∂_y this case is simpler than the preceding one.

Lemma 5.3. *Consider the system*

$$K_1 z \equiv (\partial_{xxy} + p_1 \partial_{yyy} + p_2 \partial_{xx} + p_3 \partial_{xy} + p_4 \partial_{yy} + p_5 \partial_x + p_6 \partial_y + p_7)z = R_1,$$

$$K_2 z \equiv (\partial_{xyy} + q_1 \partial_{yyy} + q_2 \partial_{xx} + q_3 \partial_{xy} + q_4 \partial_{yy} + q_5 \partial_x + q_6 \partial_y + q_7)z = R_2.$$
$$(5.33)$$

It is assumed that the coefficients p_i and q_j satisfy the coherence conditions of Lemma 2.3 such that K_1 and K_2 generate an ideal $\mathbb{J}_{xxy}$. Consistency requires that the right-hand sides R_1 and R_2 obey

$$R_{1,y} - R_{2,x} + q_1 R_{2,y} - (p_2 + q_1 q_2 - q_3)R_1 - (p_3 + q_1 q_3 - q_4)R_2 = 0. \quad (5.34)$$

Proof. Due to the coherence of the left-hand sides, the single integrability condition reduces to the above constraint. $\square$

It is assumed now that the ideal $\langle K_1, K_2 \rangle$ generated by the operators defined in the above lemma is the intersection of first-order operators with leading derivative ∂_x and ∂_y as described in Theorem 2.3 on page 51; similar as in the preceding case this has the consequence that a differential fundamental system for the corresponding homogeneous equations is known. The following theorem provides the means for obtaining a special solution of the inhomogeneous system (5.33).

Theorem 5.2. *Let the system (5.33) be given as in the above lemma. Furthermore, let both K_1 and K_2 allow first-order right factors $l_1 \equiv \partial_x + a_1 \partial_y + b_1$ and $l_2 \equiv \partial_y + b_2$; $\mathscr{E}_1(x, y)$ is defined by (B.3). Then a special solution z_0 is given by*

$$z_0 = \mathscr{E}_1(x, y) \int \frac{r(x, y)}{\mathscr{E}_1(x, y)}\Big|_{y=\psi_1(x,\bar{y})} dx \Big|_{\bar{y}=\varphi_1(x,y)}. \quad (5.35)$$

The inhomogeneity $r(x, y)$ is determined by the system

$$r_{xy} + b_2 r_x - (a_{1,y} - p_3 - q_3 a_1 + a_1 b_2 + b_1)r_y$$

$$-(b_{1,y} - p_5 - q_5 a_1 + b_1 b_2)r = R_1 + a_1 R_2, \quad (5.36)$$

$$r_{yy} + q_3 r_y + q_5 r = R_2.$$

Proof. The functions $r_i(x, y)$ and $z_i(x, y)$ are defined by $l_1 z_1 = r_1$ and $l_2 z_2 = r_2$ for $i = 1, 2$. Because both l_1 and l_2 are divisors of K_1 and K_2, reduction of K_1 and K_2 w.r.t. these first-order operators lead to the following linear system for the inhomogeneities.

$$r_{1,xy} + r_{2,xx} + a_1 r_{2,xy} + b_2 r_{1,x} - (a_{1,y} - p_3 - q_3 a_1 + a_1 b_2 + b_1) r_{1,y}$$
$$+ (p_3 + q_3 a_1 - a_1 b_2) r_{2,x} + (p_4 + q_4 a_1) r_{2,y} - (b_{1,y} - p_5 - q_2 a_1 + b_1 b_2) r_1$$
$$- (b_{2,x} a_1 + p_4 b_2 - p_6 + q_4 a_1 b_2 - q_6 a_1) r_2 = R_1 + a_1 R_2,$$
$$r_{1,yy} + r_{2,xy} + a_1 r_{2,yy} + q_3 r_{1,y} + (q_3 - b_2) r_{2,x}$$
$$+ (q_4 - a_1 b_2) r_{2,y} + q_5 r_1 - (b_{2,x} + 2b_{2,y} a_1 + q_4 b_2 - q_6 - a_1 b_2^2) r_2 = R_2.$$

The special choice $r_2 = 0$, $r_1 = r$ yields (5.36); a special solution of this system leads to the special solution (5.35).   □

Similar as for Theorem 5.1 a special solution of the system (5.36) is guaranteed to be Liouvillian over the base field.

Example 5.22. Consider the system

$$z_{xxy} - xz_{xx} + 2yz_{xy} - 2(xy + 1)z_x + y^2 z_y - y(xy + 2)z = \exp(xy),$$
$$z_{xyy} - 2xz_{xy} + yz_{yy} + x^2 z_x - 2(xy - 1)z_y + x(xy - 2)z = \exp(-xy). \tag{5.37}$$

Two first-order right factors yield the equations $z_x + yz = 0$ and $z_y - xz = 0$; correspondingly a fundamental system for the homogeneous equations is $z_1(x, y) = F(y) \exp(-xy)$ and $z_2(x, y) = G(x) \exp(xy)$. The system (5.36) is

$$r_{xy} - xr_x + yr_y - (xy + 3)r = \exp(xy), \quad r_{yy} - 2xr_y + x^2 r = \exp(-xy);$$

by Proposition 5.6 a special solution is $r(x, y) = -\frac{1}{2} \exp(xy) + \frac{1}{4x^2} \exp(-xy)$. Substitution into (5.35) yields finally the special solution

$$z_0(x, y) = -\frac{1}{4y} \exp(xy) - \frac{1}{4x} \exp(-xy)$$

of the system (5.37).   □

5.4 Transformation Theory of Second Order Linear PDE's

In the preceding subsections operators with leading derivatives ∂_{xx} and ∂_{xy} have been considered. There arises the question how these two forms are related, and how this relation may be expressed in any concrete case. In general, when dealing with differential operators and differential equations, variable transformations are an important tool for simplifying a problem. The type of a transformation and the properties of the possible result are the most important distinguishing features for obtaining a classification among the infinity of possibilities. Transformations of the operator variables, usually x and y, are considered first. They correspond to the independent variables of the corresponding differential equation, leaving the dependent variable unchanged.

Lemma 5.4. *Any second-order linear pde*

$$z_{xx} + A_1 z_{xy} + A_2 z_{yy} + A_3 z_x + A_4 z_y + A_5 z = 0 \qquad (5.38)$$

is equivalent to one of the following two normal forms by means of a transformation of the independent variables x and y; k_1 and k_2 are the roots of $k^2 + A_1 k + A_2 = 0$. The transformation functions are defined by $u = \varphi(x, y)$ and $v = \psi(x, y)$; they obey $\varphi_x \psi_y - \varphi_y \psi_x \neq 0$.

(i) $w_{uv} + B_1 w_u + B_2 w_v + B_3 w = 0$ *if* $A_1^2 - 4A_2 \neq 0$, *i.e.* $k_1 \neq k_2$; $\varphi(x, y)$ *is a first integral of* $\dfrac{dy}{dx} = -k_1$, *and* $\psi(x, y)$ *is a first integral of* $\dfrac{dy}{dx} = -k_2$.

(ii) $w_{vv} + B_1 w_u + B_2 w_v + B_3 w = 0$ *if* $A_1^2 - 4A_2 = 0$, *i.e.* $k_1 = k_2$; $\varphi(x, y)$ *is a first integral of* $\dfrac{dy}{dx} = -k_1$, $\psi(x, y)$ *is an undetermined function.*

Proof. Substituting $z(x, y) = w\big(u \equiv \varphi(x, y), v \equiv \psi(x, y)\big)$ into (5.3) yields

$$
\begin{aligned}
z_{xx} + A_1 z_{xy} &+ A_2 z_{yy} + A_3 z_x + A_4 z_y + A_5 z = \\
&[\varphi_x^2 + A_1 \varphi_x \varphi_y + A_2 \varphi_y^2] w_{uu} + [\psi_x^2 + A_1 \psi_x \psi_y + A_2 \psi_y^2] w_{vv} \\
&+ [2\varphi_x \psi_x + A_1(\varphi_x \psi_y + \varphi_y \psi_x) + 2A_2 \varphi_y \psi_y] w_{uv} \qquad (5.39) \\
&+ [\varphi_{xx} + A_1 \varphi_{xy} + A_2 \varphi_{yy} + A_3 \varphi_x + A_4 \varphi_y] w_u \\
&+ [\psi_{xx} + A_1 \psi_{xy} + A_2 \psi_{yy} + A_3 \psi_x + A_4 \psi_y] w_v + A_5 w = 0.
\end{aligned}
$$

Dividing the coefficient of w_{uu} by φ_y^2 and defining $\dfrac{\varphi_x}{\varphi_y} \equiv k$ yields the above given second order polynomial in k with roots k_1 and k_2. Assume first $k_1 \neq k_2$. If φ is chosen such that $\varphi_x - k_1 \varphi_y = 0$, i.e. if φ is a first integral of $\dfrac{dy}{dx} = -k_1$, the coefficient of w_{uu} vanishes. Similarly, choosing $\psi_x - k_2 \psi_y = 0$ with ψ a first integral of $\dfrac{dy}{dx} = -k_2$, the coefficient of w_{vv} vanishes. Upon substitution of φ and ψ the coefficient of w_{uv} becomes $-(k_1 - k_2)^2 \varphi_y \psi_y$. Because $\varphi_y \neq 0$ and $\psi_y \neq 0$ it is different from zero. This proves case (i).

If $k_1 = k_2$, then $A_1 = -2k_1$ and $A_2 = k_1^2$. The same reasoning as above may be applied either for the coefficient of w_{uu} and the transformation function φ, leaving ψ undetermined; or the coefficient of w_{vv} and the transformation function ψ, leaving φ undetermined. In either case, the coefficient of w_{uv} vanishes. Choosing the latter alternative, case (ii) is obtained. $\square$

A remarkable simplification of the normal form equation occurs in case (*ii*) of the above lemma if $B_1 = 0$; then the transformed equation is a linear ode. It is discussed next.

Corollary 5.3. *If for case (ii) of Lemma 5.4 in addition the coefficients satisfy*

$$A_{1,x} + \tfrac{1}{2} A_1 A_{1,y} + A_1 A_3 - 2A_4 = 0,$$

the resulting normal form equation is a linear ode, i.e. $B_1 = 0$ and the normal form equation is $w_{vv} + B_2 w_v + B_3 w = 0$. If (5.3) is absolutely irreducible, an ordinary normal form equation does not exist.

Proof. For case (*ii*) there follows $A_2 = \tfrac{1}{4} A_1^2$; furthermore

$$\varphi_x = -\tfrac{1}{2} A_1 \varphi_y, \varphi_{xx} = \tfrac{1}{4} A_1^2 \varphi_{yy} + \left(\tfrac{1}{4} A_1 A_{1,y} - \tfrac{1}{2} A_{1,x}\right) \varphi_y, \varphi_{xy}$$

$$= -\tfrac{1}{2} A_1 \varphi_{yy} - \tfrac{1}{2} A_{1,y} \varphi_y.$$

Substitution into (5.39) yields the following expression for the coefficient of w_u.

$$\left(-\tfrac{1}{2} A_{1,x} - \tfrac{1}{4} A_1 A_{1,y} - \tfrac{1}{2} A_1 A_3 + A_4\right) \varphi_y.$$

It vanishes if the above condition is satisfied. If not, (5.3) is absolutely irreducible because this condition agrees with (4.2) that decides absolute irreducibility.   □

The subsequent examples illustrate possible relations between reducibility and the existence of a normal form. It may occur that a given equation does not allow a factor, yet its normal form is reducible and may be solved as the next example shows.

Example 5.23. The equation $Lz = \left(\partial_{xx} - \partial_{yy} - \dfrac{2}{x}\partial_x\right)z = 0$ has been considered by Forsyth [17], vol. VI, pages 14 and 81. According to Proposition 4.1 a first order factor of L does not exist. By case (*i*) of Lemma 5.4, L may be transformed into $M \equiv \partial_{uv} - \dfrac{1}{u+v}(\partial_u + \partial_v)$ where $u = x - y$ and $v = x + y$. By Propositions 4.2 and 2.3 there are the Laplace divisors $\langle \partial_{uu}, M \rangle$ with solution $w_1 = G(v) - \tfrac{1}{2}(u + v)G'(v)$, and $\langle M, \partial_{vv} \rangle$ with solution $w_2 = F(u) - \tfrac{1}{2}(u + v)F'(u)$; F and G are undetermined functions. Returning to the original variables x and y, a differential fundamental system of $Lz = 0$ is

$$z_1(x, y) = F(x+y) - xF'(x+y), \quad z_2(x, y) = G(x-y) - xG'(x-y).   □$$

On the other hand, the reducible equation considered in Example 5.2 has a normal form with Airy function coefficients and therefore does not facilitate the solution procedure.

Example 5.24. For the operator L considered in Example 4.3 the coefficients are $A_1 = y^2 - 2x$ and $A_2 = x(x - y^2)$. Case (*i*) of Lemma 5.4 yields $k_1 = x$ and $k_2 = x - y^2$. The latter root leads to the equation $\frac{dy}{dx} = y^2 - x$. It does not have a rational solution, its general solution may be expressed in terms of Airy functions. Therefore the operator L cannot be transformed into an operator with leading derivative ∂_{uv} and rational coefficients.   □

Any equation with decomposition type $\mathcal{L}_{xx}^3$ is easily solved according to Proposition 5.1. Alternatively, the transformation into a linear ode yields the same answer as shown next.

Example 5.25. Miller's example has been considered in Example 4.6. The Equation $k^2 + \frac{2y}{x}k + \frac{y^2}{x^2} = 0$ has the double root $k_1 = k_2 = -\frac{y}{x}$. The equation $\frac{dy}{dx} = \frac{y}{x}$ leads to the first integral $\varphi(x, y) \equiv \frac{y}{x} = C$. Choosing $\psi(x, y) = xy$, i.e. $u = \frac{y}{x}$, $v = xy$ yields the transformed equation

$$w_{vv} + \frac{1}{v}w_v - \frac{1}{4v^2}w = 0 \tag{5.40}$$

for $w(u, v)$. Its Loewy decomposition $Lclm\left(\partial_v - \frac{1}{v+C} + \frac{1}{2v}\right) = 0$ yields the general solution $w = C_1(u)\sqrt{v} + C_2(u)\frac{1}{\sqrt{v}}$. Replacing the undetermined functions C_1 and C_2 by $C_1(u) \equiv \sqrt{\frac{x}{y}}F\left(\frac{y}{x}\right)$ and $C_2(u) \equiv \sqrt{\frac{y}{x}}G\left(\frac{y}{x}\right)$ finally yields $z_1(x, y) = xF\left(\frac{y}{x}\right)$ and $z_2(x, y) = \frac{1}{x}G\left(\frac{y}{x}\right)$, i.e. the same answer as in Example 5.5.   □

Example 5.26. Consider the equation

$$z_{xx} + \frac{2y}{x}z_{xy} + \frac{y^2}{x^2}z_{yy} + \frac{1}{x}z_x + \frac{y}{x^2}z_y + \frac{4y^2}{x^4}z = 0.$$

The transformation functions $\varphi(x, y)$ and $\psi(x, y)$ are the same as in the preceding example. However, by Proposition 4.1 a first-order right factor does not exist. The normal form equation $w_{vv} + \frac{1}{u}w_v + w = 0$ is a Bessel equation, i.e. its solutions may be expressed in terms of Bessel functions.   □

The following example shows that the normal form equation may have rather complicated coefficients in a field extension of the base field. That means, any theory of linear pde's that is based on a normal form with rational function coefficients may have only a limited range of application.

Example 5.27. Consider the equation

$$z_{xx} - 2\left(\frac{x}{y} + 2\right)z_{xy} + \left(\frac{x}{y} + 2\right)^2 z_{yy} + \frac{1}{x}z_x + \frac{y}{x^2}z_y - \frac{1}{x^2}z = 0.$$

According to Lemma 5.4, case (ii), the double root is $k_1 = k_2 = \frac{x}{y} + 2$. The equation $\frac{dy}{dx} = -\left(\frac{x}{y} + 2\right)$ has the first integral $u \equiv \varphi(x, y) = (x + y)\exp\left(\frac{x}{x+y}\right)$. Choosing $v \equiv \psi(x, y) = x + y$ leads to the transformed equation

$$w_{vv} + \frac{\left(\log\frac{u}{v} - 1\right)^2 + \left(\log\frac{u}{v}\right)^2 u}{\left(\log\frac{u}{v} - 2\right)^2 \left(\log\frac{u}{v}\right)^2 v^2} w_u + \frac{\left(\log\frac{u}{v} - 1\right)^2}{\left(\log\frac{u}{v} - 2\right)^2 \left(\log\frac{u}{v}\right)^2 v^2} w_v - \frac{1}{\left(\log\frac{u}{v}\right)^2} w = 0$$

for the function $w(u, v)$. $\square$

Additional normal forms of (5.38) may be obtained if the dependent variable is transformed according to $z(x, y) = \varphi(x, y)w(x, y)$. A few possibilities are considered next.

Corollary 5.4. *For the two cases of Lemma 5.4 further simplifications may be obtained.*

(i) *If in equation $z_{xx} + Az_x + Bz_y + Cz = 0$ the new dependent variable w is introduced by $z = \exp\left(-\frac{1}{2}\int A dx\right)w$, it takes the shape*

$$w_{xx} + Bw_y - \left(\frac{1}{2}A_x + \frac{1}{4}A^2 - C + \frac{1}{2}B\int A_y dx\right)w = 0.$$

(ii) *If in equation $z_{xy} + Az_x + Bz_y + Cz = 0$ the new dependent variable w is introduced by $z = \exp\left(-\int A dy\right)w$, it takes the shape*

$$w_{xy} + \left(B - \int A_x dy\right)w_y + (C - AB - A_x)w = 0.$$

The proof is considered in Exercise 5.13. The transformations described in this corollary may always be performed. However, in general the coefficient field of the resulting equation is Liouvillian over the base field.

In Exercise 1.1 it has been shown that a linear second-order ode may always be rationally transformed into an equation with no first-order derivative. The subsequent lemma generalizes this result to pde's if its coefficients satisfy certain constraints.

Lemma 5.5. *A linear second-order pde*

$$z_{xx} + A_1 z_{xy} + A_2 z_{yy} + A_3 z_x + A_4 z_y + A_5 z = 0 \tag{5.41}$$

may be transformed into

$$w_{xx} + A_1 w_{xy} + A_2 w_{yy} + rw = 0 \tag{5.42}$$

without first-order derivatives by a transformation $z = \varphi w$ if one of the following cases applies.

(i) $A_1^2 - 4A_2 \neq 0$. *Define*

$$p \equiv \frac{A_1 A_4 - 2A_2 A_3}{A_1^2 - 4A_2} \quad and \quad q \equiv \frac{A_1 A_3 - 2A_4}{A_1^2 - 4A_2};$$

if $p_y = q_x$, (5.41) is transformed into (5.42) with

$$r = p_x - p^2 + A_1(p_y - pq) + A_2(q_y - q^2) - A_3 p - A_4 q + A_5;$$

the coefficient r is in the base field of (5.41).

(ii) $A_1^2 - 4A_2 = 0$. *There must hold $2A_4 - A_1 A_3 = 0$. The first integral $\varphi(x, y)$ of $\frac{dy}{dx} = \frac{1}{2} A_1(x, y)$ has to be determined. Define $\bar{y} \equiv \psi(x, y)$ and $y = \bar{\psi}(x, \bar{y})$. Then the transformation function $\varphi(x, y)$ is given by*

$$\varphi = \Phi(\bar{y}) \exp\left(-\frac{1}{2} \int A_3(x, \bar{y}) dx\right)\Big|_{\bar{y}=\varphi(x,y)}$$

with Φ an undetermined function of its argument.

The proof of this lemma is considered in Exercise 5.14.

In Exercise 5.15 the equivalent result for second-order equations with mixed leading derivative is obtained.

5.5 Exercises

Exercise 5.1. Discuss the equation

$$z_{xx} - (2x - y^2)z_{xy} + (x^2 - xy^2)z_{yy} - (2xy + y^2 + 1)z_y - z = 0$$

and compare the result with the equation considered in Example 5.2 on page 93.

Exercise 5.2. Determine the solution of the inhomogeneous equation

$$z_{xy} + \frac{2}{x - y}z_x - \frac{2}{x - y}z_y - \frac{4}{(x - y)^2}z = xy.$$

The corresponding homogeneous equation has been considered in Example 4.9; see also Example 5.18.

Exercise 5.3. Explain the relation between the special solutions $z_0(x, y)$ given in Example 5.19 on page 107.

Exercise 5.4. Determine a special solution for the following inhomogeneous versions of Imschentzky's equation.

$$z_{xy} + xyz_x - 2yz = \frac{x^2 + 1}{y} \quad \text{and} \quad z_{xy} + xyz_x - 2yz = \frac{1}{xy}.$$

Exercise 5.5. In Example 5.23 on page 114 the operator $L \equiv \partial_{xx} - \partial_{yy} - \frac{2}{x}\partial_x$ has been considered. Represent the ideal $\langle L \rangle$ as the intersection of two ideals of the type $\langle \partial_{xx}, \partial_{xy} \rangle_{LT}$. Does this support solving the equation $Lz = 0$?

Exercise 5.6 (*Forsyth VI, page 260, Ex. 1*). Solve the equation

$$z_{xx} - \frac{1 + xy}{x}z_{xy} + \frac{y}{x}z_{yy} + \frac{1 - x}{x(1 - xy)}z_x + \frac{1 - y}{x(1 - xy)}z_y = 0.$$

Exercise 5.7 (*Forsyth VI, page 264, Ex. 2*). Solve the inhomogeneous equation

$$z_{xx} + \left(1 - \frac{y}{x}\right)z_{xy} - \frac{y}{x}z_{yy} - \frac{1}{x}z_x - \frac{1}{x}z_y = -xy.$$

Exercise 5.8 (*Forsyth VI, page 360, Ex. 4*). Solve the equation

$$z_{xx} + z_{yy} + \frac{8z}{(1 + x^2 + y^2)^2} = 0.$$

Exercise 5.9. Assume that for the equation considered in Example 5.11 on page 99 the Laplace divisor has not been found. How does the solution procedure proceed? Compare the result with the solution given in the above example.

Exercise 5.10. Solve the equation $Lz \equiv z_{xy} - \frac{2}{y}z_x - yz_y = 0$.

Exercise 5.11. Solve the equation $Lz \equiv z_{xy} + (xy - 1)z_y + 2xz = 0$.

Exercise 5.12. Solve the equation $Lz \equiv z_{xy} + \frac{1}{(xy + 1)y}z_x - \frac{y}{xy + 1}z_y = 0$.

Exercise 5.13. Prove Corollary 5.4; consider also the case where both first order derivatives in the normal form equation vanish.

Exercise 5.14. Prove Lemma 5.5.

Exercise 5.15. Prove an equivalent result as in Lemma 5.5 for an equation $z_{xy} + A_1 z_x + A_2 z_y + A_3 z = 0$.

Chapter 6
Decomposition of Third-Order Operators

Abstract Similar to second order operators and equations considered in preceding chapters, third order operators are characterized in the first place by their leading derivative. If invariance under permutations is taken into account, three cases with leading derivative ∂_{xxx}, ∂_{xxy} or ∂_{xyy} are distinguished. The corresponding ideals are of differential dimension $(1, 3)$.

6.1   Operators with Leading Derivative ∂_{xxx}

This case is particularly interesting for historical reasons because an operator of this kind was the first third order partial differential operator for which factorizations were considered in Blumberg's thesis [5]; it is discussed in detail in Example 6.9.

Proposition 6.1. *Let the third order partial differential operator*

$$L \equiv \partial_{xxx} + A_1\partial_{xxy} + A_2\partial_{xyy} + A_3\partial_{yyy}$$
$$+ A_4\partial_{xx} + A_5\partial_{xy} + A_6\partial_{yy} + A_7\partial_x + A_8\partial_y + A_9 \tag{6.1}$$

be given with $A_i \in \mathbb{Q}(x, y)$ for all i. Any first order right factor $\partial_x + a\partial_y + b$ with $a, b \in \mathbb{Q}(x, y)$ is essentially determined by the roots a_1, a_2 and a_3 of the equation $a^3 - A_1a^2 + A_2a - A_3 = 0$. The following alternatives may occur.

(i) *If $a_i \neq a_j$ for $i \neq j$ are three pairwise different rational roots and the corresponding b_i are determined by (6.9), each pair a_i, b_i satisfying (6.10) and (6.11) yields a factor $l_i = \partial_x + a_i\partial_y + b_i$. If there are three factors, the operator is completely reducible, and $L = Lclm(l_1, l_2, l_3)$; if there are two factors, their intersection may or may not be principal according to Theorem 2.2; there may be a single factor or no factor at all.*

(ii) *If $a_1 = a_2$ is a twofold rational root and $a_3 \neq a_1$ a simple one, the following factors may exist. For $a = a_3 \in \mathbb{Q}(x, y)$, the value of $b = b_3$ is determined*

F. Schwarz, *Loewy Decomposition of Linear Differential Equations*, Texts & Monographs   119
in Symbolic Computation, DOI 10.1007/978-3-7091-1286-1_6,
© Springer-Verlag/Wien 2012

by (6.9); if the pair (a_3, b_3) satisfies (6.10) and (6.11), there is a factor $\partial_x + a_3 \partial_y + b_3$.

For the double root $a = a_1 = a_2$, a necessary condition for a factor to exist is

$$(A_1 - 3a)a_x + (3a^2 - 3A_1a + 2A_2)a_y - A_4a^2 + A_5a = A_6. \qquad (6.2)$$

The type of solutions for $b = b_1 = b_2$ of the system comprising (6.10) and (6.11) determines the possible factors. The following alternatives may occur; r and r_i are undetermined functions of the respective arguments.

$$\partial_x + a_1\partial_y + r(x, y, c_1, c_2), \quad c_1 \text{ and } c_2 \text{ constants};$$

$$\partial_x + a_1\partial_y + r(x, y, c), \quad c \text{ constant};$$

$$\partial_x + a_1\partial_y + r_i(x, y), \quad i = 1 \text{ or } i = 1, 2;$$

(iii) If $a_1 = a_2 = a_3 = \frac{1}{3}A_1$ is a threefold rational solution, the following relation must be valid

$$A_1^2 A_4 - 3A_1 A_5 + 9A_6 = 0 \qquad (6.3)$$

in order for a factor to exist. The following subcases may occur.

(a) If the coefficient of b in

$$\left(A_{1,x} + \tfrac{1}{3}A_1 A_{1,y} + \tfrac{2}{3}A_1 A_4 - A_5\right) b = \tfrac{1}{3}A_{1,xx} + \tfrac{2}{9}A_1 A_{1,xy} + \tfrac{1}{27}A_1^2 A_{1,yy}$$

$$-\tfrac{2}{9}A_{1,x}A_{1,y} + \tfrac{1}{3}A_4 A_{1,x} - \tfrac{2}{27}A_1 A_{1,y}^2 - \tfrac{1}{9}A_1 A_4 A_{1,y}$$

$$+\tfrac{1}{3}A_5 A_{1,y} + \tfrac{1}{3}A_1 A_7 - A_8 = 0 \qquad (6.4)$$

does not vanish, b may be determined uniquely from this equation. A factor does exist if the constraint

$$b_{xx} + \tfrac{2}{3}A_1 b_{xy} + \tfrac{1}{9}A_1^2 b_{yy} - 3bb_x + A_4 b_x - A_1 bb_y$$

$$- \left(\tfrac{2}{3}A_{1,x} + \tfrac{2}{9}A_1 A_{1,y} + \tfrac{1}{3}A_1 A_4 - A_5\right) b_y + b^3$$

$$-A_4 b^2 + A_7 b - A_9 = 0. \qquad (6.5)$$

is satisfied.

(b) If the coefficient of b in (6.4) vanishes and the two conditions

$$A_{1,x} + \tfrac{1}{3}A_1 A_{1,y} + \tfrac{2}{3}A_1 A_4 - A_5 = 0,$$

$$A_{1,xx} + \tfrac{2}{3}A_1 A_{1,xy} + \tfrac{1}{9}A_1^2 A_{1,yy} - \tfrac{2}{3}A_{1,x}A_{1,y} + A_4 A_{1,x}$$

$$-\tfrac{2}{9}A_1 A_{1,y}^2 - \tfrac{1}{3}A_1 A_4 A_{1,y} + A_5 A_{1,y} + A_1 A_7 - 3A_8 = 0 \qquad (6.6)$$

are satisfied, then b has to be determined from

$$b_{xx} + \tfrac{2}{3}A_1 b_{xy} + \tfrac{1}{9}A_1^2 b_{yy} - 3bb_x + A_4 b_x - A_1 bb_y$$

$$+ \tfrac{1}{3}\left(A_{1,x} + \tfrac{1}{3}A_1 A_{1,y} + A_1 A_4\right) b_y + b^3 - A_4 b^2 + A_7 b - A_9 = 0. \quad (6.7)$$

Proof. Dividing the operator (6.1) by $\partial_x + a\partial_y + b$, the requirement that this division be exact leads to the following set of equations between the coefficients.

$$a^3 - A_1 a^2 + A_2 a - A_3 = 0, \qquad\qquad (6.8)$$

$$(A_1 - 3a)a_x + (3a^2 - 3A_1 a + 2A_2)a_y - A_4 a^2 + A_5 a + (3a^2 - 2A_1 a + A_2)b = A_6, \qquad (6.9)$$

$$(A_1 - 3a)b_x + (3a^2 - 3A_1 a + 2A_2)b_y - (A_1 - 3a)b^2$$
$$+ (A_5 - 2A_4 a - 2A_1 a_y + 3aa_y - 3a_x)b$$
$$+ a_{xx} + (A_1 - a)a_{xy} + (a^2 - A_1 a + A_2)a_{yy} \qquad\qquad (6.10)$$
$$- 2a_x a_y + A_4 a_x - (A_1 - a)a_y^2 - (A_4 a - A_5)a_y + A_7 a - A_8 = 0,$$

$$b_{xx} + (A_1 - a)b_{xy} + (a^2 - A_1 a + A_2)b_{yy} - (2a_x + (A_1 - a)a_y + A_4 a - A_5)b_y$$
$$+ (A_4 - 3b)b_x + (3a - 2A_1)bb_y + b^3 - A_4 b^2 + A_7 b - A_9 = 0. \qquad (6.11)$$

The algebraic Eq. (6.8) determines a. The following discussion is organized by the type of its roots.

Case (i). Assume at first that (6.8) has three simple roots a_1, a_2, and a_3. None of them may be rational, there may be a single rational solution, or all three roots may be rational. For none of these roots the coefficient of b in (6.9) does vanish; this follows because it is the derivative of the left hand side of (6.8) that does not vanish for simple roots. Therefore for each a_i, Eq. (6.9) determines the corresponding value b_i. For those pairs a_i, b_i which satisfy the constraints (6.10) and (6.11), a factor $l_i \equiv \partial_x + a_i \partial_y + b_i$ exists. If there are three right factors, by Proposition 2.6 L is completely reducible and $L = Lclm(l_1, l_2, l_3)$.

Case (ii). Assume now (6.8) has a twofold rational root $a_1 = a_2$ and a simple one $a_3 \neq a_1$. This is assured if its coefficients satisfy

$$A_1^2 A_2^2 - 4A_2^3 + 18A_1 A_2 A_3 - 27A_3^2 - 4A_1^3 A_3 = 0 \qquad\qquad (6.12)$$

and $A_1^2 - 3A_2 \neq 0$. There follows

$$a_1 = a_2 = \frac{1}{2} \frac{A_1 A_2 - 9A_3}{A_1^2 - 3A_2}, \quad a_3 = \frac{A_1^3 - 4A_1 A_2 + 9A_3}{A_1^2 - 3A_2}.$$

For the root a_3, the coefficient b_3 follows from (6.9). The existence of a factor corresponding to a_3 and b_3 depends on whether these values satisfy the constraints (6.10) and (6.11).

The double root $a_1 = a_2$ is one of the roots of $3a^2 - 2A_1 a + A_2 = 0$. Hence the coefficient of b in (6.9) vanishes for this value of a; the remaining part of (6.9) becomes the constraint (6.2). If it is not obeyed, a factor originating from a_1 does not exist. If it is obeyed, the corresponding value for b has to be determined from the system comprising (6.10) and (6.11) with $a = a_1$. Because $A_1 \neq 3a_1$, reducing (6.11) w.r.t. (6.10) yields a system of the type

$$b_x + O(b_y) = 0, \quad b_{yy} + O(b_y) = 0 \qquad (6.13)$$

if *lex* term order with $x \succ y$ is applied. If this autoreduced system forms a Janet basis, the Riccati equation (6.10) has to be solved applying Lemma B.4. If its rational solution contains an undetermined function, it has to be adjusted such that it satisfies the second equation (6.13). Any rational solution without undetermined elements is only retained if it satisfies this equation. In any case, the final result may be a rational function $r(x, y, c_1, c_2)$ involving two constants; it may be a rational function $r(x, y, c)$ involving a single constant; or there may be one or two rational solutions $r_i(x, y)$ containing no constant; or there may be no rational solution at all. If the autoreduced system (6.13) is not a Janet basis, its integrability conditions have to be included and autoreduction has to be applied again; possibly this procedure has to be repeated several times. It cannot be described for generic coefficients $A_1, \ldots, A_9$ because the resulting expressions become too large. The final result may be a system of the type $b_x + o(b) = 0$, $b_y + o(b) = 0$ the general solution of which contains a constant; it may be an algebraic equation for b with one or two solutions; or it may turn out to be inconsistent. The respective solutions are subsumed among those described above.

Case (*iii*). Finally assume there is a threefold solution $a_1 = a_2 = a_3 = \frac{1}{3} A_1$ of (6.8). This is assured if $A_2 = \frac{1}{3} A_1^2$ and $A_3 = \frac{1}{27} A_1^3$. The coefficient of b in (6.9) vanishes again; the remaining part becomes the constraint (6.3). Equation (6.10) simplifies to (6.4); if the coefficient of b does not vanish, b may be determined from it. In order for a factor to exist, in addition (6.5) must be satisfied which originates from (6.11). This is subcase (*a*). In the exceptional case that the coefficient of b in (6.4) vanishes, it reduces to constraints (6.6); b has to be determined from (6.7) which is obtained from (6.11) by simplification. This is subcase (*b*). □

In order to apply the above proposition for solving concrete problems the question arises to what extent the various factors may be determined algorithmically. The answer may be summarized as follows.

Corollary 6.1. *Any first-order factor corresponding to a simple root of the symbol polynomial of the operator (6.1) may be determined algorithmically. This is not possible in general for factors corresponding to a double or triple root. In more detail the answer is as follows.*

 (i) Simple root of symbol polynomial. Any factor may be determined.
(ii) Double or triple root of symbol polynomial. In general it is not possible to determine the right factors over the base field. The existence of factors in a universal field may always be decided.

Proof. If there are three simple roots a_i, $i = 1, 2, 3$, the b_i, may be determined from the algebraic system (6.8) and (6.9); the constraints (6.10) and (6.11) require only arithmetic and differentiations in the base field. These operations may always be performed. The same arguments apply for the simple root in case (ii). For the double root in case (ii) the corresponding value of b has to be determined from (6.13); it may lead to a partial Riccati equation, an ordinary Riccati equation, an algebraic equation, or turn out to be inconsistent. For the first alternative, rational solutions may not be determined in general whereas this is possible in the remaining cases. If there is a threefold root of (6.8) it may occur that b has to be determined from Eq. (6.7); in general there is no solution algorithm available for solving it.   □

The term *absolutely irreducible* has been avoided deliberately, although the last sentence of the above theorem may suggest it; however, this would exclude also second-order factors that have not been taken into account.

After the possible factorizations of an operator (6.1) have been determined, a listing of its various decomposition types involving first-order principal factors may be set up as follows.

Theorem 6.1. *Let the differential operator L be given by*

$$L \equiv \partial_{xxx} + A_1\partial_{xxy} + A_2\partial_{xyy} + A_3\partial_{yyy}$$
$$+ A_4\partial_{xx} + A_5\partial_{xy} + A_6\partial_{yy} + A_7\partial_x + A_8\partial_y + A_9, \tag{6.14}$$

with $A_1, \ldots, A_9 \in \mathbb{Q}(x, y)$. Moreover let $l_i \equiv \partial_x + a_i\partial_y + b_i$ for $i = 1, 2, 3$ and $l(\Phi) \equiv \partial_x + a\partial_y + b(\Phi)$ be first order operators with $a_i, b_i, a \in \mathbb{Q}(x, y)$; Φ is an undetermined function of a single argument. Then L may be decomposed according to one of the following types involving first-order principal divisors.

$$\mathscr{L}_{xxx}^1 : L = l_3l_2l_1; \quad \mathscr{L}_{xxx}^2 : L = Lclm(l_3, l_2)l_1; \quad \mathscr{L}_{xxx}^3 : L = Lclm(l(\Phi))l_1;$$
$$\mathscr{L}_{xxx}^4 : L = l_3Lclm(l_2, l_1); \quad \mathscr{L}_{xxx}^5 : L = l_3Lclm(l(\Phi));$$
$$\mathscr{L}_{xxx}^6 : L = Lclm(l_3, l_2, l_1); \quad \mathscr{L}_{xxx}^7 : L = Lclm(l(\Phi), l_1).$$

If none of these alternatives applies, the decomposition type is defined to be $\mathscr{L}_{xxx}^0$.

Proof. It is based on Proposition 6.1. In the separable case (i) there may be three first-order factors with a principal intersection, this yields type $\mathscr{L}^6_{xxx}$. If there are two factors with a principal intersection they lead to a type $\mathscr{L}^4_{xxx}$ decomposition. If there is a single right factor it is divided out and a second-order left factor with leading derivative ∂_{xx} is obtained. It may be decomposed according to Theorem 4.1; if it is not irreducible it may yield the type $\mathscr{L}^1_{xxx}$, $\mathscr{L}^2_{xxx}$ or $\mathscr{L}^3_{xxx}$ decomposition respectively.

If in case (ii) only a single factor is allowed, the same reasoning as above leads to a decomposition of type $\mathscr{L}^1_{xxx}$ or $\mathscr{L}^2_{xxx}$.

In case (iii), subcase (a), a single factor may exist leading again to type $\mathscr{L}^1_{xxx}$ or type $\mathscr{L}^2_{xxx}$ decompositions as above. $\qquad\qquad\square$

A similar remark as for ordinary operators on page 7 applies here. By definition of a Loewy divisor, a decomposition type $\mathscr{L}^1_{xxx}$ for example implies that there is no decomposition type $\mathscr{L}^i_{xxx}$ with $i > 1$. This is a consequence of the fact that a Loewy factor is defined to comprise *all* irreducible right factors. Similar remarks apply to the other decomposition types.

These results are illustrated by several examples now. They show that each decomposition type actually does exist. The equations corresponding to these operators and its solutions are discussed in Chap. 7.

Example 6.1. The operator

$$L \equiv \partial_{xxx} + y\partial_{xxy} - \left(1 - \frac{1}{x}\right)\partial_{xx} - \left(y - \frac{y}{x}\right)\partial_{xy} - \left(\frac{1}{x} + \frac{1}{x^2}\right)\partial_x - \frac{y}{x}\partial_y + \frac{1}{x^2}$$

has the symbol equation $a^2(a - y) = 0$; therefore case (ii) of Proposition 6.1 applies. There is a double root $a_1 = a_2 = 0$ and a simple root $a_3 = y$. The latter yields $b_3 = 0$. Because the pair a_3, b_3 violates constraints (6.10) and (6.11) it does not yield a factor. On the other hand, the double root leads to the equation $b_x - b^2 - (1 - \frac{1}{x})b + \frac{1}{x} = 0$ for b. Its single rational solution is $b = -1$; it yields the factor $\partial_x - 1$. Dividing it out the operator $\partial_{xx} + y\partial_{xy} + \frac{1}{x}\partial_x + \frac{y}{x}\partial_y - \frac{1}{x^2}$ is obtained. By Proposition 4.1 it has the right factor $\partial_x + \frac{1}{x}$. Altogether the decomposition

$$L = (\partial_x + y\partial_y)\left(\partial_x + \frac{1}{x}\right)(\partial_x - 1)$$

of type $\mathscr{L}^1_{xxx}$ follows. Continued in Example 7.1.

Example 6.2. Consider the operator

$$L \equiv \partial_{xxx} + (x + y - 1)\partial_{xxy} - (x + y)\partial_{xyy} - (x - y - 1)\partial_{xx} - (x^2 + xy - x + 1)\partial_{xy}$$
$$- (x + y)\partial_{yy} - (xy + x - y + 1)\partial_x - (x^2 + xy + y)\partial_y - xy - 1.$$

Equation (6.8) reads $a^3 - (x + y - 1)a^2 - (x + y)a = 0$ with roots $a_1 = 0$, $a_2 = -1$ and $a_3 = x + y$, i.e. case (i) of Proposition 6.1 applies. Equation (6.9) yields the corresponding values

$$b_1 = 1, \quad b_2 = -x \quad \text{and} \quad b_3 = -\frac{2x^2 + xy - x - y^2 - y + 2}{4x + 5y - 3}.$$

Only a_1 and b_1 satisfy the constraints (6.3) and (6.5). Hence there is a single first order factor $l_1 \equiv \partial_x + 1$. Dividing L by l_1, the completely reducible operator

$$\partial_{xx} + (x + y - 1)\partial_{xy} - (x + y)\partial_{yy} - (x - y)\partial_x - (x^2 + xy + y)\partial_y - xy - 1$$
$$= Lclm(\partial_x + (x + y)\partial_y + y, \ \partial_x - \partial_y - x)$$

follows; L has the type $\mathcal{L}^2_{xxx}$ decomposition

$$L = Lclm(\partial_x + (x + y)\partial_y + y, \ \partial_x - \partial_y - x)(\partial_x + 1).$$

Continued as Example 7.2.   □

Example 6.3. Consider the operator

$$L \equiv \partial_{xxx} - \frac{2y}{x}\partial_{xxy} + \frac{y^2}{x^2}\partial_{xyy} - \frac{x-3}{x}\partial_{xx} + \frac{(2x-1)y}{x^2}\partial_{xy} - \frac{y^2}{x^2}\partial_{yy} - \frac{3}{x}\partial_x + \frac{y}{x^2}\partial_y.$$

Equation (6.8) is $\left(a + \frac{y}{x}\right)^2 a = 0$; its roots are $a_1 = a_2 = -\frac{y}{x}$ and $a_3 = 0$, i.e. case (ii) of Proposition 6.1 applies. From the simple root there follows $b_3 = -1$; the pair a_3, b_3 satisfies the constraints given there; therefore a factor $l_1 \equiv \partial_x - 1$ exists. It turns out that the double root $a_1 = a_2$ does not lead to a right factor; therefore the factor l_1 is divided out with the result

$$\partial_{xx} - \frac{2y}{x}\partial_{xy} + \frac{y^2}{x^2}\partial_{yy} + \frac{3}{x}\partial_x - \frac{y}{x^2}\partial_y.$$

According to Theorem 4.1 it has a type $\mathcal{L}^3_{xx}$ decomposition involving an undetermined function. Putting these findings together the type $\mathcal{L}^3_{xxx}$ decomposition

$$L \equiv \left(\partial_{xx} - \frac{2y}{x}\partial_{xy} + \frac{y^2}{x^2}\partial_{yy} + \frac{3}{x}\partial_x - \frac{y}{x^2}\partial_y\right)(\partial_x - 1)$$
$$= Lclm\left(\partial_x - \frac{y}{x}\partial_y + \frac{2}{x}\frac{\Phi(xy)}{\Phi(xy) + x^2}\right)(\partial_x - 1)$$

for L is obtained. Continued as Example 7.3.   □

Example 6.4. For the operator

$$L \equiv \partial_{xxx} + (y+1)\partial_{xxy} + \left(1 - \frac{1}{x}\right)\partial_{xx} + \left(1 - \frac{1}{x}\right)(y+1)\partial_{xy}$$
$$- \frac{1}{x}\partial_x - \frac{1}{x}(y+1)\partial_y$$

Eq. (6.8) reads $a^2(a - y - 1) = 0$ with double root $a_1 = a_2 = 0$ and simple root $a_3 = y + 1$, thus case (ii) applies with $a_1 = 0$, $b_1 = 1$ and $a_3 = y + 1$, $b_3 = 0$. The corresponding first order factors yield the divisor as the principal intersection

$$Lclm(\partial_x + 1, \partial_x + (y+1)\partial_y) = \partial_{xx} + (y+1)\partial_{xy} + \partial_x + (y+1)\partial_y. \quad (6.15)$$

Therefore L has the decomposition

$$L = \left(\partial_x - \frac{1}{x}\right)Lclm\left(\partial_x + 1, \partial_x + (y+1)\partial_y\right)$$

of type $\mathscr{L}^4_{xxx}$; continued in Example 7.4.   □

Although the symbol equation of an operator is the most decisive criterion for its decomposition type, the next example shows how lower-order terms may have a significant effect.

Example 6.5. The operator

$$L \equiv \partial_{xxx} - \frac{2y}{x}\partial_{xxy} + \frac{y^2}{x^2}\partial_{xyy} - \frac{x-3}{x}\partial_{xx}$$
$$+ \frac{(2x+1)y}{x^2}\partial_{xy} - \frac{(x+2)y^2}{x^3}\partial_{yy} - \frac{3(x+1)}{x^2}\partial_x + \frac{(x+2)y}{x^3}\partial_y$$

has the same symbol as the operator of the preceding example. Therefore its roots are again $a_1 = a_2 = -\frac{y}{x}$ and $a_3 = 0$ of the symbol equation. Yet a_3 does not yield a factor because a_3 and the corresponding value $b_3 = -1 - \frac{2}{x}$ do not satisfy the necessary constraints. However, for the double root (6.10) leads now to the partial Riccati equation

$$b_x - \frac{y}{x}b_y - b^2 + \frac{3}{x}b = 0$$

for b with solution $b = \dfrac{2\Phi(xy)}{x(\Phi(xy)) + x^2}$; Φ is an undetermined function of xy. It yields finally the type $\mathscr{L}^5_{xxx}$ decomposition

$$L = (\partial_x - 1)\left(\partial_{xx} - \frac{2y}{x}\partial_{xy} + \frac{y^2}{x^2}\partial_{yy} + \frac{3}{x}\partial_x - \frac{y}{x^2}\partial_y\right)$$

$$= (\partial_x - 1)Lclm\left(\partial_x - \frac{y}{x}\partial_y + \frac{2}{x}\frac{\Phi(xy)}{\Phi(xy) + x^2}\right).$$

Continued as Example 7.5. □

Example 6.6. Consider the operator

$$L \equiv \partial_{xxx} + (y+2)\partial_{xxy} + (y+1)\partial_{xyy}$$

$$+ \left(1 - \frac{1}{y}\right)\partial_{xx} + \left(y + 2 - \frac{1}{y}\right)\partial_{xy} + (y+1)\partial_{yy} - \frac{1}{y}\partial_x - \frac{1}{y}\partial_y.$$

Equation (6.8) reads $a^3 - (y+2)a^2 + (y+1)a = 0$ with the three roots $a_1 = 0$, $a_2 = 1$ and $a_3 = y+1$, i.e. case (i) of Proposition 6.1 applies. The corresponding values of b are $b_1 = 1$, $b_2 = b_3 = 0$. It turns out that all three pairs a_i, b_i satisfy conditions (6.10) and (6.11). Hence there are three right factors $l_1 = \partial_x + 1$, $l_2 = \partial_x + \partial_y$ and $l_3 = \partial_x + (y+1)\partial_y$; thus, $L = Lclm(l_3, l_2, l_1)$; the decomposition type is $\mathscr{L}^6_{xxx}$. Continued as Example 7.6. □

Example 6.7. Another operator with the same symbol equation is

$$L \equiv \partial_{xxx} - \frac{2y}{x}\partial_{xxy} + \frac{y^2}{x^2}\partial_{xyy} - \frac{xy-5}{x}\partial_{xx}$$

$$+ \frac{(2xy-3)y}{x^2}\partial_{xy} - \frac{y^3}{x^2}\partial_{yy} - \frac{3(xy-1)}{x^2}\partial_x + \frac{y^2}{x^2}\partial_y.$$

The double root $a_1 = a_2 = -\frac{y}{x}$ leads to the same right factor containing an undetermined function Φ as in the preceding example. The simple root $a_3 = 0$ yields now $b_3 = -y$ and the factor $\partial_x - y$. Altogether the type $\mathscr{L}^7_{xxx}$ decomposition

$$L = Lclm\left(\partial_x - y, \partial_x - \frac{y}{x}\partial_y + \frac{2}{x}\frac{\Phi(xy)}{\Phi(xy) + x^2}\right)$$

follows. Continued as Example 7.7. □

Example 6.8. The operator

$$L \equiv \partial_{xxx} + \partial_{xxy} + \tfrac{1}{3}\partial_{xyy} + \tfrac{1}{27}\partial_{yyy} + 3xy\partial_{xx} + xy\partial_{xy}$$

has the symbol equation $a^3 - a^2 + \tfrac{1}{3}a - \tfrac{1}{27} = 0$ with the triple root $a_1 = a_2 = a_3 = \tfrac{1}{3}$. Constraint (6.3) is satisfied and the coefficient of b in (6.4) is different from zero, i.e. case (iii), subcase (a) applies; it yields the factor $\partial_x + \frac{1}{3}\partial_y$. The second-order left factor obtained by dividing it out does not have any right factor and the Loewy decomposition

$$L = \left(\partial_{xx} + \frac{2}{3}\partial_{xy} + \frac{1}{3}\partial_{yy} + 3xy\partial_x\right)\left(\partial_x + \frac{1}{3}\partial_y\right)$$

is obtained; due to its second-order left factor it does not belong to any of the types defined in Theorem 6.1.   □

In addition to the decompositions described in Theorem 6.1 there exists a decomposition type involving a non-principal right divisor; it occurs when there are two first-order right factors with *non-principal* intersection ideal $\mathbb{J}_{xxx}$ introduced in Lemma 2.2.

Theorem 6.2. *Assume that the differential operator*

$$L \equiv \partial_{xxx} + A_1\partial_{xxy} + A_2\partial_{xyy} + A_3\partial_{yyy}$$

$$+ A_4\partial_{xx} + A_5\partial_{xy} + A_6\partial_{yy} + A_7\partial_x + A_8\partial_y + A_9 \qquad (6.16)$$

has two first-order right factors $l_i \equiv \partial_x + a_i\partial_y + b_i$; $A_1, \ldots, A_9, a_i, b_i \in \mathbb{Q}(x, y)$. *Assume further that* l_1 *and* l_2 *have the non-principal left intersection ideal* $\mathbb{J}_{xxx}$. *Then L may be decomposed as*

$$\mathscr{L}^8_{xxx} : L = Exquo(\langle L \rangle, \mathbb{J}_{xxx})\mathbb{J}_{xxx} =$$

$$\begin{pmatrix} 1, & A_1 \\ 0, \partial_x + (A_1 - q_1)\partial_y + A_{1,y} + q_3A_1 + p_3 + q_1q_3 - q_4 \end{pmatrix}\begin{pmatrix} L_1 \\ L_2 \end{pmatrix}.$$

Proof. According to Theorem 2.2 the ideal $\mathbb{J}_{xxx}$ is generated by two third-order operators $L_1 = \partial_{xxx} + o(\partial_{xyy})$ and $L_2 = \partial_{xxy} + o(\partial_{xyy})$. Hence, an operator L with two generic factors l_1 and l_2 is contained in this ideal and has the form $L = L_1 + A_1L_2$, i.e. the exact quotient in this basis is $(1, A_1)$. The exact quotient module is the sum of this quotient and the syzygy of $\mathbb{J}_{xxx}$ given in Lemma 2.2 on page 45.   □

In the following example the operator introduced in Blumberg's dissertation [5] is discussed in detail; originally it has been suggested by Landau to him. This operator is the generic case for operators that are not completely reducible allowing only two first-order right factors as may be seen from Theorem 2.2.

Example 6.9. In his dissertation Blumberg [5] considered the third order operator

$$L \equiv \partial_{xxx} + x\partial_{xxy} + 2\partial_{xx} + 2(x + 1)\partial_{xy} + \partial_x + (x + 2)\partial_y \qquad (6.17)$$

generating a principal ideal of differential dimension $(1, 3)$. He gave its factorizations

$$L = \begin{cases} (\partial_{xx} + x\partial_{xy} + \partial_x + (x+2)\partial_y)(\partial_x + 1), \\ (\partial_{xx} + 2\partial_x + 1)(\partial_x + x\partial_y). \end{cases} \tag{6.18}$$

This result may be obtained by Proposition 6.1 as follows. Equation (6.8) is $a^3 - xa^2 = a^2(a - x) = 0$ with the double root $a_1 = a_2 = 0$, and the simple root $a_3 = x$. The latter yields $b_3 = 0$. For the double root $a_1 = 0$, the system (6.10) and (6.11) has the form

$$b_x - b^2 + \left(2 + \frac{2}{x}\right)b - 1 - \frac{2}{x} = 0,$$

$$b_{xx} + xb_{xy} - 3bb_x + 2b_x - 2xbb_y + 2(x+1)b_y + b^3 - 2b^2 + b = 0.$$

It yields the Janet basis $b - 1 = 0$ and the factor $l_1 \equiv \partial_x + 1$. Because a_3 and b_3 satisfy (6.10) and (6.11), there is a second factor $l_2 \equiv \partial_x + x\partial_y$, i.e. case (iv) of Theorem 6.1 applies.

The second order left factor in the first line at the right hand side of (6.18) is absolutely irreducible, whereas the second order factor in the second line is the left intersection of two first order factors, thus (6.18) may be further decomposed into irreducibles as

$$L = \begin{cases} (\partial_{xx} + x\partial_{xy} + \partial_x + (x+2)\partial_y)(l_1 = \partial_x + 1), \\ Lclm\left(\partial_x + 1, \partial_x + 1 - \frac{1}{x}\right)(l_2 = \partial_x + x\partial_y). \end{cases} \tag{6.19}$$

The intersection ideal of l_1 and l_2 is not principal, by Theorem 2.2 it is

$Lclm(l_2, l_1) =$

$$\Big\langle\!\Big\langle L_1 \equiv \partial_{xxx} - x^2\partial_{xyy} + 3\partial_{xx} + (2x+3)\partial_{xy} - x^2\partial_{yy} + 2\partial_x + (2x+3)\partial_y,$$

$$L_2 \equiv \partial_{xxy} + x\partial_{xyy} - \frac{1}{x}\partial_{xx} - \frac{1}{x}\partial_{xy} + x\partial_{yy} - \frac{1}{x}\partial_x - \left(1+\frac{1}{x}\right)\partial_y \Big\rangle\!\Big\rangle \tag{6.20}$$

with differential dimension $(1, 2)$; therefore the type $\mathscr{L}_{xxx}^8$ decomposition of (6.17) is

$$L = \begin{pmatrix} 1 & x \\ 0 & \partial_x + 1 + \frac{1}{x} \end{pmatrix}\begin{pmatrix} L_1 \\ L_2 \end{pmatrix};$$

there follows $L = L_1 + xL_2$. L_1 as well as L_2 have the divisors l_1 and l_2 as will be explicitly shown in Exercise 6.2. Completed in Example 7.8.   □

6.2 Operators with Leading Derivative ∂_{xxy}

If an equation does not contain a derivative ∂_{xxx} but ∂_{yyy} instead, permuting x and y leads to an equation of the form (6.1) such that the above theorem may be applied. If there is neither a term ∂_{xxx} or ∂_{yyy}, the general third order operator contains the mixed third-order derivatives ∂_{xxy} and ∂_{xyy}. Its possible factorizations into first order factors are the topic of this subsection.

Proposition 6.2. *Let the third order partial differential operator*

$$L \equiv \partial_{xxy} + A_1\partial_{xyy} + A_2\partial_{xx} + A_3\partial_{xy} + A_4\partial_{yy} + A_5\partial_x + A_6\partial_y + A_7 \quad (6.21)$$

be given with $A_i \in \mathbb{Q}(x, y)$ for all i. Any first-order right factor $\partial_x + a\partial_y + b$ with $a, b \in \mathbb{Q}(x, y)$ is essentially determined by the roots a_1 and a_2 of the equation $a^2 - A_1a = 0$. The following alternatives for a_1 and a_2 may occur.

(i) *If $A_1 \neq 0$, the roots are $a_1 = 0$ and $a_2 = A_1$; b_1 and b_2 may be determined from (6.25). A first order right factor $\partial_x + a_i\partial_y + b_i$ exists if the pair a_i, b_i for $i = 1$ or $i = 2$ satisfies the constraints (6.26) and (6.27).*

(ii) *If $A_1 = A_4 = 0$, there follows $a = 0$. Two subcases are distinguished.*

(a) *A factor $\partial_x + b$ exists with $b = \dfrac{A_{6,y} + A_2A_6 - A_7}{A_{3,y} + A_2A_3 - A_5}$ if*

$$A_{3,y} + A_2A_3 - A_5 \neq 0, \qquad b_x - b^2 + A_3b - A_6 = 0.$$

(b) *If the two constraints*

$$A_{6,y} + A_2A_6 - A_7 = 0, \quad A_{3,y} + A_2A_3 - A_5 = 0 \quad (6.22)$$

are satisfied, there may be a right factor of the form $\partial_x + R(x, y, \Phi(y))$, where R is the general rational solution of (6.28) involving an undetermined function $\Phi(y)$. There may be two factors $\partial_x + r_i(x, y)$ where $r_i(x, y)$ are special rational solutions of (6.28); there may be a single such factor or no factor at all.

(iii) *A right factor $\partial_y + A_2$ exists if*

$$A_5 = A_2A_3 + 2A_{2,x} + A_{2,y}A_1 - A_2^2A_1,$$

$$A_7 = A_2A_6 + A_{2,y}A_4 - A_4A_2^2 + A_3A_{2,x} \quad (6.23)$$

$$+ A_{2,xx} + A_{2,xy}A_1 - 2A_{2,x}A_2A_1.$$

(iv) *There may exist a Laplace divisor $\mathbb{L}_{y^n}(L)$ for $n \geq 2$.*

(v) *There may exist a Laplace divisor $\mathbb{L}_{x^m}(L)$ for $m \geq 3$.*

(vi) *There may exist both Laplace divisors $\mathbb{L}_{x^m}(L)$ and $\mathbb{L}_{y^n}(L)$. In this case L is completely reducible; L is the left intersection of two Laplace divisors.*

Proof. Dividing the operator (6.21) by $\partial_x + a\partial_y + b$, the requirement that this division be exact leads to the following system of equations between the coefficients.

$$a^2 - A_1 a = 0, \tag{6.24}$$

$$a_x - (3a - 2A_1)a_y - A_2 a^2 + A_3 a - (2a - A_1)b = A_4, \tag{6.25}$$

$$b_x - (3a - 2A_1)b_y - b^2 - (2a_y + 2A_2 a - A_3)b + a_{xy} - (a - A_1)a_{yy}$$
$$+ A_2 a_x - a_y^2 - (A_2 a - A_3)a_y + A_5 a = A_6, \tag{6.26}$$

$$b_{xy} + (A_1 - a)b_{yy} + A_2 b_x - (2b + a_y + A_2 a - A_3)b_y - A_2 b^2 + A_5 b = A_7. \tag{6.27}$$

Case (i). If $A_1 \neq 0$ the first Eq. (6.24) determines the two roots $a_1 = 0$ and $a_2 = A_1$. The corresponding values b_1 and b_2 follow uniquely from (6.25). If the two constraints (6.26) and (6.27) are satisfied for a pair of values a_i and b_i, $i = 1$ or 2, a factor $\partial_x + a_i\partial_y + b_i$ exists.

Case (ii). If $A_1 = 0$, there follows $a = 0$ and Eq. (6.25) simplifies to $A_4 = 0$. Simplification and autoreduction of Eqs. (6.26) and (6.27) yields the system

$$b_x - b^2 + A_3 b - A_6 = 0, \tag{6.28}$$

$$A_{6,y} + A_2 A_6 - A_7 - (A_{3,y} + A_2 A_3 - A_5)b = 0 \tag{6.29}$$

for b. The following alternatives are distinguished. If the coefficient of b in (6.29) does not vanish, b may be determined uniquely from it and (6.28) becomes a constraint; this is subcase (a). If the coefficient of b in (6.29) vanishes and the two constraints (6.22) are satisfied, b is determined by the Riccati equation (6.28); its rational solutions determine a possible factorization.

Case (iii). Dividing the operator (6.21) by $\partial_y + c$, the requirement that this division be exact yields the equations $c - A_2 = 0$ and

$$c_x + \frac{1}{2}c_y A_1 - \frac{1}{2}c^2 A_1 + \frac{1}{2}cA_3 - \frac{1}{2}A_5 = 0,$$

$$c_{xx} + c_{xy}A_1 - 2c_x cA_1 + c_x A_3 + c_y A_4 - c^2 A_4 + cA_6 - A_7 = 0. \tag{6.30}$$

Transformation into a Janet basis yields (6.23).

The possible existence of the Laplace divisors in cases (iv) to (vi) is a consequence of Proposition 2.4 and the constructive proof given there.   □

In order to solve concrete problems it is important to know to what extent the factorizations described in this section may be determined algorithmically. The answer to this question is given in the following corollary.

Corollary 6.2. *Any first-order right factor in the base field may be determined algorithmically. Existence of a factor in a field extension may always be decided. Laplace divisors of fixed order may be determined; a bound for the order of a Laplace divisor is not known.*

Proof. In cases (i) and (iii), and subcase (a) of case (ii) of Theorem 6.2 the statement is obvious. If in subcase (b) of case (ii) the conditions (6.22) are satisfied, the rational solutions of the ordinary Riccati equation (6.28) have to be determined; this is always algorithmically possible according to Lemma B.1. For any Laplace divisor of fixed order this follows from the constructive proof of Proposition 2.3 and Corollary 2.1.   □

After the possible factorizations of an operator (6.21) have been determined, a listing of its various decomposition types involving first-order principal factors is given next.

Theorem 6.3. *Let the differential operator L be defined by*

$$L \equiv \partial_{xxy} + A_1\partial_{xyy} + A_2\partial_{xx} + A_3\partial_{xy} + A_4\partial_{yy} + A_5\partial_x + A_6\partial_y + A_7; \quad (6.31)$$

$A_i \in \mathbb{Q}(x, y)$ *for all* i. *Let* $l_i \equiv \partial_x + a_i\partial_y + b_i$ *for* $i = 1, 2$; $l(\Phi) \equiv \partial_x + a\partial_y + b(\Phi)$; $k \equiv \partial_y + c$; $a_i, b_i, c \in \mathbb{Q}(x, y)$ *such that* $a_1a_2 = 0$. *The following decomposition types* $\mathscr{L}^1_{xxy}, \dots, \mathscr{L}^{11}_{xxy}$ *involving principal right divisors may be distinguished.*

$$\mathscr{L}^1_{xxy} : L = kl_2l_1; \mathscr{L}^2_{xxy} : L = l_2kl_1; \mathscr{L}^3_{xxy} : L = l_2l_1k;$$

$$\mathscr{L}^4_{xxy} : L = Lclm(l_2, k)l_1; \mathscr{L}^5_{xxy} : L = Lclm(l_2, l_1)k; \mathscr{L}^6_{xxy} : L = Lclm\big(l(\Phi)\big)k;$$

$$\mathscr{L}^7_{xxy} : L = l_2Lclm(l_1, k); \mathscr{L}^8_{xxy} : L = kLclm(l_2, l_1); \mathscr{L}^9_{xxy} : L = kLclm\big(l(\Phi)\big);$$

$$\mathscr{L}^{10}_{xxy} : L = Lclm(l_2, l_1, k); \mathscr{L}^{11}_{xxy} : L = Lclm\big(l(\Phi), k\big).$$

Proof. It is based on Proposition 6.2. In case (i) there may be two factors with a principal intersection yielding a decomposition of type $\mathscr{L}^7_{xxy}$; a non-principal intersection ideal as described in Theorem 2.2 is excluded because the highest derivative ∂_{xxx} of its first generator does not occur in L. If there is a single factor, it is divided out and a second-order operator with leading derivative ∂_{xy} is obtained. It has to be decomposed by Theorem 4.2; if it is not irreducible the possible decomposition types are $\mathscr{L}^1_{xxy}$, $\mathscr{L}^2_{xxy}$ or $\mathscr{L}^4_{xxy}$.

Case (ii), subcase (a), may lead to a single factor; the same decompositions as for a single factor in case (i) may occur. The same is true if subcase (b) yields a single factor. If there are two factors, they generate a principal intersection because $a_1 = a_2$ and the decomposition type is $\mathscr{L}^7_{xxy}$ or $\mathscr{L}^9_{xxy}$.

If in case (iii) there is a factor k, it is divided out; the resulting second-order left factor may be decomposed according to Theorem 4.1. The possible decomposition types are $\mathscr{L}^3_{xxy}$, $\mathscr{L}^5_{xxy}$ and $\mathscr{L}^6_{xxy}$.

It may occur that factors exist originating from case (iii) and in addition from case (i) or case (ii). The various alternatives may be described as follows. If there is a single additional factor due to case (i) with a principal intersection, it may be divided out and decomposition type $\mathscr{L}^8_{xxy}$ results. If there are two additional factors, their combined intersection is principal and generates the ideal of the given operator. If there is a single additional factor originating from case (ii), its intersection is principal due to Theorem 2.3 and the decomposition type is again $\mathscr{L}^8_{xxy}$. Furthermore, in case (ii), subcase (b), there may be two factors generating a principal intersection of a type $\mathscr{L}^{10}_{xxy}$ decomposition; or there is a factor containing an undetermined function in its coefficient generating a decomposition of type $\mathscr{L}^{11}_{xxy}$.   □

The various alternatives for decomposing an operator (6.21) into first-order components are illustrated now by some examples. The first three cases have identical symbol equations but different decompositon types, albeit they differ only by the order of the factors.

Example 6.10.   Consider the operator

$$L \equiv \partial_{xxy} + x\partial_{xyy} + \frac{1}{x}\partial_{xx} + (y+1)\partial_{xy} + (xy+1)\partial_{yy}$$

$$+ \left(1 + \frac{y}{x}\right)\partial_x + \left(x + y + \frac{1}{x}\right)\partial_y.$$

Because $A_1 \neq 0$, case (i) of Proposition 6.2 applies; there follows $a_1 = 0$, $b_1 = y + \frac{1}{x}$ and $a_2 = x$, $b_2 = 0$. Only the latter pair satisfies the constraints (6.26) and (6.27); thus there is a factor $l \equiv \partial_x + x\partial_y$. Because the constraints for case (iii) are violated, there is no factor with leading derivative ∂_y. Dividing out l the second-order operator $\partial_{xy} + \frac{1}{x}\partial_x + y\partial_y + 1 + \frac{y}{x}$ is obtained. By Theorem 4.3 it has a type $\mathscr{L}^1_{xy}$ decomposition. Altogether this yields the type $\mathscr{L}^1_{xxy}$ decomposition

$$L = \left(\partial_y + \frac{1}{x}\right)(\partial_x + y)(\partial_x + x\partial_y).$$

Continued in Example 7.10.   □

Example 6.11.   The operator

$$L \equiv \partial_{xxy} + x\partial_{xyy} + \frac{1}{x}\partial_{xx} + (y+1)\partial_{xy} + (xy+1)\partial_{yy} + \left(\frac{y}{x} - \frac{1}{x^2}\right)\partial_x + y\partial_y$$

differs from the preceding one only in the first-order terms; hence there is the same right factor $l \equiv \partial_x + x\partial_y$. Because the constraints for case (iii) are violated, there is no factor with leading derivative ∂_y. Dividing out l the second-order operator $\partial_{xy} + \frac{1}{x}\partial_x + y\partial_y + 1 + \frac{y}{x} - \frac{1}{x^2}$ is obtained now. By Theorem 4.3 is has a type $\mathscr{L}^2_{xy}$ decomposition, and for L the type $\mathscr{L}^2_{xxy}$ decomposition

$$L = (\partial_x + y)\left(\partial_y + \frac{1}{x}\right)(\partial_x + x\partial_y)$$

is obtained. Continued in Example 7.10.   □

Example 6.12. The operator

$$L \equiv \partial_{xxy} + x\partial_{xyy} + \frac{1}{x}\partial_{xx} + (y+1)\partial_{xy} + (xy+1)\partial_{yy}$$

$$+ \left(\frac{y}{x} - \frac{2}{x^2}\right)\partial_x + y\partial_y - \frac{y}{x^2} + \frac{2}{x^3}$$

differs from the preceding two examples again only in the first-order terms; the
same values for the pairs a_1, b_1 and a_2, b_2 are obtained. However none satisfies the
constraints (6.26) and (6.27), i.e. there is no factor with leading derivative ∂_x.

Case (ii) is excluded due to $A_1 \neq 0$. However, the conditions (6.30) are satisfied;
thus, there is a factor $k = \partial_y + \frac{1}{x}$. Dividing it out yields the second-order operator
$\partial_{xx} + x\partial_{xy} + y\partial_x + (xy+1)\partial_y$. By Theorem 4.1 is has a type $\mathcal{L}_{xx}^2$ decomposition
with a right factor $\partial_x + x\partial_y$. Summarizing these results the type $\mathcal{L}_{xxy}^3$ decomposition

$$L = (\partial_x + y)(\partial_x + x\partial_y)\left(\partial_y + \frac{1}{x}\right)$$

follows. Continued in Example 7.10.   □

In the next chapter it will be investigated how the order of the factors in
the preceeding three examples determines the structure of the solutions of the
corresponding pde's.

Example 6.13. Consider the operator

$$L \equiv \partial_{xxy} - \partial_{xyy} + (x+1)\partial_{xx} - (x-y+1)\partial_{xy} - y\partial_{yy} + (xy+y+1)\partial_x - (xy+y+1)\partial_y.$$

Because $A_1 = -1$ case (i) of Proposition 6.2 applies; there follows $a_1 = 0$ and
$a_2 = 1$. The corresponding values of b are $b_1 = -y$ and $b_2 = 0$. Only the pair
a_2, b_2 satisfies the constraints (6.26) and (6.27) leading to the factor $l \equiv \partial_x - \partial_y$.
The conditions for case (iii) are violated, an additional right factor does not exist.
Dividing out l, the second-order operator $\partial_{xy} + \partial_{xy} + (x+1)\partial_x - y\partial_y + xy + 1$
follows. According to Theorem 4.1 it has a type $\mathcal{L}_{xx}^2$ decomposition, and the type
$\mathcal{L}_{xxy}^4$ decomposition

$$L = Lclm(\partial_x + y, \partial_y + x + 1)(\partial_x - \partial_y)$$

is obtained. Continued in Example 7.11.   □

Example 6.14. Consider the operator

$$L \equiv \partial_{xxy} + \partial_{xyy} + x\partial_{xx} + (x - y - 1)\partial_{xy} - \partial_{yy}$$
$$-(xy + x - 2)\partial_x - (x - y - 1)\partial_y + xy - y - 1.$$

Because $A_1 = 1$ case (i) of Proposition 6.2 applies; it follows $a_1 = 0$ and $a_2 = 1$. The corresponding values of b are $b_1 = -1$ and $b_2 = -y$. Both pairs a_i, b_i do not satisfy the constraints (6.26) and (6.27), i.e. a factor with leading derivative ∂_x does not exist. However, the conditions of case (iii) are obeyed and there is a factor $\partial_y + x$. Dividing it out, the second-order operator $\partial_{xx} + \partial_{xy} - (y + 1)\partial_x - \partial_y + y$ follows. According to Theorem 4.1 it has a type $\mathcal{L}_{xx}^2$ decomposition. Putting these results together the type $\mathcal{L}_{xxy}^5$ decomposition

$$L = Lclm(\partial_x + \partial_y - y, \partial_x - 1)(\partial_y + x)$$

is obtained. Continued in Example 7.12.   □

Example 6.15. Consider the operator

$$L \equiv \partial_{xxy} + \left(x + \frac{y}{x}\right)\partial_{xx} + \frac{2}{x}\partial_{xy} + 4\partial_x + \frac{2}{x}.$$

Because $A_1 = A_4 = 0$ and $A_{3,y} + A_2A_3 - A_5 \neq 0$, case (ii), subcase (a) of Proposition 6.2 applies; $b = \dfrac{x}{x^2 - y}$ is obtained. This value does not satisfy the Riccati equation $b_x - b^2 + \frac{2}{x}b = 0$, i.e. a factor with leading derivative ∂_x does not exist. However, the constraints (6.23) of case (iii) are satisfied and a factor $\partial_y + x + \frac{y}{x}$ is obtained. Dividing it out yields the second order factor $\partial_{xx} + \frac{2}{x}\partial_x$ with type $\mathcal{L}_{xx}^3$ decomposition. Putting these results together the $\mathcal{L}_{xxy}^6$ decomposition

$$L = \left(\partial_{xx} + \frac{2}{x}\partial_x\right)\left(\partial_y + x + \frac{y}{x}\right) = Lclm\left(\partial_x + \frac{1}{x} - \frac{1}{x + \Phi}\right)\left(\partial_y + x + \frac{y}{x}\right)$$

follows. Continued in Example 7.13.   □

Example 6.16. Consider the operator

$$L \equiv \partial_{xxy} + (x + 1)\partial_{xx} + (y - 1)\partial_{xy} + (xy - x + y + 1)\partial_x - y\partial_y - xy - 1.$$

By case (i) of Proposition 6.2 the factor $l_1 \equiv \partial_x + y$ exists. In addition there is the factor $k \equiv \partial_y + x + 1$ originating from case (iii). They have the principal intersection

$$Lclm(l_1, k) = \partial_{xy} + (x + 1)\partial_x + y\partial_y + xy + y + 1$$

and L has the type $\mathcal{L}_{xxy}^7$ decomposition

$$L = (\partial_x - 1)Lclm(\partial_x + y, \partial_y + x + 1).$$

Continued in Example 7.14.   □

Example 6.17. Consider the operator

$$L \equiv \partial_{xxy} + \partial_{xyy} - (y + 1)\partial_{xy} - \partial_{yy} - \partial_x + y\partial_y + 1.$$

By case (i) of Proposition 6.2 the two factors $l_1 \equiv \partial_x - 1$ and $l_2 \equiv \partial_x + \partial_y - y$ exist. They have a principal left intersection $\partial_{xx} + \partial_{xy} - (y + 1)\partial_x - \partial_y + y$, and L has the type $\mathscr{L}^8_{xxy}$ decomposition

$$L = \partial_y Lclm(\partial_x + \partial_y - y, \partial_x - 1).$$

Continued in Example 7.15.   □

Example 6.18. Consider the operator

$$L \equiv \partial_{xxy} + (x + y)\partial_{xx} + \frac{2}{x}\partial_{xy} + 2\left(1 + \frac{y}{x}\right)\partial_x.$$

Because $A_1 = A_4 = 0$ and $A_{3,y} + A_2 A_3 - A_5 = 0$, case (ii), subcase (b) of Proposition 6.2 applies. The Riccati equation for b is $b_x - b^2 + \frac{2}{x}b = 0$, it has the general solution $b = \frac{1}{x} - \frac{1}{x + \Phi(y)}$ where Φ is an undetermined function of y; it yields the factor $l \equiv \partial_x + \frac{1}{x} - \frac{1}{x + \Phi(y)}$. The constraints of case (iii) are not satisfied, i.e. there is no factor with leading derivative ∂_y. Dividing out the factor l the type $\mathscr{L}^9_{xxy}$ decomposition

$$L = (\partial_y + x + y)\left(\partial_{xx} + \frac{2}{x}\right) = (\partial_y + x + y)Lclm\left(\partial_x + \frac{1}{x} - \frac{1}{x + \Phi}\right)$$

follows. Continued in Example 7.16.   □

Example 6.19. Consider the operator

$$L \equiv \partial_{xxy} + \partial_{xyy} - y\partial_{xx} - (2y + 1)\partial_{xy} - \partial_{yy} + (y^2 + y - 1)\partial_x + 2y\partial_y - y^2 + 1.$$

Because $A_1 = 1$ case (i) of Proposition 6.2 applies; it follows $a_1 = 0$ and $a_2 = 1$. The corresponding values of b are $b_1 = -1$ and $b_2 = -y$. Both pairs a_i, b_i satisfy the constraints (6.26) and (6.27), i.e. the factors $\partial_x - 1$ and $\partial_x + \partial_y - y$ exist. Because the conditions of case (iii) are satisfied there is the additional factor $\partial_y - y$, and L has the type $\mathscr{L}^{10}_{xxy}$ decomposition

$$L = Lclm(\partial_x + \partial_y - y, \partial_x - 1, \partial_y - y).$$

Continued in Example 7.17.   □

Example 6.20. Consider the operator

$$L \equiv \partial_{xxy} - y\partial_{xx} + \frac{2}{x}\partial_{xy} - \frac{2y}{x}\partial_x.$$

Because $A_1 = A_4 = 0$ and $A_{3,y} + A_2 A_3 - A_5 = 0$, case (ii), subcase (b) of Proposition 6.2 applies. The Riccati equation for b is $b_x - b^2 + \frac{2}{x}b = 0$, it has the general solution $b = \frac{1}{x} - \frac{1}{x + \Phi(y)}$ where Φ is an undetermined function of y; it yields the factor $l \equiv \partial_x + \frac{1}{x} - \frac{1}{x + \Phi(y)}$. The constraints of case (iii) are satisfied now, i.e. there is another factor $\partial_y - y$. Altogether the type $\mathscr{L}^{11}_{xxy}$ decomposition

$$L = Lclm\left(\partial_y + x + y, \partial_{xx} + \frac{2}{x}\partial_x\right) = Lclm\left(\partial_x + \frac{1}{x} - \frac{1}{x + \Phi}, \partial_y - y\right)$$

is obtained. Continued in Example 7.18.   □

In addition to the decompositions into principal first-order components there are four decompositions types of an operator with leading derivative ∂_{xxy} involving non-principal right divisors; they are considered next.

Theorem 6.4. *Let the differential operator L be defined by*

$$L \equiv \partial_{xxy} + A_1\partial_{xyy} + A_2\partial_{xx} + A_3\partial_{xy} + A_4\partial_{yy} + A_5\partial_x + A_6\partial_y + A_7 \quad (6.32)$$

such that $A_i \in \mathbb{Q}(x, y)$ for all i. The following decomposition types $\mathscr{L}^{12}_{xxy}, \ldots, \mathscr{L}^{15}_{xxy}$ involving non-principal right divisors may be distinguished; $m \geq 3$ and $n \geq 2$. The ideal $\mathbb{J}_{xxy}$ is defined on page 45.

$$\mathscr{L}^{12}_{xxy} : L = Lclm\big(\mathbb{L}_{x^m}(L), \mathbb{L}_{y^n}(L)\big);$$

$$\mathscr{L}^{13}_{xxy} : L = Exquo\big(L, \mathbb{L}_{x^m}(L)\big)\mathbb{L}_{x^m}(L) = \begin{pmatrix} 1 & 0 \\ 0 & \partial_y + A_2 \end{pmatrix}\begin{pmatrix} L \\ l_m \end{pmatrix};$$

$$\mathscr{L}^{14}_{xxy} : L = Exquo\big(L, \mathbb{L}_{y^n}(L)\big)\mathbb{L}_{y^n}(L) =$$

$$\begin{pmatrix} 1 & 0 \\ 0 & \partial_{xx} + A_1\partial_{xy} + (A_{1,y} - A_1 A_2 + A_3)\partial_x + A_4\partial_y + A_{4,y} - A_2 A_4 + A_6 \end{pmatrix}\begin{pmatrix} L \\ l_n \end{pmatrix};$$

$$\mathscr{L}^{15}_{xxy} : L = Exquo\big(\langle L\rangle, \mathbb{J}_{xxy}\big)\mathbb{J}_{xxy} =$$

$$\begin{pmatrix} 1, & A_1 \\ 0, \ \partial_x + (A_1 - q_1)\partial_y + p_3 + q_1 q_3 - q_4 - (p_2 + q_1 q_2 - q_3)A_1 + A_{1,y} \end{pmatrix}\begin{pmatrix} K_1 \\ K_2 \end{pmatrix}.$$

Proof. The conditions for the existence of Laplace divisors $\mathbb{L}_{x^m}(L)$ and $\mathbb{L}_{y^n}(L)$ for an operator (6.32) have been proved in Proposition 2.4 on page 39. If there are two divisors, decomposition type $\mathscr{L}_{xxy}^{12}$ follows. If there is a single one, dividing it out yields decomposition type $\mathscr{L}_{xxy}^{13}$ or $\mathscr{L}_{xxy}^{14}$.

According to Proposition 6.2, for any given operator there may exist factors simultaneously by case (i) or (ii) and case (iii). By Theorem 2.3 on page 50 their generic intersection is an ideal $\mathbb{J}_{xxy}$ yielding a type $\mathscr{L}_{xxy}^{15}$ decomposition.   □

Example 6.21. Consider the operator

$$L \equiv \partial_{xxy} + x\partial_{xyy} - \frac{1}{y+1}\partial_{xx} + \partial_{xy} + \partial_{yy} - \frac{1}{y+1}\partial_x - \frac{1}{y+1}\partial_y.$$

According to case (i) of Proposition 6.2 there are two pairs $a_1 = 0$, $b_1 = \frac{1}{x}$ and $a_2 = x$, $b_2 = \frac{x}{y+1} + 1$; neither satisfies the constraints (6.26) and (6.27), i.e. there is no first-order factor with leading derivative ∂_x. The same is true for a factor with leading derivative ∂_y because the second constraint (6.23) is violated. However, there is a Laplace divisor $\mathbb{L}_{y^2}(L) = \langle\!\langle L, \partial_{yy} \rangle\!\rangle$ leading to decomposition

$$L = \begin{pmatrix} 1 & 0 \\ 0 \; \partial_{xx} + x\partial_{xy} + \dfrac{x+y+1}{y+1}\partial_x + \partial_y \end{pmatrix} \begin{pmatrix} L \\ \partial_{yy} \end{pmatrix}$$

of type $\mathscr{L}_{xxy}^{14}$. Continued in Example 7.19.   □

Example 6.22. For the operator

$$L \equiv \partial_{xxy} + \frac{y}{x}\partial_{xyy} - x\partial_{xx} + \frac{y^2}{x}\partial_{yy} - (xy+2)\partial_x - \frac{y}{x}(xy-2)\partial_y - 4y$$

$A_1 = \frac{y}{x} \neq 0$; thus $a_1 = 0$ and $a_2 = \frac{y}{x}$. By (6.25) there follows $b_1 = y$ and $b_2 = \frac{x}{y^2}(xy+2) + x - 2$. Only the first pair a_1, b_1 satisfies the constraints of case (i) and leads to a factor $l \equiv \partial_x + y$. However, the conditions of case (iii) are obeyed and yield a factor $k \equiv \partial_y - x$. The intersection ideal of l and k is not principal; by Theorem 2.3 it is

$$Lclm(k,l) = \langle\!\langle K_1 \equiv \partial_{xxy} - x\partial_{xx} + 2y\partial_{xy} - 2(xy+1)\partial_x + y^2\partial_y - xy^2 - 2y,$$

$$K_2 \equiv \partial_{xyy} - 2x\partial_{xy} + y\partial_{yy} + x^2\partial_x - 2(xy-1)\partial_y + x^2y - 2x \rangle\!\rangle,$$

i.e. L has the type $\mathscr{L}_{xxy}^{15}$ decomposition

$$L = \begin{pmatrix} 1 & \dfrac{y}{x} \\ 0 \; \partial_x + \dfrac{y}{x}\partial_y + \dfrac{1}{x} \end{pmatrix} \begin{pmatrix} K_1 \\ K_2 \end{pmatrix};$$

it may be represented as $L = K_1 + \frac{y}{x}K_2$. Continued in Example 7.20.   □

6.3 Operators with Leading Derivative ∂_{xyy}

If an equation contains no unmixed third-order derivative, but only a single mixed third-order one, it is assumed to be ∂_{xyy}; otherwise x and y may be exchanged or, what amounts to the same thing, the term order may be changed to $y \succ x$. Although this case may be obtained from Proposition 6.2 by specialization if $A_1 = 0$ is assumed, it is treated as a separate alternative because solving the corresponding differential equations will be simpler than in the general case.

Proposition 6.3. *Let the third order partial differential operator*

$$L \equiv \partial_{xyy} + A_1 \partial_{xy} + A_2 \partial_{yy} + A_3 \partial_x + A_4 \partial_y + A_5 \qquad (6.33)$$

be given with $A_i \in \mathbb{Q}(x, y)$ for all i. The following first order right factors may occur.

(i) A right factor $\partial_x + A_2$ exists if

$$A_4 = 2A_{2,y} + A_1 A_2,$$
$$A_5 = A_{2,yy} + A_{2,y} A_1 + A_2 A_3. \qquad (6.34)$$

(ii) Right factors $\partial_y + c$ with $c \in \mathbb{Q}(x, y)$. Define

$$P \equiv A_{3,x} + A_2 A_3 - A_5 \quad and \quad Q = A_{1,x} + A_1 A_2 - A_4. \qquad (6.35)$$

The following two subcases have to be distinguished.

(a) If $P = Q = 0$, c must be a rational solution of $c_y - c^2 + A_1 c - A_3 = 0$.
(b) If $Q \neq 0$ and $P_y Q - P(Q_y + P) + A_1 PQ - A_3 Q^2 = 0$, there follows

$$c = \frac{A_{3,x} + A_2 A_3 - A_5}{A_{1,x} + A_1 A_2 - A_4}.$$

(iii) There may exist a Laplace divisor $\mathbb{L}_{y^n}(L)$ for $n \geq 3$.
(iv) There may exist a Laplace divisor $\mathbb{L}_{x^m}(L)$ for $m \geq 2$.
(v) There may exist both Laplace divisors $\mathbb{L}_{x^m}(L)$ and $\mathbb{L}_{y^n}(L)$. In this case L is completely reducible; L is the left intersection of two Laplace divisors.

Proof. Case (i). Reducing (6.33) w.r.t. $\partial_x + a \partial_y + b$, the requirement that this division be exact leads to the constraints $a = 0$ and

$$2a_y + A_1 a - A_2 + b = 0,$$
$$a_{yy} + A_1 a_y + 2b_y + A_1 b + A_3 a - A_4 = 0,$$
$$b_{yy} + A_1 b_y + A_3 b - A_5 = 0.$$

Transforming this system of pde's into a Janet basis yields the system (6.34).

Case (ii). The requirement that $\partial_y + c$ divides (6.33) leads to the constraints

$$c_y - c^2 + A_1c - A_3 = 0,$$

$$c_{xy} + A_1c_x - 2cc_x + A_2c_y - A_2c^2 + A_4c - A_5 = 0. \qquad (6.36)$$

Reduction of the second equation w.r.t. the first one yields

$$(A_{1,x} + A_1A_2 - A_4)c - A_{3,x} - A_2A_3 + A_5 = 0. \qquad (6.37)$$

If the coefficient of c vanishes, there are the two constraints $P = 0$ and $Q = 0$; c has to be determined from the first equation of (6.36). This is subcase (a).

Subcase (b) is obtained if the coefficient of c in (6.37) does not vanish. Then c is uniquely determined by this equation. Substituting it into the first equation of (6.36) yields the relation for P and Q given above.

The possible existence of the Laplace divisors in cases (iii) to (v) is a consequence of Proposition 2.5 and the constructive proof given there. Furthermore, the existence of a first-order factor in addition to a Laplace divisor is not excluded.

□

What concerns algorithmic methods for obtaining the various decompositions, the answer is the same as for operators (6.21), i.e. Corollary 6.2 applies here as well.

Using this result, a listing of the possible decomposition types of an operator (6.33) involving first-order principal factors only is given in the subsequent theorem; decompositions involving non-principal divisors are discussed later in this subsection.

Theorem 6.5. *Let the differential operator L be defined by*

$$L \equiv \partial_{xyy} + A_1\partial_{xy} + A_2\partial_{yy} + A_3\partial_x + A_4\partial_y + A_5 \qquad (6.38)$$

such that $A_i \in \mathbb{Q}(x, y)$ for all i. Let $l \equiv \partial_x + b$, $k_i \equiv \partial_y + c_i$ for $k = 1, 2$ and $k(\Phi) \equiv \partial_y + c(\Phi)$; $b, c_i \in \mathbb{Q}(x, y)$. The following decomposition types $\mathscr{L}^1_{xyy}, \ldots, \mathscr{L}^{11}_{xyy}$ involving principal first-order right divisors may be distinguished.

$\mathscr{L}^1_{xyy} : L = k_2k_1l; \mathscr{L}^2_{xyy} : L = k_2lk_1; \mathscr{L}^3_{xyy} : L = lk_2k_1;$

$\mathscr{L}^4_{xyy} : L = Lclm(l, k_2)k_1; \mathscr{L}^5_{xyy} : L = Lclm(k_2, k_1)l; \mathscr{L}^6_{xyy} : L = Lclm(k(\Phi))l;$

$\mathscr{L}^7_{xyy} : L = k_2Lclm(k_1, l); \mathscr{L}^8_{xyy} : L = lLclm(k_2, k_1); \mathscr{L}^9_{xyy} : L = lLclm(k(\Phi));$

$\mathscr{L}^{10}_{xyy} : L = Lclm(l, k_2, k_1); \mathscr{L}^{11}_{xyy} : L = Lclm(l, k(\Phi)).$

Proof. It is based on Proposition 6.3. According to case (i), a single right factor $\partial_x + A_2$ exists if the conditions (6.34) are satisfied; then L factors into

$$L = (\partial_{yy} + A_1\partial_y + A_3)(\partial_x + A_2).$$

The second-order left factor is a generic second-order ordinary operator in y. The coefficient b in a possible factor $\partial_y + b$ is determined by an ordinary Riccati equation. Therefore all decomposition types listed in Theorem 4.1 may occur; they lead to decomposition types $\mathscr{L}^1_{xyy}$, $\mathscr{L}^4_{xyy}$ and $\mathscr{L}^6_{xyy}$.

Any right factor $\partial_y + c$ satisfies

$$L = (\partial_{xy} + (A_1 - c)\partial_x + A_2\partial_y + A_4 - c_x - A_2c)(\partial_y + c).$$

If in case (ii), subcase (a), there is a single rational solution, or subcase (b) applies, the resulting decomposition type is $\mathscr{L}^2_{xyy}$, $\mathscr{L}^3_{xyy}$ or $\mathscr{L}^5_{xyy}$. If there is more than a single rational solution in subcase (a), the decomposition type is $\mathscr{L}^7_{xyy}$ or $\mathscr{L}^8_{xyy}$.

There may be a factor l originating from case (i), and in addition a single factor k, or two factors k_1 and k_2 originating from case (ii); these alternatives yield decompositions of type $\mathscr{L}^7_{xyy}$ or type $\mathscr{L}^{10}_{xyy}$.   □

Subsequently examples are given for the various decomposition types; they show that each decomposition does in fact exist. As usual, these examples are continued in the next chapter where solutions of the corresponding equations are given.

Example 6.23. For the operator

$$L \equiv \partial_{xyy} - (x + y)\partial_{xy} + y\partial_{yy} + xy\partial_x - (xy + y^2 - 2)\partial_y + xy^2 - x - y$$

$A_1 = -(x + y)$, $A_2 = y$, $A_3 = xy$, $A_4 = -(xy + y^2 - 2)$, and $A_5 = xy^2 - x - y$. The conditions (6.34) are satisfied, there is a factor $l \equiv \partial_x + y$. Because $P \neq 0$ and $Q \neq 0$, case (ii), subcase (a), does not apply; the same is true for subcase (b) because the conditions for P and Q are not satisfied, i.e. l is the only first-order right factor. Dividing it out, the second-order left factor $\partial_{yy} - (x + y)\partial_y + xy\partial_y$ is obtained. By Theorem 4.1 it factors uniquely into $k_1 \equiv \partial_y - x$ and $k_2 \equiv \partial_y - y$ and the type $\mathscr{L}^1_{xyy}$ decomposition

$$L = (\partial_y - y)(\partial_y - x)(\partial_x + y)$$

for L follows. Continued in Example 7.21.   □

Example 6.24. For the operator

$$L \equiv \partial_{xyy} - (x + y)\partial_{xy} + y\partial_{yy} + xy\partial_x - (xy + y^2)\partial_y + xy^2 - x + y$$

$A_1 = -(x + y)$, $A_2 = y$, $A_3 = xy$, $A_4 = -(xy + y^2)$, and $A_5 = xy^2 - x + y$. The conditions (6.34) are violated, i.e. there is no factor with leading derivative ∂_x.

Because $P \neq 0$ and $Q \neq 0$, case (ii), subcase (a), does not apply; however the conditions for subcase (b) are satisfied, there is only the first-order right factor $k_1 = \partial_y - x$. Dividing it out, the second-order left factor $\partial_{xy} - x\partial_x + y\partial_y - xy - 1$ is obtained. By Theorem 4.2 it factors uniquely into $k_1 \equiv \partial_y - x$ and $k_2 \equiv \partial_y - y$; L has the type $\mathscr{L}^2_{xyy}$ decomposition

$$L = (\partial_y - y)(\partial_x + y)(\partial_y - x).$$

Continued in Example 7.22.   □

Example 6.25. For the operator

$$L \equiv \partial_{xyy} - (x + y)\partial_{xy} + y\partial_{yy} + xy\partial_x - (xy + y^2 + 1)\partial_y + (xy + 1)y$$

$A_1 = -(x + y)$, $A_2 = y$, $A_3 = xy$, $A_4 = -(xy + y^2 + 1)$, and $A_5 = (xy + 1)y$. The conditions (6.34) are violated, i.e. a factor with leading derivative ∂_x does not exist. However, because $P = Q = 0$, the Riccati equation for the coefficient c is $c_y - c^2 - (x + y)c - xy = 0$, it has the single rational solution $c = -x$, there is the factor $k_1 \equiv \partial_y - x$. Dividing it out, the second-order left factor is $\partial_{xy} - y\partial_x + y\partial_y - y^2$. According to Proposition 4.2 it factors into $l \equiv \partial_x + y$ and $k_2 \equiv \partial_y - y$; therefore the full type $\mathscr{L}^3_{xyy}$ factorization is

$$L = (\partial_x + y)(\partial_y - y)(\partial_y - x).$$

Continued in Example 7.23.   □

The factorizations of the three preceding examples differ only be the order of the factors; therefore the remarks following Example 6.12 on page 134 apply for these examples as well.

Example 6.26. Consider the operator

$$L \equiv \partial_{xyy} + \left(x + 1 + \frac{1}{y}\right)\partial_{xy} + y\partial_{yy} + \left(x + \frac{1}{y}\right)\partial_x + (xy + y + 2)\partial_y + xy + 2.$$

Because the conditions (6.34) are violated there is no factor with leading derivative ∂_x. There follows $P = Q = 0$, subcase (a) of case (ii) applies. The Riccati equation

$$c_y - c^2 + \left(x + 1 + \frac{1}{y}\right)c - x - \frac{1}{y}$$

for c is obtained. Its only rational solution is $c = 1$; it yields the right factor $\partial_y + 1$. Dividing it out the second-order operator $\partial_{xy} + (x + \frac{1}{y})\partial_x + y\partial_y + xy + 2$ follows. According to Proposition 4.2 it has a type $\mathscr{L}^3_{xy}$ decomposition. Finally the type $\mathscr{L}^4_{xyy}$ decomposition

$$L = Lclm\left(\partial_x + y, \partial_y + x + \frac{1}{y}\right)(\partial_y + 1)$$

is obtained. Continued in Example 7.24.   □

Example 6.27. Consider the operator

$$L \equiv \partial_{xyy} - \frac{x^2y^2 - 2}{(xy+1)y}\partial_{xy} - \frac{y}{x}\partial_{yy} - \frac{(xy+2)x}{(xy+1)y}\partial_x$$
$$+ \frac{x^2y^2 - 2xy - 4}{(xy+1)x}\partial_y + \frac{2(x^2y^2 + xy - 1)}{xy(xy+1)}.$$

Conditions (6.34) are satisfied, there is a factor $l \equiv \partial_x - \frac{y}{x}$. Because the conditions for subcase (b) of case (ii) are violated, this is the only right factor. Dividing L by l, the operator

$$\partial_{yy} - \frac{x^2y^2 - 2}{(xy+1)y}\partial_y - \frac{(xy+2)x}{xy+1}$$

follows. Exchangining x and y, Theorem 4.1 may be applied and the type $\mathcal{L}_{xx}^2$ decomposition follows; L has the type $\mathcal{L}_{xyy}^5$ decomposition

$$L = Lclm\left(\partial_y + \frac{1}{y}, \partial_y - x\right)\left(\partial_x - \frac{y}{x}\right).$$

Continued in Example 7.25.   □

Example 6.28. For the operator

$$L \equiv \partial_{xyy} + \frac{2}{y}\partial_{xy} + (x + y^2)\partial_{yy} + \frac{2x + 6y^2}{y}\partial_y + 6$$

the conditions (6.34) are satisfied and the factor $l \equiv \partial_x + x + y^2$ exists. However, the conditions for subcase (b) of case (ii) are violated, there is no additional right factor. Dividing L by l, the operator $\partial_{yy} + \frac{2}{y}\partial_y$ follows. According to Theorem 4.1 it has a type $\mathcal{L}_{xx}^2$ decomposition involving an undetermined function Φ. Therefore L has the type $\mathcal{L}_{xyy}^6$ decomposition

$$L = \left(\partial_{yy} + \frac{2}{y}\partial_y\right)(\partial_x + x + y^2) = Lclm\left(\partial_y + \frac{1}{y} - \frac{1}{\Phi + y}\right)(\partial_x + x + y^2).$$

Continued in Example 7.26.   □

Example 6.29. For the operator

$$L \equiv \partial_{xyy} + \left(x + 1 + \frac{1}{y}\right)\partial_{xy} + y\partial_{yy} + \left(x + \frac{1}{y} - \frac{1}{y^2}\right)\partial_x$$
$$+ (xy + y + 3)\partial_y + xy + x + 2$$

by case (*i*) and case (*ii*), subcase (*b*) of Proposition 6.3 the type $\mathscr{L}^7_{xyy}$ decomposition

$$L = (\partial_y + 1)Lclm\left(\partial_x + y, \partial_y + x + \frac{1}{y}\right)$$
$$= (\partial_y + 1)\left(\partial_{xy} + \left(x + \frac{1}{y}\right)\partial_x + y\partial_y + xy + 2\right)$$

follows. Continued in Example 7.27.   □

Example 6.30. For the operator

$$L \equiv \partial_{xyy} + \frac{y^2 - 2}{y - 1}\partial_{xy} + \frac{y}{x}\partial_{yy} + \frac{y^2 - y - 1}{y - 1}\partial_x + \frac{y(y^2 - 2)}{x(y - 1)}\partial_y + \frac{y(y^2 - y - 1)}{x(y - 1)}$$

conditions (6.34) are not satisfied, i.e. a factor $\partial_x + A_3$ does not exist. However, $P = Q = 0$, and case (*ii*), subcase (*b*) applies. The Riccati equation

$$c_y - c^2 + \frac{y^2 - 2}{y + 1}c + \frac{y^2 + y - 1}{y + 1} = 0$$

has the rational solutions $c = y$ and $c = -1$. They yield the factors $\partial_y + y$ and $\partial_y - 1$ and the type $\mathscr{L}^8_{xyy}$ decomposition

$$L = \left(\partial_x + \frac{y}{x}\right)\left(\partial_{yy} + \frac{y^2 - 2}{y - 1}\partial_y + \frac{y^2 - y - 1}{y - 1}\right)$$
$$= \left(\partial_x + \frac{y}{x}\right)Lclm(\partial_y + y, \partial_y + 1).$$

Continued in Example 7.28.   □

Example 6.31. Consider the operator

$$L \equiv \partial_{xyy} + \frac{2}{y}\partial_{xy} + (x + y^2)\partial_{yy} + \frac{2x + 2y^2}{y}\partial_y.$$

Conditions (6.34) are not satisfied, i.e. a factor with leading derivative ∂_x does not exist. Because $P = Q = 0$ the possible factors $\partial_y + c$ are determined by the Riccati equation $c_y - c^2 + \frac{2}{y}c = 0$. Its general solution is $c = \frac{1}{y} - \frac{1}{\Phi + y}$ where Φ is an undetermined function of x. Therefore L has the type $\mathcal{L}^9_{xyy}$ decomposition

$$L = (\partial_x + x + y^2)\left(\partial_{yy} + \frac{2}{y}\partial_y\right) = (\partial_x + x)Lclm\left(\partial_y + \frac{1}{y} - \frac{1}{\Phi + y}\right).$$

Continued in Example 7.29. □

Example 6.32. Let the operator L be defined by

$$L \equiv \partial_{xyy} + \left(2x + 1 - \frac{1}{y-1} + \frac{2}{y}\right)\partial_{xy} + y\partial_{yy}$$

$$+ \left(x(x+1) - \frac{x+1}{y-1} + 2\frac{x+1}{y}\right)\partial_x + \left((2x+1)y + 3 - \frac{1}{y-1}\right)\partial_y$$

$$+ x(x+1)y + 3x + 2 - \frac{x+2}{y-1} + \frac{2}{y}.$$

Its coefficients satisfy conditions (6.34) of Proposition 6.3, i.e. there is a factor $\partial_y + x$. Furthermore, $P = Q = 0$ such that case (ii), subcase (a) applies. The Riccati equation for c has the two rational solutions $c = x + 1$ and $c = x + \frac{1}{y}$; they yield the factors $\partial_y + x + 1$ and $\partial_y + x + \frac{1}{y}$; the type $\mathcal{L}^{10}_{xyy}$ decomposition

$$L = Lclm\left(\partial_x + y, \partial_y + x + 1, \partial_y + x + \frac{1}{y}\right)$$

follows. Continued in Example 7.30. □

Example 6.33. As in the preceding example the operator

$$L \equiv \partial_{xyy} + \frac{2}{y}\partial_{xy} + x\partial_{yy} + \frac{2x}{y}\partial_y$$

has a first-order factor $\partial_x + x$. Because $P = Q = 0$ the possible factors $\partial_y + c$ are determined by the Riccati equation $c_y - c^2 + \frac{2}{y}c = 0$. Its general solution is $c = \frac{1}{y} - \frac{1}{\Phi + y}$ where Φ is an undetermined function of x; therefore L has the type $\mathcal{L}^{11}_{xyy}$ decomposition

$$L = Lclm(\partial_x + x, \partial_{yy} + \frac{2}{y}\partial_y) = Lclm\left(\partial_x + x, \partial_y + \frac{1}{y} - \frac{1}{\Phi + y}\right).$$

Continued in Example 7.31. □

In addition to the decompositions into first-order principal divisors, there may exist Laplace divisors and divisors originating from the non-principal intersection of first-order operators described in case (ii) of Theorem 2.3 on page 51. They are the subject of the subsequent theorem.

Theorem 6.6. *Let the differential operator L be defined by*

$$L \equiv \partial_{xyy} + A_1 \partial_{xy} + A_2 \partial_{yy} + A_3 \partial_x + A_4 \partial_y + A_5 \qquad (6.39)$$

such that $A_i \in \mathbb{Q}(x, y)$ for all i. The following decomposition types $\mathcal{L}^{12}_{xyy}, \ldots, \mathcal{L}^{15}_{xyy}$ involving non-principal divisors may be distinguished; $m \geq 2$ and $n \geq 3$. The ideal $\mathbb{J}_{xyy}$ is defined on page 45.

$$\mathcal{L}^{12}_{xyy} : L = Lclm\big(\mathbb{L}_{x^m}(L), \mathbb{L}_{y^n}(L)\big);$$

$$\mathcal{L}^{13}_{xyy} : L = Exquo\big(L, \mathbb{L}_{y^n}(L)\big)\mathbb{L}_{y^n}(L) = \begin{pmatrix} 1 & 0 \\ 0 & \partial_x + A_3 \end{pmatrix} \begin{pmatrix} L \\ \ell_n \end{pmatrix};$$

$$\mathcal{L}^{14}_{xyy} : L = Exquo\big(L, \mathbb{L}_{x^m}(L)\big)\mathbb{L}_{x^m}(L) = \begin{pmatrix} 1 & 0 \\ 0 & \partial_{yy} + A_1 \partial_y + A_3 \end{pmatrix} \begin{pmatrix} L \\ \ell_m \end{pmatrix};$$

$$\mathcal{L}^{15}_{xyy} : L = Exquo\big(\langle L \rangle, \mathbb{J}_{xxy}\big)\mathbb{J}_{xxy} = \begin{pmatrix} 0, & 1 \\ \partial_y - p_2 - q_1 q_2 + q_3, & 0 \end{pmatrix} \begin{pmatrix} K_1 \\ K_2 \end{pmatrix}.$$

Proof. The existence of Laplace divisors $\mathbb{L}_{x^m}(L)$ and $\mathbb{L}_{y^n}(L)$ for an operator (6.39) has been proved in Proposition 2.5 on page 41; their differential dimension is $(1, 1)$ and $(1, 2)$ respectively. If there are two divisors, by case (iii) of Proposition 2.5 the representation of $\mathcal{L}^{12}_{xyy}$ follows. If there is a single divisor $\mathbb{L}_{y^n}(L)$, the exact quotient module of $\langle L \rangle$ and $\mathbb{L}_{y^n}(L)$ is constructed. It is generated by $(1, 0)$ and the syzygies determined in Corollary 2.5. Transformation into a Janet basis yields the representation given above for the $\mathcal{L}^{13}_{xyy}$. The discussion for decomposition type $\mathcal{L}^{14}_{xyy}$ is similar. Finally, a Janet basis for the module generated by the exact quotient $(0, 1)$ and the syzygy (2.29) given in Lemma 2.3 yields the decomposition for type $\mathcal{L}^{15}_{xyy}$. □

Example 6.34. Let the third order operator

$$L \equiv \partial_{xyy} + (x + y)\partial_{xy} + (x + y)\partial_x - 2\partial_y - 2$$

be given. By Proposition 6.3 it does not have a first order right factor. By Proposition 2.3 a divisor $\mathbb{L}_{x^m}(L)$ for $m \leq 2$, or a divisor $\mathbb{L}_{y^n}(L)$ for $n \leq 5$ does not exist. However, there is a divisor $\mathbb{L}_{x^3}(L) = \langle\!\langle L, \partial_{xxx} \rangle\!\rangle$, i.e. L has the decomposition

$$L = \begin{pmatrix} 1 & 0 \\ 0 & \partial_{yy} + (x + y)\partial_y + x + y \end{pmatrix} \begin{pmatrix} L \\ \partial_{xxx} \end{pmatrix}$$

of type $\mathcal{L}^{14}_{xyy}$. Continued in Example 7.32. □

Example 6.35. Consider the operator

$$L \equiv \partial_{xyy} - 2x\partial_{xy} + y\partial_{yy} + x^2\partial_x - 2(xy - 1)\partial_y + x(xy - 2).$$

By case (i) and case (ii), subcase (b) of Proposition 6.3 the factorizations

$$L = \begin{cases} (\partial_{yy} - 2x\partial_y + x^2)(l_1 \equiv \partial_x + y) \\ (\partial_{xy} - x\partial_x + y\partial_y - xy + 3)(l_2 \equiv \partial_y - x). \end{cases}$$

are obtained. The intersection ideal $\mathbb{J}_{xxy}$ of l_1 and l_2 is

$$Lclm(l_1, l_2) = \langle\!\langle K_1 \equiv \partial_{xxy} - x\partial_{xx} + 2y\partial_{xy} - 2(xy + 1)\partial_x + y^2\partial_y - y(xy + 2),$$
$$K_2 \equiv \partial_{xyy} - 2x\partial_{xy} + y\partial_{yy} + x^2\partial_x - 2(xy - 1)\partial_y + x(xy - 2)\rangle\!\rangle$$

with differential dimension $(1, 2)$. It is not principal, L has the decomposition

$$L = \begin{pmatrix} 0, & 1 \\ \partial_y - x, 0 \end{pmatrix} \begin{pmatrix} K_1 \\ K_2 \end{pmatrix}.$$

of type $\mathcal{L}^{15}_{xyy}$. Continued in Example 7.33. □

6.4 Exercises

Exercise 6.1. Derive the two generators of the *Lclm* in Eq. (6.20) on page 129 as follows. Start from the solution $z = F(y - \frac{1}{2}x^2) + G(y)e^{-x}$ given in Example 7.8, derive it up to third order and eliminate the undetermined functions F and G.

Exercise 6.2. Show that both l_1 and l_2 divide L_1 and L_2 in Eq. (6.20).

Exercise 6.3. Determine the Loewy decomposition of

$$L \equiv \partial_{xxx} + \partial_{xxy} - x(x-1)\partial_{xyy} + \left(3 - \frac{1}{x}\right)\partial_{xx} + \left(2x + 3 - \frac{1}{x}\right)\partial_{xy} + x(x-1)\partial_{yy}$$
$$+ \left(2 - \frac{1}{x}\right)\partial_x + \left(2x + 2 - \frac{1}{x}\right)\partial_y$$

and discuss the result.

Exercise 6.4. Determine the Loewy decomposition of

$$L \equiv \partial_{xyy} + x\partial_{xy} + \left(1 + \frac{1}{x}\right)\partial_{yy} + \partial_x + x\partial_y + \left(1 + \frac{1}{x}\right).$$

Chapter 7
Solving Homogeneous Third-Order Equations

Abstract The operators corresponding to third-order equations considered in this chapter generate ideals of differential dimension $(1, 3)$. Therefore, by Kolchin's Theorem 2.1, these equations have a differential fundamental system containing three undetermined functions of a single argument. The remarks on the structure of the solutions of linear pde's on page 91 apply here as well. Similar as for second-order equations, various cases differing by leading derivatives are distinguished. As opposed to second-order equations, third-order equations have virtually never been treated in the literature before.

7.1 Equations with Leading Derivative z_{xxx}

The topic of this subsection are equations with leading derivatives z_{xxx} of an undetermined function z. Equations corresponding to decompositions involving only principal divisors are considered first.

Proposition 7.1. *Let a third-order equation*

$$Lz \equiv (\partial_{xxx} + A_1\partial_{xxy} + A_2\partial_{xyy} + A_3\partial_{yyy}$$

$$+A_4\partial_{xx} + A_5\partial_{xy} + A_6\partial_{yy} + A_7\partial_x + A_8\partial_y + A_9)z = 0$$

be given with $A_1, \ldots, A_9 \in \mathbb{Q}(x, y)$. *Define* $l_i \equiv \partial_x + a_i\partial_y + b_i$, $a_i, b_i \in \mathbb{Q}(x, y)$ *for* $i = 1, 2, 3$; $\varphi_i(x, y) = const$ *is a rational first integral of* $\dfrac{dy}{dx} = a_i(x, y)$; $\bar{y} \equiv \varphi_i(x, y)$ *and the inverse* $y \equiv \psi_i(x, \bar{y})$; *both* φ_i *and* ψ_i *are assumed to exist;* F_1, F_2, *and* F_3 *are undetermined functions of a single argument. Furthermore, define*

$$\mathscr{E}_i(x, y) \equiv \exp\left(-\int b_i(x, y)|_{y=\psi_i(x,\bar{y})}dx\right)\bigg|_{\bar{y}=\varphi_i(x,y)} \tag{7.1}$$

F. Schwarz, *Loewy Decomposition of Linear Differential Equations*, Texts & Monographs in Symbolic Computation, DOI 10.1007/978-3-7091-1286-1_7,
© Springer-Verlag/Wien 2012

for $i = 1, 2, 3$. For decomposition types $\mathscr{L}_{xxx}^1, \ldots, \mathscr{L}_{xxx}^7$ involving only principal divisors a differential fundamental system has the following structure.

$$\mathscr{L}_{xxx}^1 : \begin{cases} z_1 = \mathscr{E}_1(x, y) F_1(\varphi_1), \\[2mm] z_2 = \mathscr{E}_1(x, y) \int \dfrac{\mathscr{E}_2(x, y)}{\mathscr{E}_1(x, y)} F_2(\varphi_2(x, y)) \Big|_{y = \psi_1(x, \bar{y})} dx \Big|_{\bar{y} = \varphi_1(x, y)}, \\[3mm] z_3 = \mathscr{E}_1(x, y) \int \dfrac{r(x, y)}{\mathscr{E}_1(x, y)} \Big|_{y = \psi_1(x, \bar{y})} dx \Big|_{\bar{y} = \varphi_1(x, y)}, \\[3mm] r(x, y) = \mathscr{E}_2(x, y) \int \dfrac{\mathscr{E}_3(x, y)}{\mathscr{E}_2(x, y)} F_3(\varphi_3(x, y)) \Big|_{y = \psi_2(x, \bar{y})} dx \Big|_{\bar{y} = \varphi_2(x, y)}; \end{cases}$$

$$\mathscr{L}_{xxx}^2 : \begin{cases} z_1 = \mathscr{E}_1(x, y) F_1(\varphi_1(x, y)), \\[2mm] z_i = \mathscr{E}_1(x, y) \int \dfrac{\mathscr{E}_i(x, y)}{\mathscr{E}_1(x, y)} F_i(\varphi_i(x, y)) \Big|_{y = \psi_1(x, \bar{y})} dx \Big|_{\bar{y} = \varphi_1(x, y)}, \quad i = 2, 3; \end{cases}$$

$\mathscr{L}_{xxx}^3$: *The same as preceding case except that $\varphi_2 = \varphi_3 = \varphi$, $\psi_2 = \psi_3 = \psi$;*

$$\mathscr{L}_{xxx}^4 : \begin{cases} z_i = \mathscr{E}_i(x, y) F_i(\varphi_i(x, y)), \quad i = 1, 2, \\[2mm] z_3 = \mathscr{E}_1(x, y) \int \dfrac{r(x, y)}{\mathscr{E}_1(x, y)} \Big|_{y = \psi_1(x, \bar{y})} dx \Big|_{\bar{y} = \varphi_1(x, y)} \\[3mm] \qquad\qquad - \mathscr{E}_2(x, y) \int \dfrac{r(x, y)}{\mathscr{E}_2(x, y)} \Big|_{y = \psi_2(x, \bar{y})} dx \Big|_{\bar{y} = \varphi_2(x, y)}, \\[3mm] r(x, y) = r_0 \int \dfrac{\mathscr{E}_3(x, y)}{a_2 - a_1} \dfrac{F_3(\varphi_3)}{r_0} dy, \quad r_0 = \exp\left(- \int \dfrac{b_2 - b_1}{a_2 - a_1} dy \right); \end{cases}$$

$$\mathscr{L}_{xxx}^5 : \begin{cases} \text{The same as preceding case except that } r(x, y) = \dfrac{\mathscr{E}_3(x, y)}{b_2 - b_1} F_3(\varphi_3) \\[2mm] \text{and } \varphi_1 = \varphi_2, \ \psi_1 = \psi_2; \end{cases}$$

$\mathscr{L}_{xxx}^6$: $z_i = \mathscr{E}_i(x, y) F_i(\varphi_i(x, y)), \ i = 1, 2, 3;$
$\mathscr{L}_{xxx}^7$: *The same as preceding case except that $\varphi_2 = \varphi_3$ and $\psi_2 = \psi_3$.*

$F_i, \ i = 1, 2, 3$ are undetermined functions of a single argument; f, g, h, φ_i, ψ_i, φ, and ψ are determined by the coefficients $A_1, \ldots, A_9$ of the given equation.

Proof. It is based on Theorem 6.1 and Lemma B.3. For a decomposition $L = l_3 l_2 l_1$ of type $\mathscr{L}_{xxx}^1$, Eq. (B.5) applied to the factor l_1 yields the above given $z_1(x, y)$. The left factor equation $l_2 w = 0$ yields $w_2 = \mathscr{E}_2(x, y) F_2(\varphi_2)$. Taking it as inhomogeneity for the right factor equation, by (B.4) the given expression for $z_2(x, y)$ is obtained. Finally, the equation $l_3 w = 0$ yields the solution $w_3 = \mathscr{E}_3(x, y) F_3(\varphi_3(x, y))$. By Proposition 5.4, case (ii), the solution of $l_2 l_1 z = w_3$ yields the solution $z_3(x, y)$.

For decomposition $L = Lclm(l_3, l_2)l_1$ of type $\mathcal{L}^2_{xxx}$, the right factor equation yields $z_1(x, y)$ as above. The two left factor equations $l_i w = 0$ for $i = 2, 3$ have the solutions $w_i = \mathcal{E}_i(x, y) F_i(\varphi_i(x, y))$. They lead to the inhomogeneous equations $l_i z = w_i$ with the above solutions.

For decomposition $L = Lclm(l(\Phi))l_1$ of type $\mathcal{L}^3_{xxx}$, the arguments of F_2 and F_3 are the same according to Proposition 5.1.

For decomposition $L = l_3 Lclm(l_2, l_1)$ of type $\mathcal{L}^4_{xxx}$, the first-order right factors l_1 and l_2 yield z_1 and z_2 as above. The left factor equation $l_3 w = 0$ has the solution $w_3 = \mathcal{E}_3(x, y) F_3(\varphi_3)$; it yields the inhomogeneous equation $Lclm(l_2, l_1)z = w_3$. It has been considered in Proposition 5.4, case (ii). The solution z_3 follows from (5.9) upon substitution of R by w_3.

For decomposition $L = l_3 Lclm(l(\Phi))$ of type $\mathcal{L}^5_{xxx}$, by the same arguments as for type $\mathcal{L}^4_{xxx}$ the transformation functions are identical to each other for $i = 1$ and $i = 2$.

For a decomposition $L = Lclm(l_3, l_2, l_1)$ of type $\mathcal{L}^6_{xxx}$, the right-factor equations $l_i z = 0$, $i = 1, 2, 3$, lead to the above solutions $z_i(x, y)$.

For decomposition $L = Lclm(l(\Phi), l_1)$ of type $\mathcal{L}^7_{xxx}$, z_1 follows from the factor l_1. The remaining solutions z_2 and z_3 result from the type $\mathcal{L}^4_{xx}$ decomposition of $l(\Phi)$ according to Proposition 5.1.   □

In the subsequent examples the results given in the above proposition are applied for determining the solutions of the corresponding equations. The reader is encouraged to verify them by substitution.

Example 7.1. The three factors of the type $\mathcal{L}^1_{xxx}$ decomposition in Example 6.1 yield $\varphi_1 = \varphi_2 = y$, $\varphi_3 = y e^{-x}$, $\psi_1 = \psi_2 = \bar{y}$, and $\psi_3 = \bar{y} e^x$. Furthermore, $\mathcal{E}_1 = e^x$, $\mathcal{E}_2 = \frac{1}{x}$, and $\mathcal{E}_3 = 1$. Substituting these values into the expressions given in Proposition 7.1 leads to

$$z_1(x, y) = F(y)e^x, \quad z_2(x, y) = G(y)Ei(-x)e^x,$$

$$z_3(x, y) = e^x \int \frac{e^{-x}}{x} \int x H(y e^{-x}) dx dx.$$

F, G and H are undetermined functions.   □

Example 7.2. The first-order right factor in the type $\mathcal{L}^2_{xxx}$ decomposition of Example 6.2 yields $z_1 = e^{-x} F(y)$. The homogeneous equations corresponding to the two left factors and its solutions are

$$w_x + (x + y)w_y + yw = 0 \;\; with\; solution \;\; w = G\big((x + y + 1)e^{-x}\big) \exp\big(\tfrac{1}{2}x^2 - y\big)$$

and

$$w_x - w_y - xw = 0 \;\; with\; solution \;\; w = H(x + y) \exp\big(\tfrac{1}{2}x^2\big).$$

Taking them as inhomogeneities of the right factor equation, the remaining two solutions are

$$z_2 = e^{-x} \int G\big((x+y+1)e^{-x}\big) \exp\big(\tfrac{1}{2}x^2 + x - y\big)dx$$

and

$$z_3 = e^{-x} \int H(x+y) \exp\big(\tfrac{1}{2}x^2 + x\big)dx;$$

as usual, F, G and H are undetermined functions of its argument.   □

Example 7.3. The rightmost factor of the type $\mathscr{L}^3_{xxx}$ decomposition in Example 6.3 is $l_1 = \partial_x - 1$. Choosing $\Phi = 0$ and $\Phi \to \infty$ in the left factor, $l_2 = \partial_x - \tfrac{y}{x}\partial_y$ and $l_3 = \partial_x - \tfrac{y}{x}\partial_y + \tfrac{2}{x}$ are obtained. With the notation of the above proposition there follows

$$\varphi_1(x, y) = y \text{ and } \varphi_2(x, y) = \varphi_3(x, y) \equiv \varphi(x, y) = xy;$$

furthermore $\mathscr{E}_1(x, y) = e^x$, $\mathscr{E}_2(x, y) = 1$ and $\mathscr{E}_3(x, y) = \tfrac{1}{x^2}$. Using these expressions the following fundamental system

$$z_1 = F(y)e^x, \quad z_2 = e^x \int G(xy)e^{-x}dx, \quad z_3 = e^x \int H(xy)e^{-x}\frac{dx}{x^2}$$

is obtained; F, G and H are undetermined functions.   □

Example 7.4. The two first-order right factors of the type $\mathscr{L}^4_{xxx}$ decomposition in Example 6.4 yield

$$z_1 = F(y)e^{-x}, \quad z_2 = G\big((y+1)e^{-x}\big).$$

The first-order left factor leads to the equation $w_x - \tfrac{1}{x}w = 0$ with the solution $w = xH(y)$; H is an undetermined function. Taking it as inhomogeneity of the second-order equation corresponding to the right factors yields

$$z_{xx} + (y+1)z_{xy} + z_x + (y+1)z_y = xH(y). \tag{7.2}$$

Due to its $\mathscr{L}^2_{xx}$ decomposition, Proposition 5.4 applies. According to case *(i)* of Corollary 5.1, $r_1 = 0$, $r_{2,x} + r_2 = xH(y)$ leads to $r_2 = (x-1)H(y)$. Therefore the desired special solution of (7.2) satisfies $z_{3,x} + (y+1)z_{3,y} = (x-1)H(y)$. The result is

$$z_3 = \int (x-1)H(y)\big|_{y=\psi(x,\bar{y})}dx\big|_{\bar{y}=\varphi(x,y)}; \tag{7.3}$$

there follows $\varphi(x, y) = \log(y+1) - x$ and $\psi(x, \bar{y}) = \exp(\bar{y} + x) - 1$; F, G and H are undetermined functions. In Exercise 7.1 this solution will be verified. □

Example 7.5. Choosing again $\varPhi = 0$ and $\varPhi \to \infty$, two right factors of the type $\mathscr{L}^5_{xxx}$ decomposition of Example 6.5 are $l_1 = \partial_x - \frac{y}{x}\partial_y$ and $l_2 = \partial_x - \frac{y}{x}\partial_y + \frac{2}{x}$. There follows $\varphi_i(x, y) = xy$, $\psi_i(x, \bar{y}) = \frac{\bar{y}}{x}$ for $i = 1, 2$ and $\mathscr{E}_1(x, y) = 1$, $\mathscr{E}_2(x, y) = \frac{1}{x^2}$. This yields $z_1(x, y) = F(xy)$ and $z_2(x, y) = G(xy)\frac{1}{x^2}$. Dividing out the left intersection generator

$$Lclm(l_2, l_1) = \partial_{xx} - \frac{2y}{x}\partial_{xy} + \frac{y^2}{x^2}\partial_{yy} + \frac{3}{x}\partial_x - \frac{y}{x^2}\partial_y$$

the third factor $l_3 = \partial_x - 1$ is obtained. It yields $\varphi_3 = y$, $\psi_3 = \bar{y}$, and $\mathscr{E}_3(x, y) = \exp(x)$. Substitution into the respective expression given in Proposition 7.1 the third solution

$$z_3(x, y) = \int xe^{-xy}H(y)\Big|_{y=\psi_1(x,\bar{y})}dx\Big|_{\bar{y}=\varphi_1(x,y)}$$
$$-\frac{1}{x^2}\int x^3 e^{-xy}H(y)\Big|_{y=\psi_1(x,\bar{y})}dx\Big|_{\bar{y}=\varphi_1(x,y)}$$

follows; F G and H are undetermined functions. □

Example 7.6. The three first-order right factors of the type $\mathscr{L}^6_{xxx}$ decomposition of Example 6.6 yield the solutions z_i, $i = 1, 2, 3$, by means of three independent integrations with the result

$$z_1 = F(y)e^{-x}, \quad z_2 = G(x - y), \quad z_3 = H((y + 1)e^{-x});$$

F, G and H are undetermined functions. □

Example 7.7. Choosing $\varPhi = 0$ and $\varPhi \to \infty$ in the type $\mathscr{L}^7_{xxx}$ decomposition of Example 6.7 as above yields the fundamental system

$$z_1 = F(xy), \quad z_2 = G(xy)\frac{1}{x^2}, \quad \text{and} \quad z_3 = H(y)\exp(xy);$$

F, G and H are undetermined functions. □

According to Theorem 6.2 on page 128 there is one more decomposition of operators with leading derivative ∂_{xxx} involving a non-principal divisor. The subsequent proposition shows how it may be applied for solving the corresponding equation.

Proposition 7.2. *Let a third-order equation*

$$Lz \equiv (\partial_{xxx} + A_1\partial_{xxy} + A_2\partial_{xyy} + A_3\partial_{yyy}$$
$$+A_4\partial_{xx} + A_5\partial_{xy} + A_6\partial_{yy} + A_7\partial_x + A_8\partial_y + A_9)z = 0$$

be given with $A_1, \ldots, A_9 \in \mathbb{Q}(x, y)$; *assume it has two first-order right factors* $l_i \equiv \partial_x + a_1 \partial_y + b_i$, $i = 1, 2$ *generating a non-principal divisor* $\mathbb{J}_{xxx} = \langle L_1, L_2 \rangle = Lclm(l_1, l_2)$. *A fundamental system may be obtained as follows.*

$$\mathcal{L}_{xxx}^8 : \begin{cases} z_i(x, y) = \mathcal{E}_i(x, y) F_i(\varphi_i(x, y)), \ i = 1, 2; \\ z_3(x, y) \ \text{is a special solution of} \ L_1 z = w_1, \ L_2 z = w_2, \\ w_1 \ \text{and} \ w_2 \ \text{are given by (7.5) below.} \end{cases}$$

Proof. The first two members z_i follow immediately as solutions of $l_i z = 0$. In order to obtain the third member $z_3(x, y)$ of a fundamental system the exact quotient module

$$\begin{aligned} Exquo(\langle L \rangle, \langle L_1, L_2 \rangle) &= \langle (1, A_1), (\partial_y + q_3, -\partial_x + q_1 \partial_y - p_3 - q_1 q_3 + q_4) \rangle \\ &= \langle (1, A_1), (0, \partial_x + (A_1 - q_1)\partial_y + A_{1,y} + q_3 A_1 + p_3 + q_1 q_3 - q_4) \rangle \end{aligned}$$

is constructed. The first generator of the module at the right hand side in the first line follows from the division, i.e. from $L = L_1 + A_1 L_2$; the second generator represents the single syzygy (2.28) given in Lemma 2.2. In the last line the generators have been transformed into a a Janet basis. Introducing the new differential indeterminates w_1 and w_2 the equations

$$\begin{aligned} w_1 + A_1 w_2 &= 0, \\ w_{2,x} + (A_1 - q_1)w_{2,y} + (A_{1,y} + q_3 A_1 + p_3 + q_1 q_3 - q_4)w_2 &= 0 \end{aligned} \tag{7.4}$$

are obtained. According to Corollary B.1 on page 207 its solutions are

$$w_1(x, y) = -A_1 w_2(x, y),$$

$$w_2(x, y) = \Phi(\varphi) \exp \left(-\int (A_{1,y} + q_3 A_1 + p_3 + q_1 q_3 - q_4) \Big|_{y = \psi(x, \bar{y})} dx \right) \Big|_{\bar{y} = \varphi(x, y)} ; \tag{7.5}$$

here $\varphi(x, y)$ is a first integral of $\frac{dy}{dx} = A_1 - q_1$, $\bar{y} = \varphi(x, y)$, and $y = \psi(x, \bar{y})$. Then $z_3(x, y)$ is a special solution of $L_1 z = w_1$, $L_2 z = w_2$; its solution has been described in Theorem 5.1.   □

Apparently it is not meaningful to describe $z_3(x, y)$ more explicitly as in the preceding proposition because several alternatives may occur due to the special structure of the problem at hand. This will become clear in the next example.

Example 7.8. Blumberg's Example 6.9 has a type $\mathcal{L}_{xxx}^8$ decomposition. The two first-order factors yield $z_1(x, y) = F(y - \frac{1}{2}x^2)$ and $z_2(x, y) = G(y)e^{-x}$. The coefficients are $A_1 = x$, $p_3 = 3$, $q_1 = x$, and $q_3 = q_4 = -\frac{1}{x}$; the system (7.4) becomes $w_1 + xw_2 = 0$ and $w_{2,x} + (1 + \frac{1}{x})w_2 = 0$. Its solutions are $w_1 = -H(y)e^{-x}$ and $w_2 = H(y)\frac{1}{x}e^{-x}$; H is an undetermined function of y. Thus z_3 is a special solution of

$$L_1z = -H(y)e^{-x}, \quad L_2z = H(y)\frac{1}{x}e^{-x}.$$

It turns out that case (i) of Corollary 5.2 of Theorem 5.1 is most suitable for solving it because the second-order ordinary equation obtained for r is homogeneous. Defining $r_2 \equiv r$, system (5.31) is

$$r_{xy} - \frac{1}{x}r_x + \left(1 - \frac{1}{x}\right)r_y - \frac{1}{x}r = -\frac{1}{x}e^{-x}H(y), \quad r_{xx} + 2r_x + r = 0. \quad (7.6)$$

Systems of this type have been considered in Proposition 5.6. A special solution is $r(x, y) = -xe^{-x}H(y)$. The remaining member z_3 of a generalized fundamental system is obtained from $l_2z \equiv z_x + xz_y = -xe^{-x}H(y)$ with the result

$$z_3(x, y) = \int xe^{-x}H\left(\bar{y} + \tfrac{1}{2}x^2\right)dx \Big|_{\bar{y}=y-\frac{1}{2}x^2}. \quad (7.7)$$

H is an undetermined function. □

It is instructive to compare expression (7.7) with the solutions obtained by taking into account only a single first-order right factor. The two possibilities corresponding to the factorizations in the second and the first line of (6.18) are discussed in Exercise 7.2 and Exercise 7.3 respectively.

The above discussion shows in which respect Blumberg's example is very special; because $A_1 - q_1 = 0$, the coefficient of $w_{2,y}$ in the equation for w_2 in the second equation (7.4) becomes an ordinary equation for w_2.

Another extreme is shown in the next example.

Example 7.9. The equation

$$Lz = \left(\partial_{xxx} + (2x + 1)\partial_{xxy} + x(x + 1)\partial_{xyy} + \left(1 - \tfrac{1}{x}\right)\partial_{xx}\right.$$
$$\left. + \left(2x + 1 - \tfrac{1}{x}\right)\partial_{xy} + x(x + 1)\partial_{yy} - \tfrac{1}{x}\partial_x - \tfrac{1}{x}\partial_y\right)z = 0$$

has the same right factors as Blumberg's equation in the preceding example. Therefore $z_1(x, y)$ and $z_2(x, y)$ are the same as given there. Now the coefficient of w_2 in the second equation (7.5) vanishes, it simplifies to $w_{2,x} + (x + 1)w_{2,y} = 0$ with the solution $w_2(x, y) = H(x^2 + 2x - 2y)$; H is an undetermined function of its single argument. It leads to

$$L_1z = -(2x + 1)H(x^2 + 2x - 2y), \quad L_2z = H(x^2 + 2x - 2y).$$

The system (7.6) reads now

$$r_{xy} - \frac{1}{x}r_x + \left(1 - \frac{1}{x}\right)r_y - \frac{1}{x}r = H(x^2 + 2x - 2y),$$

$$r_{xx} + 2r_x + r = -(2x + 1)H(x^2 + 2x - 2y).$$

Applying again Proposition 5.6 the special solution

$$r(x, y) = e^{-x}\left(1 + \int (x + 1)xe^x H(x^2 + 2x - 2y)dx\right.$$

$$\left. -x\int (x + 1)e^x H(x^2 + 2x - 2y)dx\right)$$

follows. The desired third element z_3 of a fundamental system is a special solution of $l_2 z \equiv z_x + xz_y = r$; the result is

$$z_3(x, y) = \int \int (x + 1)xe^x H(x^2 + 2x - 2y)dx\Big|_{y=\psi(x,\bar{y})}dx\Big|_{\bar{y}=\varphi(x,y)}$$

$$- \int x\int (x + 1)xe^x H(x^2 + 2x - 2y)dx\Big|_{y=\psi(x,\bar{y})}dx\Big|_{\bar{y}=\varphi(x,y)} - e^{-x};$$

here $\varphi(x, y) = y - \frac{1}{2}x^2$ and $\psi(x, \bar{y}) = \bar{y} + \frac{1}{2}x^2$.   □

The two preceding examples show how sensitive the solution in a $\mathcal{L}^8_{xxx}$ decomposition is to small changes in the coefficients of the given equation, despite the fact that they have been chosen such that they lead to very special equations (7.4) for w_2.

7.2   Equations with Leading Derivative z_{xxy}

The equations considered now contain only mixed third-order derivatives, i.e. z_{xxy} and z_{xyy}. As a consequence, two first-order factors with leading derivative ∂_x cannot contain a derivative ∂_y simultaneously; this constraint entails some simplifications in the structure of its solutions.

Proposition 7.3. *Let a third-order equation*

$$Lz \equiv (\partial_{xxy} + A_1\partial_{xyy} + A_2\partial_{xx} + A_3\partial_{xy} + A_4\partial_{yy} + A_5\partial_x + A_6\partial_y + A_7)z = 0$$

be given with $A_1, \ldots, A_7 \in \mathbb{Q}(x, y)$. Define $l_i \equiv \partial_x + a_i\partial_y + b_i$ for $i = 1, 2$; $k \equiv \partial_y + c$; $a_i, b_i, c \in \mathbb{Q}(x, y)$; $\varphi_i(x, y) = const$ is a rational first integral of $\frac{dy}{dx} = a_i(x, y)$; $\bar{y} \equiv \varphi_i(x, y)$ and the inverse $y = \psi_i(x, \bar{y})$; both φ_i and ψ_i are assumed to exist; F_1, F_2 and G are undetermined functions of a single argument.

Furthermore define

$$\varepsilon_k(x, y) \equiv \exp\left(-\int c(x, y)dy\right) \text{ and}$$

$$\mathscr{E}_i(x, y) \equiv \exp\left(-\int b_i(x, y)|_{y=\psi_i(x,\bar{y})}dx\right)\Big|_{\bar{y}=\varphi_i(x,y)} \tag{7.8}$$

for $i = 1, 2$. For decomposition types $\mathscr{L}_{xxy}^1, \ldots, \mathscr{L}_{xxy}^{11}$ into first-order components a fundamental system has the following structure.

$$\mathscr{L}_{xxy}^1 : \begin{cases} z_1(x, y) = F_1(\varphi_1)\mathscr{E}_1(x, y), \\[2mm] z_2(x, y) = \mathscr{E}_1(x, y) \int \frac{\mathscr{E}_2(x, y)}{\mathscr{E}_1(x, y)} F_2(\varphi_2)\Big|_{y=\psi_1(x,\bar{y})} dx\Big|_{\bar{y}=\varphi_1(x,y)}, \\[2mm] z_3(x, y) = \mathscr{E}_1(x, y) \int \frac{r(x, y)}{\mathscr{E}_1(x, y)}\Big|_{y=\psi_1(x,\bar{y})} dx\Big|_{\bar{y}=\varphi_1(x,y)}, \\[2mm] r(x, y) = \mathscr{E}_2(x, y) \int \frac{\varepsilon_k(x, y)}{\mathscr{E}_2(x, y)}\Big|_{y=\psi_2(x,\bar{y})} G(x)dx\Big|_{\bar{y}=\varphi_2(x,y)}; \end{cases}$$

$$\mathscr{L}_{xxy}^2 : \begin{cases} z_1(x, y) = F_1(\varphi_1)\mathscr{E}_1(x, y), \\[2mm] z_2(x, y) = \mathscr{E}_1(x, y) \int \frac{\varepsilon_k(x, y)}{\mathscr{E}_1(x, y)} G(x)\Big|_{y=\psi_1(x,\bar{y})} dx\Big|_{\bar{y}=\varphi_1(x,y)}, \\[2mm] z_3(x, y) = \mathscr{E}_1(x, y) \int \frac{\varepsilon_k(x, y)}{\mathscr{E}_1(x, y)} \\[2mm] \qquad\qquad \times \int \frac{\mathscr{E}_2(x, y)}{\varepsilon_k(x, y)} F_2(\varphi_2)dy\Big|_{y=\psi_1(x,\bar{y})} dx\Big|_{\bar{y}=\varphi_1(x,y)}; \end{cases}$$

$$\mathscr{L}_{xxy}^3 : \begin{cases} z_1(x, y) = \varepsilon_k(x, y)G(x), \quad z_2(x, y) = \varepsilon_k(x, y) \int \frac{\mathscr{E}_1(x, y)}{\varepsilon_k(x, y)} F_1(\varphi_1)dy, \\[2mm] z_3(x, y) = \varepsilon_k(x, y) \int \frac{\mathscr{E}_1(x, y)}{\varepsilon_k(x, y)} \int \frac{\mathscr{E}_2(x, y)}{\mathscr{E}_1(x, y)} \\[2mm] \qquad\qquad \times F_2(\varphi_2)\Big|_{y=\psi_1(x,\bar{y})} dx\Big|_{\bar{y}=\varphi_1(x,y)} dy; \end{cases}$$

$$\mathscr{L}_{xxy}^4 : \begin{cases} z_1(x, y) = F_1(\varphi_1)\mathscr{E}_1(x, y), \\[2mm] z_2(x, y) = \mathscr{E}_1(x, y) \int \frac{\mathscr{E}_2(x, y)}{\mathscr{E}_1(x, y)} F_2(\varphi_2)\Big|_{y=\psi_1(x,\bar{y})} dx\Big|_{\bar{y}=\varphi_1(x,y)}, \\[2mm] z_3(x, y) = \mathscr{E}_1(x, y) \int \frac{\varepsilon_k(x, y)}{\mathscr{E}_1(x, y)}\Big|_{y=\psi_1(x,\bar{y})} G(x)dx\Big|_{\bar{y}=\varphi_1(x,y)}; \end{cases}$$

$$\mathscr{L}_{xxy}^5 : z_1(x, y) = \varepsilon(x, y)G(x), \quad z_{i+1}(x, y) = \varepsilon_k(x, y) \int \frac{\mathscr{E}_i(x, y)}{\varepsilon_k(x, y)} F_i(\varphi_i)dy,$$
$$i = 1, 2;$$

$\mathscr{L}_{xxy}^6$: *The same as preceding case except that $\varphi_1 = \varphi_2 = \varphi$, $\psi_1 = \psi_2 = \psi$.*

$$\mathscr{L}_{xxy}^7 : \begin{cases} z_1(x, y) = F_1(\varphi_1)\mathscr{E}_1(x, y), \quad z_2(x, y) = G(x)\varepsilon(x, y), \\[2mm] z_3(x, y) = \varepsilon_k(x, y) \int \dfrac{\mathscr{E}_1(x, y)}{\varepsilon_k(x, y)} \int \dfrac{\mathscr{E}_1(x, y)}{\mathscr{E}_2(x, y)} F_2(\varphi_2)dxdy, \\[2mm] \textit{or the same expression with } \varepsilon_k \textit{ and } \mathscr{E}_1, \textit{ and } dx \textit{ and } dy \textit{ interchanged;} \end{cases}$$

$$\mathscr{L}_{xxy}^8 : \begin{cases} z_i(x, y) = F_i(\varphi_i)\mathscr{E}_i(x, y), \quad i = 1, 2, \\[2mm] z_3(x, y) = \mathscr{E}_1(x, y) \int \left. \dfrac{r(x, y)}{\mathscr{E}_1(x, y)} \right|_{y=\psi_1(x,\bar{y})} dx \Big|_{\bar{y}=\varphi_1(x,y)} \\[2mm] \qquad\qquad -\mathscr{E}_2(x, y) \int \left. \dfrac{r(x, y)}{\mathscr{E}_2(x, y)} \right|_{y=\psi_2(x,\bar{y})} dx \Big|_{\bar{y}=\varphi_2(x,y)}, \\[2mm] r(x, y) = r_0 F_3(x) \int \dfrac{\varepsilon_k(x, y)}{a_2 - a_1} \dfrac{dy}{r_0}, \quad r_0 = \exp\left(-\int \dfrac{b_2 - b_1}{a_2 - a_1} dy\right); \end{cases}$$

$\mathscr{L}_{xxy}^9$: *The same expressions for $z_i(x, y)$, $i = 1, 2, 3$ as in the preceding case except that $\varphi_1 = \varphi_2 = \varphi$, $\psi_1 = \psi_2 = \psi$ and $r(x, y) = F_3(y) \dfrac{\varepsilon(x, y)}{b_2 - b_1}$.*

$\mathscr{L}_{xxy}^{10} : z_i(x, y) = \mathscr{E}_i(x, y) F_i(\varphi_i)$, $i = 1, 2$, $z_3(x, y) = \varepsilon_k(x, y) F_3(x)$;

$\mathscr{L}_{xxy}^{11}$: *The same as preceding case except that $\varphi_1 = \varphi_2 = \varphi$, $\psi_1 = \psi_2 = \psi$.*

Proof. It based on Theorem 6.3 and Lemma B.3. The reasoning how the various elements of a fundamental system are obtained are very similar as in the proof of Proposition 7.1 and therefore are not repeated in detail. Two particularities should be mentioned. Due to the condition $a_1 a_2 = 0$ in Theorem 6.3 it is always possible to simplify one of the functions $\mathscr{E}_i$ to ε_i; correspondingly $\psi(x, \bar{y}) = \bar{y}$ and $\varphi(x, y) = y$ in such a case and the shifted integral disappears. Moreover, in decomposition type $\mathscr{L}_{xxy}^7$ the right Loewy factor may have the form $Lclm(\partial_y + c, \partial_x + a_1\partial_y + b_1)$ and case (ii) of Proposition 5.6 does not apply. In such a case the term order may be changed to $y \succ x$ and the solution is obtained by case (ii) of Proposition 5.4. $\square$

Taking these remarks into account all possible cases that may occur in the decomposition of an operator in the above proposition are covered. They are illustrated by the subsequent examples.

Example 7.10. The decompositions of Examples 6.10, 6.11, or 6.12 differ only by the order of its factors. For all three cases $\varphi_1(x, y) = y - \frac{1}{2}x^2$, $\psi_1(x, \bar{y}) = \bar{y} + \frac{1}{2}x^2$, $\varphi_2(x, y) = y$, $\psi_2(x, \bar{y}) = \bar{y}$, $\varepsilon_k(x, y) = \exp\left(-\frac{y}{x}\right)$, $\mathscr{E}_1(x, y) = 1$, and $\mathscr{E}_2(x, y) = \exp(-xy)$. Substitution into the expressions given in Proposition 7.3 yields the following results.

For Example 6.10, decomposition type $\mathcal{L}^1_{xxy}$, a fundamental system is

$$z_1(x, y) = F\left(y - \tfrac{1}{2}x^2\right), \quad z_2(x, y) = \int \exp(-xy)G(y)\Big|_{y=\psi_1(x,\bar{y})}dx\Big|_{\bar{y}=\varphi_1(x,y)},$$

$$z_3(x, y) = \int \exp(-xy)\int \exp\left(xy + \frac{y}{x}\right)H(x)dx\Big|_{y=\psi_1(x,\bar{y})}dx\Big|_{\bar{y}=\varphi_1(x,y)}.$$

For Example 6.11, decomposition type $\mathcal{L}^2_{xxy}$, a fundamental system is

$$z_1(x, y) = F_1\left(y - \tfrac{1}{2}x^2\right), \quad z_2(x, y) = \int \exp\left(-\frac{y}{x}\right)G(x)\Big|_{y=\psi_1(x,\bar{y})}dx\Big|_{\bar{y}=\varphi_1(x,y)},$$

$$z_3(x, y) = \int \exp\left(-\frac{y}{x}\right)\int \exp\left(\frac{y}{x} - xy\right)F_2(y)dy\Big|_{y=\psi_1(x,\bar{y})}dx\Big|_{\bar{y}=\varphi_1(x,y)}.$$

Finally, for Example 6.12, decomposition type $\mathcal{L}^3_{xxy}$, the fundamental system

$$z_1(x, y) = F_1(x)\exp\left(-\frac{y}{x}\right), \quad z_2(x, y) = \exp\left(-\frac{y}{x}\right)\int \exp\left(\frac{y}{x}\right)G\left(y - \tfrac{1}{2}x^2\right)dy,$$

$$z_3(x, y) = \exp\left(\frac{y}{x}\right)\int \exp\left(-\frac{y}{x}\right)\int \exp(-xy)F_2(y)\Big|_{y=\psi_1(x,\bar{y})}dx\Big|_{\bar{y}=\varphi_1(x,y)}$$

is obtained.   □

The different structure of the solutions in the preceding example originating from the different order of the first order factors should be observed. There is always one member with no integrals involved; a second member contains an integral over an undetermined function, possibly with a shifted argument.

Example 7.11. The factors in the type $\mathcal{L}^4_{xxy}$ decomposition of Example 6.13 yield $\varphi_1(x, y) = y + x$, $\psi_1(x, y) = \bar{y} - x$, $\mathcal{E}_1(x, y) = 1$, $\mathcal{E}_2(x, y) = \exp(-xy)$, and $\varepsilon_k(x, y) = \exp(-xy - y)$. Substitution into the expressions given in Proposition 7.3 yields

$$z_1(x, y) = F(x + y), \quad z_2(x, y) = \int \exp(-xy)G(y)\Big|_{y=\bar{y}-x}dx\Big|_{\bar{y}=y+x},$$

$$z_3(x, y) = \int \exp(-xy - y)\Big|_{y=\bar{y}-x}H(x)dx\Big|_{\bar{y}=y+x};$$

F, G and H are undetermined functions.   □

Example 7.12. The factors in the type $\mathcal{L}^5_{xxy}$ decomposition of Example 6.14 yield $\varepsilon_k(x, y) = \exp(-xy)$, $\mathcal{E}_1(x, y) = \exp(x)$, and $\mathcal{E}_2(x, y) = \exp\left(xy - \tfrac{1}{2}x^2\right)$. The fundamental system is

$$z_1(x, y) = F(x) \exp(-xy), \quad z_2(x, y) = \exp(x(1 - y)) \int \exp(xy) G(y) dy,$$

$$z_3(x, y) = \exp\left(-xy - \tfrac{1}{2}x^2\right) \int \exp(2xy) H(x - y) dy;$$

F, G and H are undetermined functions.   □

Example 7.13. The factors in the type $\mathscr{L}^6_{xxy}$ decomposition of Example 6.15 yield

$\varphi_1 = \varphi_2 = y$, $\psi_1 = \psi_2 = \bar{y}$, $\varepsilon_k(x, y) = \exp\left(-xy - \dfrac{y^2}{2x}\right)$, $\mathscr{E}_1(x, y) = 1$, and

$\mathscr{E}_2(x, y) = \dfrac{1}{x}$. Substitution into the above expressions leads to the fundamental system

$$z_1(x, y) = F(x) \exp\left(-xy - \tfrac{y^2}{2x}\right),$$

$$z_2(x, y) = \exp\left(-xy - \tfrac{y^2}{2x}\right) \int \exp\left(xy + \tfrac{y^2}{2x}\right) G(y) dy,$$

$$z_3(x, y) = \exp\left(-xy - \tfrac{y^2}{2x}\right) \exp\left(xy + \tfrac{y^2}{2x}\right) H(y) \tfrac{dy}{x};$$

F, G and H are undetermined functions.   □

Example 7.14. The two arguments of the $Lclm$ in the type $\mathscr{L}^7_{xxy}$ decomposition of Example 6.16 yield $\varepsilon_k(x, y) = \exp(-xy - y)$ and $\mathscr{E}_1(x, y) = \exp(-xy)$; from the left factor l_2 there follows $\mathscr{E}_2(x, y) = \exp(x)$. The fundamental system

$$z_1(x, y) = F(y) \exp(-xy), \quad z_2(x, y) = G(x) \exp(-(x + 1)y),$$

$$z_3(x, y) = \exp(-xy - y) \int \exp(x) \int \exp(xy + y) H(y) dy dx$$

is obtained; F, G and H are undetermined functions.   □

Example 7.15. The two arguments of the $Lclm$ in the type $\mathscr{L}^8_{xxy}$ decomposition of Example 6.17 yield $\varphi_1 = y$, $\psi_1 = \bar{y}$, $\varphi_2 = x - y$, and $\psi_2 = x - \bar{y}$. There follows $\mathscr{E}_1(x, y) = \exp(x)$ and $\mathscr{E}_2(x, y) = \exp\left(xy - \tfrac{1}{2}x^2\right)$. Two elements of a fundamental system are

$$z_1 = F(y) \exp(x) \quad \text{and} \quad z_2 = G(x - y) \exp\left(xy - \tfrac{1}{2}x^2\right).$$

From the left factor $k = \partial_y$ there follows $\varepsilon_k(x, y) = 1$ and

$$r(x, y) = H(x) \exp\left(\tfrac{1}{2}y^2 - y\right) \int \exp\left(y - \tfrac{1}{2}y^2\right) dy.$$

Substitution into the expression for $z_3(x, y)$ finally leads to

$$z_3 = e^x \int e^{-x} r(x, y) dx$$
$$- \exp\left(xy - \tfrac{1}{2}x^2\right) \int \exp\left(\tfrac{1}{2}x^2 - xy\right) r(x, y)\Big|_{y=x-\bar{y}} dx\Big|_{\bar{y}=x-y};$$

F, G and H are undetermined functions.   □

Example 7.16. Choosing $\Phi = 0$ and $\Phi \to \infty$ in the argument of the $Lclm$ in the type $\mathcal{L}^9_{xxy}$ decomposition of Example 6.18 yields $l_1 = \partial_x + \frac{1}{x}$ and $l_2 = \partial_x$; there follows $\varphi_1 = \varphi_2 = y$, $\psi_1 = \psi_2 = \bar{y}$, $\mathscr{E}_1 = 1$, and $\mathscr{E}_2 = \frac{1}{x}$. Substitution into the respective expressions for z_1 and z_2 of decomposition type $\mathcal{L}^9_{xxy}$ yields $z_1(x, y) = F(y)\frac{1}{x}$ and $z_2(x, y) = G(y)$. The factor $k = \partial_y + x + y$ leads to $\varepsilon_k(x, y) = \exp\left(-xy - \tfrac{1}{2}y^2\right)$. Substitution into the expression given in Proposition 7.3 and some simplification yield the third element of a fundamental system

$$z_3(x, y) = H(y)\left(\int x \exp\left(-xy - \tfrac{1}{2}y^2\right) dx + \frac{1}{x} \int x^2 \exp\left(-xy - \tfrac{1}{2}y^2\right) dx \right);$$

F, G and H are undetermined functions.   □

Example 7.17. The three arguments in the $Lclm$ of the type $\mathcal{L}^{10}_{xxy}$ decomposition of Example 6.19 yield the three solutions

$$z_1 = F(x) \exp\left(\tfrac{1}{2}y^2\right), \quad z_2(x, y) = G(y)e^x, \quad z_3 = H(x - y) \exp\left(xy - \tfrac{1}{2}x^2\right);$$

F, G and H are undetermined functions.   □

Example 7.18. The first two elements $z_1(x, y)$ and $z_2(x, y)$ of a fundamental system following from the $\mathcal{L}^{11}_{xxy}$ decomposition in Example 6.20 are the same as in the above Example 7.16. The third element corresponding to the factor $k = \partial_y - y$ is $z_3(x, y) = H(x) \exp\left(\tfrac{1}{2}y^2\right)$. H is an undetermined function.   □

According to Theorem 6.6 on page 146 there are four additional decompositions of operators with leading derivative ∂_{xxy} involving non-principal divisors. In the subsequent proposition it is shown how they may be utilized for solving the corresponding equation.

Proposition 7.4. *Let a third-order equation*

$$Lz \equiv (\partial_{xxy} + A_1 \partial_{xyy} + A_2 \partial_{xx} + A_3 \partial_{xy} + A_4 \partial_{yy} + A_5 \partial_x + A_6 \partial_y + A_7)z = 0$$

be given with $A_1, \ldots, A_7 \in \mathbb{Q}(x, y)$. *A differential fundamental system has the following structure for the various decomposition types involving non-principal divisors. The superscript i means the i-th derivative w.r.t. the single argument of the respective function.*

$$\mathscr{L}_{xxy}^{12} : \begin{cases} z_1(x, y) = \sum_{i=0}^{m-1} f_i(x, y) F^{(i)}(y), \quad z_2(x, y) = \sum_{i=0}^{m-1} g_i(x, y) G^{(i)}(y), \\ \\ \qquad\qquad z_3(x, y) = \sum_{i=0}^{n-1} h_i(x, y) H^{(i)}(x); \end{cases}$$

$$\mathscr{L}_{xxy}^{13} : \begin{cases} \qquad\qquad z_1(x, y) \ and \ z_2(x, y) \ as \ above, \\ z_3(x, y) \ is \ solution \ of \ Lz = 0, \ \mathfrak{l}_m z = H(x) \exp\left(- \int A_2 dy \right); \end{cases}$$

$$\mathscr{L}_{xxy}^{14} : \begin{cases} z_1(x, y) = \sum_{i=0}^{n-1} f_i(x, y) F^{(i)}(x), \quad z_2(x, y) \ and \ z_3(x, y) \ follow \ from \\ Lz = 0, \ and \ \mathfrak{k}_n z = w_2, where \ w_2 \ is \ solution \ of \ (7.9) \ below. \end{cases}$$

If L has two right factors $l \equiv \partial_x + a$ and $k \equiv \partial_y + b$ such that $\mathbb{J}_{xxy} = \langle K_1, K_2 \rangle = Lclm(l, k)$ is a non-principal divisor as determined in Lemma 2.3, a fundamental system may be determined as follows.

$$\mathscr{L}_{xxy}^{15} : \begin{cases} z_1(x, y) = F(y) \exp\left(- \int a \, dx \right), \quad z_2(x, y) = G(x) \exp\left(- \int b \, dy \right), \\ z_3(x, y) \ is \ a \ special \ solution \ of \ K_1 z = w_1, \quad K_2 z = w_2 \ where \\ w_1 \ and \ w_2 \ are \ given \ by \ (7.12) \ below. \end{cases}$$

Proof. Let $\mathbb{L}_{x^m}(L)$ be a Laplace divisor as determined in Proposition 2.4. The linear ode $\mathfrak{l}_m z = 0$ has the general solution $z = C_1 f_1(x, y) + \ldots + C_m f_m(x, y)$. The C_i are constants w.r.t. x and undetermined functions of y. This expression for z must also satisfy the equation $Lz = 0$. Because the Laplace divisor $\mathbb{L}_{x^m}(L)$ has differential dimension $(1, 2)$, by Kolchin's Theorem 2.1 it must be possible to express $C_1, \ldots, C_m$ in terms of two undetermined functions $F(y)$ and $G(y)$ and its derivatives up to order $m - 1$.

For the second Laplace divisor $\mathbb{L}_{y^n}(L)$ the same steps with x and y interchanged are performed. Because it has differential dimension $(1, 1)$ only a single solution is obtained.

For decomposition type $\mathscr{L}_{xxy}^{12}$, the two Laplace divisors yield the fundamental system $z_i(x, y)$, $i = 1, 2, 3$.

For the type $\mathscr{L}_{xxy}^{13}$ decomposition, the divisor $\mathbb{L}_{x^m}(L)$ yields $z_1(x, y)$ and $z_2(x, y)$ as in the previous case. In order to obtain the third solution define $w_1 \equiv Lz$ and $w_2 \equiv \mathfrak{l}_m z$. The exact quotient equations $w_1 = 0$ and $w_{2,y} + A_2 w_2 = 0$ follow from Theorem 6.4. Substituting its solutions into $Lz = 0$ and $\mathfrak{l}_m z = w_2$ the third element $z_3(x, y)$ of a fundamental system follows.

For the type $\mathscr{L}_{xxy}^{14}$ decomposition the first solution $z_1(x, y)$ follows from the Laplace divisor. According to Theorem 6.4 the exact quotient equations are now $w_1 = 0$ and

$$w_{2,xx} + A_1 w_{2,xy} + (A_{1,y} - A_1 A_2 + A_3) w_{2,x} + A_4 w_{2,y} + (A_{4,y} - A_2 A_4 + A_6) w_2 = 0.$$
$$\text{(7.9)}$$

The latter equation corresponds to an ideal of differential dimension $(1,2)$; thus substituting its solution into $Lz = 0$, $\ell_n z = w_2$ yields the remaining solutions $z_2(x, y)$ and $z_3(x, y)$.

For decomposition type $\mathscr{L}_{xxy}^{15}$ the first two members follow immediately as solutions of $lz = 0$ and $kz = 0$. In order to obtain the third member $z_3(x, y)$ the exact quotient module

$$Exquo(\langle L \rangle, \langle K_1, K_2 \rangle)$$
$$= \langle (1, A_1), (\partial_y - p_2 - q_1 q_2 + q_3, -\partial_x + q_1 \partial_y - p_3 - q_1 q_3 + q_4) \rangle$$
$$= \langle (1, A_1), (0, \partial_x + (A_1 - q_1) \partial_y + P(x, y)) \rangle$$

is constructed where

$$P(x, y) \equiv A_{1,y} - (p_2 + q_1 q_2 - q_3) A_1 + p_3 + q_1 q_3 - q_4. \qquad \text{(7.10)}$$

The first generator of the module at the right hand side in the second line follows from the division, i.e. from $L = 1 + A_1 K_2$; the second generator represents the single syzygy (2.28) given in Lemma 2.3. In the last line the generators have been transformed into a Janet basis. Introducing the new differential indeterminates w_1 and w_2 the equations

$$w_1 + A_1 w_2 = 0, \quad w_{2,x} + (A_1 - q_1) w_{2,y} + P(x, y) w_2 = 0 \qquad \text{(7.11)}$$

follow. According to Corollary B.1 on page 207 its solutions are

$$w_1(x, y) = -A_1 w_2(x, y),$$
$$\text{(7.12)}$$
$$w_2(x, y) = \Phi(\varphi) \exp\left(-\int P(x, y) \Big|_{y = \psi(x, \bar{y})} dx \right) \Big|_{\bar{y} = \varphi(x, y)};$$

here $P(x, y)$ is again defined by (7.10); $\varphi(x, y)$ is a first integral of $\frac{dy}{dx} = A_1 - q_1$, $\bar{y} = \varphi(x, y)$ and $y = \psi(x, \bar{y})$; Φ is an undetermined function. Then $z_3(x, y)$ is a special solution of $K_1 z = w_1$, $K_2 z = w_2$; this system is discussed in detail in Theorem 5.2.   □

Subsequently the above theorem is applied to the decompositions considered in the examples of the preceding chapter.

Example 7.19. The equation $z_{yy} = 0$ originating from the Laplace divisor in the type $\mathscr{L}_{xxy}^{14}$ decomposition of Example 6.21 leads to $z = C_1(x) + C_2(x) y$. Substitution into $Lz = 0$ and adjusting C_1 and C_2 yields the solution

$$z_1(x, y) = e^{-x}\big(F(x) + (y + 1)F'(x) - (y + 1)F''(x)\big);$$

F is an undetermined function of x. Equation (7.9) reads

$$w_{2,xx} + xw_{2,xy} + \Big(1 + \frac{x}{y+1}\Big)w_{2,x} + w_{2,y} = 0.$$

According to Proposition 4.1, case (i), it is absolutely irreducible and no further solution of differential type 1 may be found. In Exercise 7.5 the existence of solutions of differential type 0 is discussed.   □

Example 7.20. The two equations $lz = 0$ and $kz = 0$ of the type $\mathscr{L}^{15}_{xxy}$ decomposition in Example 6.22 have the solutions

$$z_1(x, y) = F(y)\exp(-xy) \quad \text{and} \quad z_2(x, y) = G(x)\exp(xy).$$

The Eqs. (7.11) are

$$w_1 + \frac{y}{x}w_2 = 0 \quad \text{and} \quad w_{2,x} + \frac{y}{x}w_{2,y} + \frac{1}{x}w_2 = 0.$$

Its general solution is

$$w_1(x, y) = -\frac{y}{x^2}H\Big(\frac{y}{x}\Big), \quad w_2(x, y) = \frac{1}{x}H\Big(\frac{y}{x}\Big).$$

H is an undetermined function. Finally, the system $K_1z = w_1$, $K_2z = w_2$ has to be solved. According to Theorem 5.2 the equations

$$r_{xy} - xr_x + yr_y - (xy + 3)r = -\frac{y}{x^2}H\Big(\frac{y}{x}\Big), \quad r_{yy} - 2xr_y + x^2r = \frac{1}{x}H\Big(\frac{y}{x}\Big)$$

are obtained; a special solution is

$$r_0(x, y) = \exp(xy)\frac{1}{x}\Big(y\int H\Big(\frac{y}{x}\Big)\exp(-xy)dy - \int H\Big(\frac{y}{x}\Big)y\exp(-xy)dy\Big).$$

Substitution into (5.35) finally yields the third element of a fundamental system

$$z_3(x, y) = \exp(-xy)\int r_0(x, y)\exp(xy)dx$$

with $r_0(x, y)$ as given above.   □

7.3 Equations with Leading Derivative z_{xyy}

The subject of this section are equations involving a single mixed derivative of third order; in the variable order $grlex$ with $x \succ y$ this is z_{xyy}. If it is z_{xxy}, the variable order may be changed to $y \succ x$ in order to apply the results of this subsection. The expressions for the solutions given here are simpler than in the preceding cases. At first equations corresponding to decompositions involving principal divisors are considered.

Proposition 7.5. *Let a third-order equation*

$$Lz \equiv (\partial_{xyy} + A_1\partial_{xy} + A_2\partial_{yy} + A_3\partial_x + A_4\partial_y + A_5)z = 0$$

be given with $A_1, \ldots, A_5 \in \mathbb{Q}(x, y)$. *Define* $l \equiv \partial_x + b$, $k_i \equiv \partial_y + c_i$ *for* $i = 1, 2$; $b, c_i \in \mathbb{Q}(x, y)$, $c_1 \neq c_2$. *Furthermore define*

$$\varepsilon_l(x, y) = \exp\left(-\int b(x, y)dx\right), \quad \varepsilon_i(x, y) = \exp\left(-\int c_i(x, y)dy\right) \quad (7.13)$$

for $i = 1, 2$. F *is an undetermined function of* y; G_1 *and* G_2 *are undetermined functions of* x. *For decomposition types* $\mathscr{L}^1_{xyy}, \ldots, \mathscr{L}^{11}_{xyy}$ *a generalized fundamental system has the following structure for the various decompositions into first-order components.*

$$\mathscr{L}^1_{xyy} : \begin{cases} z_1(x, y) = F(y)\varepsilon_l(x, y), \quad z_2(x, y) = \varepsilon_l(x, y)\int \dfrac{\varepsilon_1(x, y)}{\varepsilon_l(x, y)}G_1(x)dx, \\[3ex] z_3(x, y) = \varepsilon_l(x, y)\int \dfrac{\varepsilon_1(x, y)}{\varepsilon_l(x, y)}G_2(x)\int \dfrac{\varepsilon_2(x, y)}{\varepsilon_1(x, y)}dydx; \end{cases}$$

$$\mathscr{L}^2_{xyy} : \begin{cases} z_1(x, y) = G_1(x)\varepsilon_1(x, y), \quad z_2(x, y) = \varepsilon_1(x, y)\int \dfrac{\varepsilon_l(x, y)}{\varepsilon_1(x, y)}F(y)dy, \\[3ex] z_3(x, y) = \varepsilon_1(x, y)\int \dfrac{\varepsilon_l(x, y)}{\varepsilon_1(x, y)}\int \dfrac{\varepsilon_2(x, y)}{\varepsilon_l(x, y)}G_2(x)dxdy; \end{cases}$$

$$\mathscr{L}^3_{xyy} : \begin{cases} z_1(x, y) = G_1(x)\varepsilon_1(x, y), \quad z_2(x, y) = \varepsilon_1(x, y)\int \dfrac{\varepsilon_2(x, y)}{\varepsilon_1(x, y)}G_2(x)dy, \\[3ex] z_3(x, y) = \varepsilon_1(x, y)\int \dfrac{\varepsilon_2(x, y)}{\varepsilon_1(x, y)}\int \dfrac{\varepsilon_l(x, y)}{\varepsilon_2(x, y)}F(y)dydy; \end{cases}$$

$$\mathscr{L}^4_{xyy} : \begin{cases} z_1(x, y) = G_1(x)\varepsilon_1(x, y), \quad z_2(x, y) = \varepsilon_1(x, y)G_2(x)\int \dfrac{\varepsilon_2(x, y)}{\varepsilon_1(x, y)}dy, \\[3ex] z_3 = \varepsilon_1(x, y)\int \dfrac{\varepsilon_l(x, y)}{\varepsilon_1(x, y)}F(y)dy; \end{cases}$$

$$\mathscr{L}_{xyy}^5 : \begin{cases} z_1(x, y) = F(y)\varepsilon_l(x, y), \quad z_{i+1}(x, y) = \varepsilon_l(x, y) \int \dfrac{\varepsilon_i(x, y)}{\varepsilon_l(x, y)} G_i(x) dy, \\ i = 1, 2; \end{cases}$$

$\mathscr{L}_{xyy}^6$: *The same as preceding case, z_2 and z_3 in the same function field;*

$$\mathscr{L}_{xyy}^7 : \begin{cases} z_1(x, y) = F(y)\varepsilon_l(x, y), \quad z_2(x, y) = G_1(x)\varepsilon_1(x, y), \\ z_3(x, y) = \varepsilon_l(x, y) \int \dfrac{\varepsilon_1(x, y)}{\varepsilon_l(x, y)} G_2(x) \int \dfrac{\varepsilon_2(x, y)}{\varepsilon_1(x, y)} dy dx; \end{cases}$$

$$\mathscr{L}_{xyy}^8 : \begin{cases} z_i(x, y) = G_i(x)\varepsilon_i(x, y), \ i = 1, 2, \\ z_3(x, y) = \varepsilon_1(x, y) \int \dfrac{\varepsilon_1(x, y)}{\varepsilon_1(x, y)} \dfrac{F(y) dy}{c_2 - c_1} - \varepsilon_2(x, y) \int \dfrac{\varepsilon_1(x, y)}{\varepsilon_2(x, y)} \dfrac{F(y) dy}{c_2 - c_1}; \end{cases}$$

$\mathscr{L}_{xyy}^9$: *The same as preceding case, z_1 and z_2 in the same function field;*

$\mathscr{L}_{xyy}^{10}$: $z_i(x, y) = G_i(x)\varepsilon_i(x, y), \ i = 1, 2, \ z_3(x, y) = F(y)\varepsilon_l(x, y);$

$\mathscr{L}_{xyy}^{11}$: *The same as preceding case, z_1 and z_2 in the same function field.*

Proof. It is based on Theorem 6.5 and Lemma B.3. Because it is similar to the respective proofs for equations with leading derivatives ∂_{xxx} or ∂_{xxy}, only some features which are specific for the case under consideration are mentioned. Due to the absence of factors $\partial_x + a\partial_y + b$ with $a \neq 0$, shifted integrations in the solutions do not occur. In some cases this feature may be used in order to avoid undetermined functions under an integral sign, e.g. for the solution $z_3(x, y)$ of decomposition type $\mathscr{L}_{xyy}^7$; as a consequence it may be possible in special cases to execute the y−integration explicitly. □

At this point it becomes clear why equations with a single mixed third-order derivative are treated individually; its solutions never contain a shifted integral as opposed to the previously treated equations with leading derivatives z_{xxx} or z_{xxy}.

The first three examples below have identical first-order factors, albeit in different sequence; this entails the different structure of its respective solutions.

Example 7.21. The factors of the type $\mathscr{L}_{xyy}^1$ decomposition in Example 6.23 yield $\varepsilon_l = \exp(-xy)$, $\varepsilon_1 = \exp(xy)$, and $\varepsilon_2 = \exp(\frac{1}{2}y^2)$. Substitution into the respective expressions of the above proposition leads to the fundamental system

$$z_1(x, y) = F(y)\exp(-xy), \quad z_2(x, y) = \exp(-xy) \int \exp(2xy)G(x) dx;$$

$$z_3(x, y) = \exp(-xy) \int \exp(2xy)H(x) \int \exp(\tfrac{1}{2}y^2 - xy) dy dx;$$

F, G and H are undetermined functions. □

Example 7.22. The factors of the type $\mathscr{L}^2_{xyy}$ decomposition of Example 6.24 lead to the same expressions ε_i and ε_l as in the preceding example. The fundamental system is

$$z_1(x, y) = F(x) \exp(xy), \quad z_2(x, y) = \exp(-2xy) \int G(y) \exp(-2xy) dy,$$

$$z_3(x, y) = \exp(xy) \int \exp(-2xy) \int \exp\left(\tfrac{1}{2}y^2 + xy\right) H(x) dx dy;$$

F, G and H are undetermined functions. □

Example 7.23. The same remarks as in the preceding example apply for the type $\mathscr{L}^3_{xyy}$ decomposition of Example 6.25. The following fundamental system is obtained.

$$z_1(x, y) = F(x) \exp(xy), \quad z_2(x, y) = G(x) \exp(xy) \int \exp\left(\tfrac{1}{2}y^2 - xy\right) dy,$$

$$z_3(x, y) = \exp(xy) \int \exp\left(\tfrac{1}{2}y^2 - xy\right) \int \exp\left(-\tfrac{1}{2}y^2 - xy\right) H(y) dy dy;$$

F, G and H are undetermined functions. □

There are three more permutations of the factors $\partial_y - y$, $\partial_y - x$, and $\partial_x + y$ of the preceding examples; the corresponding third-order equations and the structure of its fundamental systems are discussed in Exercise 7.6.

Example 7.24. The factors of the type $\mathscr{L}^4_{xyy}$ decomposition of Example 6.26 lead to $\varepsilon_l(x, y) = \exp(-xy)$, $\varepsilon_1(x, y) = \exp(-y)$ and $\varepsilon_2(x, y) = \tfrac{1}{y} \exp(-xy)$. The following fundamental system is obtained.

$$z_1(x, y) = F(x) \exp(-y), \quad z_2(x, y) = G(x) \exp(-y) \int \exp(y - xy) \frac{dy}{y},$$

$$z_3(x, y) = \exp(-y) \int \exp(y - xy) H(y) dy;$$

F, G and H are undetermined functions. □

Example 7.25. The factors of the type $\mathscr{L}^5_{xyy}$ decomposition of Example 6.27 lead to $\varepsilon_l(x, y) = \exp(x^y)$, $\varepsilon_1(x, y) = \exp(xy)$ and $\varepsilon_2(x, y) = \tfrac{1}{y}$. Substitution into the respective expressions of Proposition 7.5 yields the fundamental system

$$z_1(x, y) = F(y)x^y, \quad z_2(x, y) = x^y \int G(x) \exp(xy) \frac{dx}{x^y},$$

$$z_3(x, y) = x^y \int H(x) \frac{dx}{yx^y};$$

F, G and H are undetermined functions. $\square$

Example 7.26. The right factor equation $z_x + xz = 0$ of the type $\mathscr{L}^6_{xyy}$ decomposition in Example 6.28 yields $z_1(x, y) = F(y) \exp\left(-\frac{1}{2}x^2 - xy^2\right)$. Choosing $\Phi = 0$ and $\Phi \to \infty$ in the argument of the *Lclm* leads to the solutions $G(x)$ and $H(x)\frac{1}{y}$. Taking them as inhomogeneities for the above right-factor equation yields

$$z_2(x, y) = \exp\left(-\tfrac{1}{2}x^2 - xy^2\right) \int G(x) \exp\left(\tfrac{1}{2}x^2 + xy^2\right) dx,$$

$$z_3(x, y) = \exp\left(-\tfrac{1}{2}x^2 - xy^2\right) \frac{1}{y} \int H(x) \exp\left(\tfrac{1}{2}x^2 + xy^2\right) dx;$$

F, G and H are undetermined functions. $\square$

Example 7.27. The two arguments of the *Lclm* of the type $\mathscr{L}^7_{xyy}$ decomposition in Example 6.29 yield the solutions $z_1 = F(y) \exp(-xy)$ and $z_2 = G(x)\frac{1}{y} \exp(-xy)$; L may be factorized as $L = (\partial_y + 1) Lclm\left(\partial_y + x + \frac{1}{y}, \partial_x + y\right)$ from which the third solution

$$z_3 = \exp\left(-(x+1)y\right) \frac{1}{y} \int H(x) \frac{xy - y - 1}{(x-1)^2} \exp(xy) dx$$

follows; F, G and H are undetermined functions. $\square$

Example 7.28. The two first-order equations corresponding to the arguments of the *Lclm* in the type $\mathscr{L}^8_{xyy}$ decomposition of Example 6.30 yield

$$z_1(x, y) = F(x) \exp(-y), \quad z_2(x, y) = G(x) \exp\left(-\tfrac{1}{2}y^2\right).$$

The last member of a generalized fundamental system is

$$z_3 = \exp\left(-\tfrac{1}{2}y^2\right) \int H(y) \exp\left(\tfrac{1}{2}y^2\right) \frac{dy}{yx^y} - \exp(-y) \int H(y) \exp(y) \frac{dy}{yx^y};$$

F, G and H are undetermined functions. $\square$

Example 7.29. Choosing $\Phi = 0$ and $\Phi \to \infty$ in the argument of the *Lclm* of the type $\mathscr{L}^9_{xyy}$ decomposition of Example 6.31 leads to the factors ∂_y and $\partial_y + \frac{1}{y}$; they yield the solutions

$$z_1(x, y) = F(x) \quad \text{and} \quad z_2(x, y) = G(x)\frac{1}{y}.$$

The first-order equation corresponding to the left factor $w_x + (x + y^2)w = 0$ has the solution $w = H(y) \exp\left(-\frac{1}{2}x^2 - xy^2\right)$. Thus the third element of a fundamental

system is obtained as a special solution of $z_{yy} + \frac{2}{y}z_y = H(y)\exp\left(-\frac{1}{2}x^2 - xy^2\right)$
with the result

$$z_3(x, y) = \int H(y)y \exp\left(-\tfrac{1}{2}x^2 - xy^2\right)dy - \frac{1}{y}\int H(y)y^2 \exp\left(-\tfrac{1}{2}x^2 - xy^2\right)dy;$$

F, G and H are undetermined functions. □

Example 7.30. The three first-order right factors in the type $\mathscr{L}_{xyy}^{10}$ decomposition of Example 6.32 yield the fundamental system

$$z_1 = F(y)\exp(-xy), \quad z_2 = G(x)\exp(-xy - y), \quad z_3 = H(x)\exp(-xy)\frac{1}{y};$$

F, G and H are undetermined functions. □

Example 7.31. Choosing $\Phi = 0$ and $\Phi \to \infty$ in the type $\mathscr{L}_{xyy}^{11}$ decomposition of Example 6.33 yields the factors ∂_y and $\partial_y + \frac{1}{y}$ respectively. The fundamental system
$$z_1 = F(x), \quad z_2 = G(x)\frac{1}{y}, \quad z_3 = H(y)\exp\left(-\tfrac{1}{2}x^2\right)$$
follows; F, G and H are undetermined functions. □

Finally equations involving decompositions with non-principal divisors are considered. According to Theorem 6.6 on page 146 there are four decomposition types to be considered.

Proposition 7.6. *Let a third-order equation*

$$Lz \equiv (\partial_{xyy} + A_1\partial_{xy} + A_2\partial_{yy} + A_3\partial_x + A_4\partial_y + A_5)z = 0$$

be given with $A_1, \ldots, A_5 \in \mathbb{Q}(x, y)$. A differential fundamental system has the following structure for the various decomposition types involving non-principal divisors. F, G and H are undetermined functions of a single argument; the superscript i means the i-th derivative w.r.t. the single argument of the respective function; f_i, g_i and h_i are determined by the given equation.

$$\mathscr{L}_{xyy}^{12}: \begin{cases} z_1(x, y) = \sum_{i=0}^{m} f_i(x, y)F^{(i)}(y), \quad z_2(x, y) = \sum_{i=0}^{n} g_i(x, y)G^{(i)}(x), \\ \qquad\qquad z_3(x, y) = \sum_{i=0}^{n} h_i(x, y)H^{(i)}(x); \end{cases}$$

$$\mathscr{L}_{xyy}^{13}: \begin{cases} z_1(x, y) \text{ and } z_2(x, y) \text{ as above,} \\ z_3(x, y) \text{ is solution of } Lz = 0, \; \mathfrak{k}_n z = H(y)\exp\left(-\int A_2 dx\right); \end{cases}$$

$$\mathscr{L}^{14}_{xyy} : \begin{cases} z_1(x, y) = \sum_{i=0}^{n-1} f_i(x, y) F^{(i)}(x), \\ z_2(x, y) \text{ and } z_3(x, y) \text{ are solutions of } Lz = 0 \text{ and } \mathfrak{l}_m z = w, \text{ where} \\ w \text{ is solution of } w_{yy} + A_1 w_y + A_3 w = 0; \end{cases}$$

Assume now further that L has two right factors $l \equiv \partial_x + a$ and $k \equiv \partial_y + b$ such that $\mathbb{J}_{xxy} = \langle K_1, K_2 \rangle = Lclm(l, k)$ is a non-principal divisor as determined in Lemma 2.3. Then a fundamental system may be described as follows.

$$\mathscr{L}^{15}_{xyy} : \begin{cases} z_1(x, y) = F(y) \exp\left(-\int a\, dx\right), \quad z_2(x, y) = G(x) \exp\left(-\int b\, dy\right), \\ z_3(x, y) \text{ is a special solution of } K_1 z = w_1, \quad K_2 z = 0 \text{ where} \\ w_1 = \exp \int (p_2 + q_1 q_2 - q_3)\, dy, \\ p_2 \text{ and the } q_i \text{ are coefficients of } K_1 \text{ and } K_2. \end{cases}$$

Proof. By similar reasoning as in the proof of Proposition 7.4 solutions corresponding to the systems $Lz = 0$, $\mathfrak{l}_m z = 0$ for $m \geq 2$ or $Lz = 0$, $\mathfrak{k}_n z = 0$, $n \geq 3$ are obtained. For decomposition type $\mathscr{L}^{12}_{xyy}$ this yields the fundamental system given above.

For decomposition type $\mathscr{L}^{13}_{xyy}$ the first two solutions $z_1(x, y)$ and $z_2(x, y)$ follow from the Laplace divisor. In order to obtain the third solution define $w_1 \equiv Lz$ and $w_2 \equiv \mathfrak{k}_n z$. The exact quotient equations $w_1 = 0$ and $w_{2,x} + A_2 w_2 = 0$ follow from Theorem 6.6. Substituting its solutions into the above two inhomogeneous equations the solution for $z_3(x, y)$ follows.

For decomposition type $\mathscr{L}^{14}_{xyy}$ the first solution $z_1(x, y)$ follows from the Laplace divisor. According to Theorem 6.6 the exact quotient equations are now $w_1 = 0$ and $(\partial_{yy} + A_1 \partial_y + A_3) w_2 = 0$. The latter equation corresponds to an ideal of differential dimension $(1, 2)$; consequently, substituting its solution into the system $Lz = 0$, $\mathfrak{l}_m z = w_2$ yields the remaining solutions $z_2(x, y)$ and $z_3(x, y)$.

For decomposition type $\mathscr{L}^{15}_{xyy}$, the first two members follow immediately as solutions of $lz = 0$ and $kz = 0$. In order to obtain the third member $z_3(x, y)$ the exact quotient module

$Exquo(\langle L \rangle, \langle K_1, K_2 \rangle)$

$$= \langle (0, 1), (\partial_y - p_2 - q_1 q_2 + q_3, -\partial_x + q_2 \partial_y - p_3 - q_1 q_3 + q_4) \rangle$$

$$= \langle (0, 1), (\partial_y - p_2 - q_1 q_2 + q_3, 0) \rangle$$

is constructed. The first generator of the module at the right hand side in the first line follows from the division, i.e. from $L = K_2$; the second generator represents the single syzygy (2.28) given in Lemma 2.3. In the last line the generators have been transformed into a a Janet basis. Introducing the new differential indeterminates w_1 and w_2 the equations

$$w_{1,y} - (p_2 + q_1 q_2 - q_3)w_1 = 0, \quad w_2 = 0 \tag{7.14}$$

are obtained. Its solutions are

$$w_1 = F(x) \exp \int (p_2 + q_1 q_2 - q_3) dy, \quad w_2 = 0; \tag{7.15}$$

$F(x)$ is an undetermined function. Then $z_3(x, y)$ is a special solution of $K_1 z = w_1$, $K_2 z = w_2$; its solution has been described in Theorem 5.2. □

The first example shows how the existence of a Laplace divisor enables the solution of a given third-order equation.

Example 7.32. The system $Lz = 0$, $\partial_{xxx}z = 0$ of the type $\mathscr{L}^{14}_{xyy}$ decomposition considered in Example 6.34 yields the solution

$$z_1(x, y) = \big((x + y)^2 - 2(x + y) + 2\big)F(y) + 2(x + y - 1)F'(y) + F''(y)$$

where F is an undetermined function of y. The equation $w_{yy} + (x+y)(w_y + w) = 0$ has the general solution

$$w = G(x)r(x, y) + H(x)r(x, y) \int s(x, y) dy$$

where

$$r(x, y) \equiv \exp\left(-\tfrac{1}{2}(x + y - 2)^2 - y\right) \quad \text{and} \quad s(x, y) = \exp\left(\tfrac{1}{2}(x + y - 2)^2\right);$$

G and H are undetermined functions of x. Substitution into $z_{xxx} = w$ yields

$$z_2(x, y) = \tfrac{1}{2} \int G(x)r(x, y)x^2 dx - x \int G(x)r(x, y)x dx + \tfrac{1}{2}x^2 \int G(x)r(x, y)dx,$$

$$z_3(x, y) = \frac{1}{2} \int H(x)r(x, y) \int s(x, y) dy x^2 dx$$

$$- x \int H(x)r(x, y) \int s(x, y) dy x dx$$

$$+ \tfrac{1}{2}x^2 \int H(x)r(x, y) \int s(x, y) dy dx;$$

G and H are undetermined functions. □

The equation in the following example may be solved because it allows two first-order right factors generating a divisor $\mathbb{J}_{xxy}$. It is emphasized again that knowing only a single factor leads to a much more involved solution procedure.

Example 7.33. The two equations $l_1 z = 0$ and $l_2 z = 0$ of the type $\mathscr{L}_{xyy}^{15}$ decomposition in Example 6.35 yield the solutions

$$z_1(x, y) = F(y) \exp(-xy) \text{ and } z_2(x, y) = G(x) \exp(xy);$$

$w_1 = H(x) \exp(xy)$ and $w_2 = 0$, i.e. $K_1 z = H(x) \exp(xy)$ and $K_2 z = 0$ have to be solved. According to Theorem 5.2 the system

$$r_{xy} - xr_x + yr_y - (xy + 3)r = H(x) \exp(xy), \quad r_{yy} - 2xr_x + x^2 r = 0$$

follows; its solution is $r(x, y) = -\frac{1}{2} H(x) \exp(xy)$. Substitution into (5.35) finally yields

$$z_3(x, y) = \exp(-xy) \int H(x) \exp(2xy) dx;$$

$H(x)$ is an undetermined function. $\square$

It is suggested that the reader verifies the solutions given in the preceding examples by substitution into the respective equation.

7.4   Transformation Theory of Third-Order Linear PDE's

Like for operators of order two, transformations of the operator variables x and y are considered first. They correspond to a change of the independent variables of the corresponding equation, leaving the dependent variable unchanged.

Proposition 7.7. *Any third order linear pde*

$$z_{xxx} + A_1 z_{xxy} + A_2 z_{xyy} + A_3 z_{yyy} + A_4 z_{xx} + A_5 z_{xy} + A_6 z_{yy} + A_7 z_x + A_8 z_y + A_9 z = 0 \tag{7.16}$$

is equivalent to one of the following normal forms by means of a transformation of the independent variables x and y; the three alternatives are determined by the roots of the symbol equation $k^3 + A_1 k^2 + A_2 k + A_3 = 0$. The transformation functions are defined by $u \equiv \varphi(x, y)$ and $v \equiv \psi(x, y)$; there must hold $\varphi_1 \psi_y - \varphi_y \psi_x \neq 0$.

- (i) $w_{uuv} + w_{uvv} + B_1 w_{uu} + B_2 w_{uv} + B_3 w_{vv} + B_4 w_u + B_5 w_v + B_6 w = 0$
  if the symbol equation has three simple roots.
- (ii) $w_{uvv} + B_1 w_{uu} + B_2 z_{uv} + B_3 z_{vv} + B_4 w_u + B_5 w_v + B_6 w = 0$
  if the symbol equation has a twofold root.
- (iii) $w_{uuu} + B_1 w_{uu} + B_2 w_{uv} + B_3 w_{vv} + B_4 w_u + B_5 w_v + B_6 w = 0$
  if the symbol equation has a threefold root.

Proof. Defining $u \equiv \varphi(x, y)$ and $v \equiv \psi(x, y)$ it follows that $z(x, y) = w(u, v)$, $z_x = w_u \varphi_x + w_v \psi_x$ and $z_y = w_u \varphi_y + w_v \psi_y$. Applying the chain rule repeatedly and substituting the resulting derivatived one obtains

$$z_{xxx} + A_1 z_{xxy} + A_2 z_{xyy} + A_3 z_{yyy} + A_4 z_{xx} + A_5 z_{xy} + A_6 z_{yy} + A_7 z_x + A_8 z_y + A_9 z$$
$$= [\varphi_x^3 + A_1 \varphi_x^2 \varphi_y + A_2 \varphi_x \varphi_y^2 + A_3 \varphi_y^3] w_{uuu}$$
$$+ [3\varphi_x^2 \psi_x + A_1 \varphi_x (\varphi_x \psi_y + 2\varphi_y \psi_x) + A_2 \varphi_y (\varphi_y \psi_x + 2\varphi_x \psi_y) + 3A_3 \varphi_y^2 \psi_y] w_{uuv}$$
$$+ [3\varphi_x \psi_x^2 + A_1 \psi_x (2\varphi_x \psi_y + \varphi_y \psi_x) + A_2 \psi_y (2\varphi_y \psi_x + \varphi_x \psi_y) + 3A_3 \varphi_y \psi_y^2] w_{uvv}$$
$$+ [\psi_x^3 + A_1 \psi_x^2 \psi_y + A_2 \psi_x \psi_y^2 + A_3 \psi_y^3] w_{vvv}$$
$$+ [3\varphi_x \varphi_{xx} + A_1 (2\varphi_x \varphi_{xy} + \varphi_{xx} \varphi_y) + A_2 (2\varphi_y \varphi_{xy} + \varphi_x \varphi_{yy}) + 3A_3 \varphi_y \varphi_{yy}$$
$$+ A_4 \varphi_x^2 + A_5 \varphi_x \varphi_y + A_6 \varphi_y^2] w_{uu}$$
$$+ [3\varphi_{xx} \psi_x + 3\varphi_x \psi_{xx} + A_1 (2\varphi_{xy} \psi_x + 2\varphi_x \psi_{xy} + \varphi_{xx} \psi_y + \varphi_y \psi_{xx})$$
$$+ A_2 (2\varphi_{xy} \psi_y + 2\varphi_y \psi_{xy} + \psi_x \varphi_{yy} + \varphi_x \psi_{yy})$$
$$+ 3A_3 (\varphi_y \psi_{yy} + \varphi_{yy} \psi_y)$$
$$+ 2A_4 \varphi_x \psi_x + A_5 (\varphi_x \psi_y + \varphi_y \psi_x) + 2A_6 \varphi_y \psi_y] w_{uv}$$

$$\text{(7.17)}$$

$$+ [3\psi_x \psi_{xx} + A_1 (2\psi_x \psi_{xy} + \psi_y \psi_{xx}) + A_2 (2\psi_y \psi_{xy} + \psi_x \psi_{yy}) + 3A_3 \psi_y \psi_{yy}$$
$$+ A_4 \psi_x^2 + A_5 \psi_x \psi_y + A_6 \psi_y^2] w_{vv}$$
$$+ [\varphi_{xxx} + A_1 \varphi_{xxy} + A_2 \varphi_{xyy} + A_3 \varphi_{yyy} + A_4 \varphi_{xx} + A_5 \varphi_{xy} + A_6 \varphi_{yy}$$
$$+ A_7 \varphi_x + A_8 \varphi_y] w_u$$
$$+ [\psi_{xxx} + A_1 \psi_{xxy} + A_2 \psi_{xyy} + A_3 \psi_{yyy} + A_4 \psi_{xx} + A_5 \psi_{xy} + A_6 \psi_{yy}$$
$$+ A_7 \psi_x + A_8 \psi_y] w_v + A_9 w = 0.$$

The coefficients of w_{uuu} and w_{vvv} are homogeneous of order 3 in the first derivatives of φ or ψ. Dividing by φ_y^3 or ψ_y^3 yields the third order polynomial $P \equiv k^3 + A_1 k^2 + A_2 k + A_3$ where $k = \varphi_x / \varphi_y$ or $k = \psi_x / \psi_y$ respectively.

Assume first that there are three different roots k_1, k_2 and k_3 of P, $k_i \neq k_j$ for $i \neq j$. Then $A_1 = -(k_1 + k_2 + k_3)$, $A_2 = k_1 k_2 + k_1 k_3 + k_2 k_3$ and $A_3 = -k_1 k_2 k_3$. If $\varphi_x = k_1 \varphi_y$ and $\psi_x = k_2 \psi_y$ are chosen, the coefficients of w_{uuu} and w_{vvv} vanish. The non-vanishing third order terms in the transformed equations are

$$(k_1 - k_2) \varphi_y \psi_y [(k_1 - k_3) \varphi_y w_{uuv} + (k_2 - k_3) \psi_y w_{uvv}].$$

If $(k_1 - k_3) \varphi_y = (k_2 - k_3) \psi_y$ is chosen, this simplifies to

$$(k_1 - k_2)(k_2 - k_3) \varphi_y \psi_y^2 (w_{uuv} + w_{uvv});$$

this is case (i).

Now let P have the double root k_1 and the simple root k_2, $k_1 \neq k_2$, and choose $\varphi_x = k_1 \varphi_y$, $\psi_x = k_2 \psi_y$. Then $A_1 = -2k_1 - k_2$, $A_2 = k_1^2 + 2k_1 k_2$, and $A_3 = -k_1^2 k_2$. Substituting these expressions into the transformed equation, the coefficients of w_{uuu}, w_{uuv}, and w_{vvv}, vanish, whereas the coefficient of w_{uvv} is proportional to $(k_1 - k_2)^3 \neq 0$; this is case (ii).

Finally, let P have the threefold root $k \equiv k_1$ and choose $\psi_x = k_1\psi_y$. Then $A_1 = -3k_1$, $A_2 = 3k_1^2$, and $A_3 = -k_1^3$. Upon substituting these expressions into the transformed equation, the coefficients of w_{uuv}, w_{uvv}, and w_{vvv} vanish. Because φ is chosen such that $\varphi_x \neq k_1\varphi_y$, the coefficient of w_{uuu} does not vanish; this is case (iii).   □

The normal forms of the above proposition require first integrals of first-order ode's. Because there is no algorithm available at present for this problem, there is no guarantee that for a given equation the normal may actually be computed. In particular it is by no means guaranteed that the coefficients of the normal form equations are in the base field of (7.16).

It is instructive to generate the normal forms given in the above proposition for various concrete examples and discuss the result w.r.t. the solution procedure.

Example 7.34. Consider again Blumberg's equation of Example 7.8. The symbol equation is $k^3 + k^2 x = k^2(k + x) = 0$ with the twofold root $k_1 = 0$ and the simple root $k_2 = -x$, i.e. case (ii) of the above proposition applies. They yield $\varphi_x = 0$ and $\psi_x + x\psi_y = 0$ with the solutions $\varphi(y)$ and $\psi(y - \frac{1}{2}x^2)$; φ is an undetermined function of y. Possible choices for the new variables are $u = y$ and $v = y - \frac{1}{2}x^2$ with the inverse $x = \sqrt{2(u - v)}$ and $y = u$. Substitution into (6.17) leads to

$$\partial_{uvv} - \left(\frac{3}{2(u-v)} + \frac{\sqrt{2}}{(u-v)^{1/2}}\right)\partial_{uv} + \left(\frac{1}{2(u-v)} + \frac{\sqrt{2}}{2(u-v)^{3/2}}\right)\partial_u = 0.$$

It has the right factors

$$\partial_u \quad \text{and} \quad \partial_v - \frac{\sqrt{2}\sqrt{u-v}+3}{3\sqrt{2}\sqrt{u-v}+2(u-v)}$$

with a rather complicated non-principal intersection ideal.   □

Example 7.35. The symbol equation for the equation considered in Example 7.6 is $k^3 + (y + 2)k^2 + (y + 1)k = 0$ with the three roots $k_1 = 0$, $k_2 = -1$ and $k_3 = -(y + 1)$. By case (i) of Proposition 7.7, $\varphi_x = 0$ and $\psi_x = -\psi_y$ may be chosen; new variables are $u = \varphi(y) = y$ and $v = \psi(x - y) = x - y$ with the inverse $x = u + v$ and $y = u$. The normal form equation is

$$w_{uuv} - \frac{u}{u+1}w_{uvv} + w_{uu} - \frac{u^2+1}{u^2+u}w_{uv} - \frac{1}{u^2+u}w_u = 0$$

with right factors ∂_u, $\partial_v + 1$ and $\partial_u - \frac{u}{u+1}$.   □

Example 7.36. The symbol equation for the equation considered in Example 6.4 is $k^3 + (y + 1)k^2 = 0$; it has the double root $k_1 = 0$ and the simple root $k_2 = y + 1$ corresponding to case (ii) of Proposition 7.7. Choosing $\varphi_x = 0$ and $\psi_x = -(y + 1)\psi_y$, new variables are $u = y$ and $v = x - \log(y + 1)$ with the inverse $x = v + \log(u + 1)$ and $y = u$. The normal form equation is

$$w_{uvv} + \frac{\log{(u+1)} + v - 1}{\log{(u+1)} + v} w_{uv} - \frac{1}{\log{(u+1)} + v} w_u = 0. \qquad \square$$

These examples show that in general the normal form equations determined in Proposition 7.7 have more complicated coefficients than the originally given equation, usually they are not in its base field. Therefore, if factorization of normal form equations is considered, elementary or Liouvillian function coefficients must be allowed which amounts to a new class of factorization problems.

7.5 Exercises

Exercise 7.1. Verify the solution of (7.2) given in Example 7.4 on page 152.

Exercise 7.2. Solve Blumberg's equation (6.17) by applying the factorization given in the second line of (6.19) on page 129. Compare the result with the solution that has been obtained in Example 7.8.

Exercise 7.3. Determine the solutions of Blumberg's equation (6.17) from the factorization in the first line of (6.19) and discuss the result.

Exercise 7.4. Solve the equation

$$z_{xxx} - x^2 z_{xyy} + 3z_{xx} + (2x + 3)z_{xy} - x^2 z_{yy} + 2z_x + (2x + 3)z_y = 0$$

and discuss the result.

Exercise 7.5. Discuss the possible existence of differential type 0 solutions of the equation of Example 7.19 on page 163.

Exercise 7.6. Determine the equations corresponding to the three permutations of the first-order factors not considered in Examples 7.21–7.23 and determine the respective fundamental systems. Discuss the result.

Exercise 7.7. Solve the equation $z_{xyy} + (x + 1)z_{xy} - z_y = 0$.

Chapter 8
Summary and Conclusions

The importance of decomposing differential equations for solving them has become obvious in this monograph. It turned out that decomposing an operator and solving the corresponding differential equation in closed form is essentially the same subject. In this way, most results known from the classical literature on solving linear pde's may be obtained in a systematic way, without heuristics or ad hoc methods.

The relevance of this monograph originates from this connection. Its main topic – generalizing Loewy's fundamental result of decomposing an ordinary differential operator into completely reducible components to partial differential operators – has been achieved to a large extent by working in an appropriate ring of partial differential operators. The main distinguishing features compared to ordinary operators are twofold. On the one hand, ideals of partial differential operators are not necessarily principal, even if the ideals to be intersected are principal. On the other hand, divisors of ideals of partial differential operators may be principal or not, independent of whether this is true for the given ideal. The clue for understanding this behavior is given in the section on the lattice structure of left ideals in rings of partial differential operators on page 47. The relevant features may be summarized as follows.

Theorem 8.1. *The left ideals in any ring of partial differential operators have the following properties.*

  (i) *The left ideals form a modular lattice.*
 (ii) *The ideals of differential type zero form a sublattice.*
(iii) *The principal ideals do not form a sublattice.*

Proof. The first property is true for any ideal in a ring. The second property is easily obtained from the solution spaces of the corresponding system of linear pde's. The third one follows from Theorems 2.2 and 2.3. □

The procedures for factorizations and more general decompositions are driven by the following questions: Does there exist any decomposition? And secondly,

F. Schwarz, *Loewy Decomposition of Linear Differential Equations*, Texts & Monographs in Symbolic Computation, DOI 10.1007/978-3-7091-1286-1_8, © Springer-Verlag/Wien 2012

may the various components be determined explicitly? In either case it has to be specified which coefficient field for the factors is admitted.

In order to obtain a satisfactory theory, and also if solving concrete problems is the main issue, the base field of the given operator or equation is usually the field of rational functions. However, there is also some interest in factorizations over a universal field that is considered next.

Theorem 8.2. *The existence of first-order principal right factors in a universal field for any second- or third-order partial differential operator in the plane may always be decided; however, in general these factors may not be determined algorithmically.*

Proof. Operators with leading derivative ∂_{xx} have been considered in Proposition 4.1. For the separable case (i) the above statement is obvious. For a double root of the symbol polynomial, condition (4.2) is necessary for any factor to exist. If it is satisfied, the solutions of the partial Riccati equation (4.3) determine the possible factors. In a sufficiently large field they always exist; hence, for a universal field (4.2) is also sufficient for the existence of a factor.

The proof for any second-order operators with leading derivative ∂_{xy} or operators of third order is similar and therefore not discussed in detail.   □

The general scheme underlying the preceding proof may be described as follows. For any operator, divisibility by a first-order factor is determined by two kinds of constraints. In the first place, there are those conditions comprising only the coefficients of the given operator; obviously they are necessary for the existence of any factor independent of its coefficient field. Secondly, there are conditions comprising also the coefficients of the desired factor. It may always be decided whether they allow solutions in a universal field, but in general they may not be determined explicitly. Proving existence of solutions in the base field however amounts to determining them explicitly right from the beginning; that cannot be performed in general.

Divisibility in a universal field is also of practical importance for factorization in the base field. For the latter the coefficient constraints for factorization in a universal field are necessary conditions that may be tested rather quickly; therefore they may be applied in order to discard a certain factor in the base field without entering the time-consuming solution algorithm for determining its coefficients explicitly.

In general the preceding theorem does not decide absolute irreducibility. On the one hand, for a third-order operator also the existence of second-order factors would have to be excluded. Secondly, an algorithm for determining Laplace divisors of any order would be required. Furthermore, in some exceptional cases like e.g. in case (iii), subcase (b) of Proposition 6.1 certain pde's have to be solved for which a solution algorithm is not available.

There is one more feature that has to be considered; it concerns the question to what extent the existence of factors or divisors of a certain type is decidable at all. In particular this applies to the existence of a Laplace divisor; right now an upper bound for any such divisor as defined in Definition 2.3 on page 34 is not known.

Its close connection to Laplace's method described in Appendix C makes it highly suspect of being not decidable at all. Similar remarks apply to the problem of finding rational first integrals as described at the end of Appendix B. To provide a sound basis for the subject of this monograph the decidability of these problems should be clarified. In case the answer is negative, the situation would be somewhat similar to solving diophatine equations; the best possible result would be to identify classes of differential equations for which factorization is decidable. On the other hand, if it is positive, more advanced methods have to be developed.

Finally, some aspects on the further development of the subject of this monograph will be discussed. The computational cost for obtaining a Loewy decomposition is very high and increases quickly with the order of the given operator and the number of variables involved. It should be more favourable to obtain the Loewy factors in a first step *without* computing their irreducible components. Only in a second step the Loewy factors should be further decomposed. Because they are of lower order, this process will be more efficient. However, at present there is no algorithm known for proceeding in this way.

For ordinary operators, the associated equations discussed on page 3 are an important tool for obtaining the coefficients of its factors. It is not obvious how to generalize them for partial differential operators, if possible at all.

There appears to be a close relation between the type of a decomposition and the structure of the solution of the corresponding differential equation as shown in Chaps. 5 and 7. It should be possible to give a more detailed description of the structure of the solution for any decomposition type.

In many applications linear pde's in three or even four independent variables occur, see e.g. the collection by Polyanin [55]. In particular this is true for the symmetry analysis of differential equations because the so-called determining system of the symmetries is a linear homogeneous system of pde's. Therefore it would be highly desirable to extend the results of this monograph to pde's in more than two variables, and possibly of higher order. A first important step into this direction is the article by Schwarz [63] on intersection ideals in three space; these results give an indication of the complexity of any such effort.

Methods of differential algebra have been applied in a more general setting by Ritt [57] and Kolchin [37], see also [66] and [27]. They consider general differential polynomials, not necessarily linear, and the solution manifolds of the corresponding equations. To this end, they decompose the radical differential ideal generated by the differential polynomials into prime components such that the former is obtained as the intersection of the latter. These components correspond to the general and the singular solutions respectively.

The algebraic methods described in this monograph turned out to be highly appropriate for linear differential equations. In order to proceed, in a first step monic quasilinear ode's and pde's might be considered. They are easier to handle than the general case, and many practical problems like e.g. Eulers equations or the Kortweg-deVries equation are of this type. A good introduction into this more general subject including a useful list of references is given by Tsarev [68].

Appendix A
Solutions to the Exercises

For many exercises extensive calculations are necessary that are difficult to perform by pencil and paper. It is recommended to apply the userfunctions provided on the website www.alltypes.de in those cases; a short description is given in Appendix E.

Chapter 1

1.1. For $y'' + a_1 y' + a_2 y = 0$ the answer is

$$a_1 = -\frac{1}{W^{(2)}} \begin{vmatrix} y_1 & y_2 \\ y_1'' & y_2'' \end{vmatrix}, \quad a_2 = -\frac{1}{W^{(2)}} \begin{vmatrix} y_1' & y_2' \\ y_1'' & y_2'' \end{vmatrix} \quad \text{where} \quad W^{(2)} = \begin{vmatrix} y_1 & y_2 \\ y_1' & y_2' \end{vmatrix}.$$

For $y''' + a_1 y'' + a_2 y' + a_1 y = 0$ the answer is

$$a_1 = -\frac{1}{W^{(3)}} \begin{vmatrix} y_1 & y_2 & y_3 \\ y_1' & y_2' & y_3' \\ y_1''' & y_2''' & y_3''' \end{vmatrix}, \quad a_2 = \frac{1}{W^{(3)}} \begin{vmatrix} y_1 & y_2 & y_3 \\ y_1'' & y_2'' & y_3'' \\ y_1''' & y_2''' & y_3''' \end{vmatrix},$$

$$a_3 = -\frac{1}{W^{(3)}} \begin{vmatrix} y_1' & y_2' & y_3' \\ y_1'' & y_2'' & y_3'' \\ y_1''' & y_2''' & y_3''' \end{vmatrix} \quad \text{where} \quad W^{(3)} = \begin{vmatrix} y_1 & y_2 & y_3 \\ y_1' & y_2' & y_3' \\ y_1'' & y_2'' & y_3'' \end{vmatrix}.$$

1.2. Go to the ALLTYPES user interface and define
```
e1:=Df(z_1,x)-z_2;
e2:=Df(z_2,x)-z_3+a_1*z_2+a_2*z_1;
e3:=Df(z_3,x)-a_3*z_1+a_1*z_3;
T==|LDFMOD(DFRATF(Q,{a_1,a_2,a_3},{x},GRLEX),
                 {z_3,z_2,z_1},{x},LEX)|;
JanetBasis({e1,e2,e3}|T|);
```
The result of Example 1.1 is returned. The term orders
```
T==|LDFMOD(DFRATF(Q,{a_1,a_2,a_3},{x},GRLEX),
                 {z_1,z_2,z_3},{x},LEX)|;
JanetBasis({e1,e2,e3}|T|);
```

F. Schwarz, *Loewy Decomposition of Linear Differential Equations*, Texts & Monographs in Symbolic Computation, DOI 10.1007/978-3-7091-1286-1,
© Springer-Verlag/Wien 2012

```
T== | LDFMOD (DFRATF (Q, {a_1, a_2, a_3}, {x}, GRLEX),
                      {z_3, z_1, z_2}, {x}, LEX) | ;
JanetBasis ({e1, e2, e3} | T | );
```

yield the associated equations for z_3 and z_2 respectively.

1.3. Let the given fourth-order equation be

$$y'''' + a_1 y''' + a_2 y'' + a_3 y' + a_4 y = 0. \tag{A.1}$$

In addition to the functions z_1, z_2, and z_3 defined in Example 1.1 the functions

$$z_4 = \begin{vmatrix} y_1 & y_2 \\ y_1''' & y_2''' \end{vmatrix}, \quad z_5 = \begin{vmatrix} y_1' & y_2' \\ y_1''' & y_2''' \end{vmatrix}, \quad z_6 = \begin{vmatrix} y_1'' & y_2'' \\ y_1''' & y_2''' \end{vmatrix}.$$

are required. They obey the system

$$z_1' = z_2, \quad z_2' = z_3 + z_4, \quad z_3' = z_5, \quad z_4' = z_5 - a_1 z_4 - a_2 z_2 + a_4 z_1,$$
$$z_5' = z_6 - a_1 z_5 - a_2 z_3 + a_4 z_1, \quad z_6' = -a_1 z_6 + a_3 a_3 + a_4 z_2.$$

A Janet basis in *lex* term order with $z_1 \succ \ldots \succ z_6$ has the form

$$z_1^{(VI)} + \sum_{k=0}^{5} r_k(a_1, \ldots, a_4) z_1^{(k)}, \quad z_2 = z_1', \quad \text{and} \quad z_i = \sum_{k=0}^{5} f_{i,k}(a_1, \ldots, a_4) z_1^{(k)}$$

for $i = 3, \ldots, 6$. The r_k and $f_{i,k}$ are differential functions of the coefficients $a_1, \ldots, a_4$. In order to determine the coefficients of a second-order factor (compare the discussion in Example 1.3) it suffices to find a solution with rational logarithmic derivative of the first equation for z_1.

1.4. Substituting $y = \phi z$ into the given equation yields

$$z'' + \left(2\frac{\phi'}{\phi} + p \right) z' + \left(\frac{\phi''}{\phi} + p\frac{\phi'}{\phi} + q \right) z = 0.$$

The coefficient of z' vanishes if $\frac{\phi'}{\phi} = -\frac{1}{2}p$; hence $\phi = C \exp\left(-\frac{1}{2}\int p\, dx\right)$, C a constant, is the most general transformation with this property. Substitution into the coefficient of z leads to $r = -\frac{1}{2} - \frac{1}{4}p^2 + q$.

1.5. The first-order right factors of a $\mathcal{L}_3^2$ type decomposition have the form $l^{(1)}(C) = D - p - \frac{r'}{r + C}$, where p and r originate from the solution of the Riccati equation (notation as in (B.1)), and C is a constant. The $Lclm$ for two operators of this form is

$$L \equiv Lclm\left(l^{(1)}(C_1), l^{(1)}(C_2)\right) = D^2 - \left(\frac{r''}{r'} + 2p\right)D + \frac{r''}{r'}p - p' + p^2.$$

By division it is shown that any operator $l^{(1)}(C)$ is a divisor of L, i.e. it is contained in the left intersection ideal generated by it.

1.6. For $\mathcal{L}_1^2$, the first solution y_1 is obtained from $(D + a_1)y = 0$, the second from $(D + a_1)y = \bar{y}_2$ with $\bar{y}_2$ a solution of $(D + a_2)y = 0$.

For $\mathcal{L}_2^2$ the two solutions are obtained from $(D + a_i)y = 0$. Linear dependence over the base field would imply a relation $q_1 y_1 + q_2 y_2 = 0$ with q_1, q_2 from the base field. Substituting the solutions this would entail $\frac{q_1}{q_2} = -\exp \int (a_2 - a_1)dx$. Due to the non-equivalence of a_1 and a_2, its difference is not a logarithmic derivative; thus the right hand side cannot be rational.

For $\mathcal{L}_3^2$ the equation $(D + a(C))y = 0$ has to be solved, then C is specialized to $\bar{C}$ and $\bar{\bar{C}}$. Substituting $a_1 = \frac{r'}{r + C} + p$ and $a_2 = \frac{r'}{r + \bar{C}} + p$ in the above quotient, the integration may be performed with the result $\frac{q_1}{q_2} = -\frac{r + \bar{\bar{C}}}{r + C}$ which is rational.

1.7. Define $\Delta \equiv \sqrt{A^2 - 4B}$. Two cases are distinguished. If $\Delta \neq 0$, two first-order right factors are $l_{1,2} = D + \frac{1}{2}A \pm \frac{1}{2}\Delta$; L has the decomposition $L = Lclm(l_1, l_2)$ of type $\mathcal{L}_2^2$; a fundamental system is $y_{1,2} = \exp\left((-\frac{1}{2}A \pm \frac{1}{2}\Delta)x\right)$. If $\Delta = 0$, the type $\mathcal{L}_3^2$ decomposition is $L = Lclm(D + \frac{1}{2}A - \frac{1}{x + C})$, C a constant; it yields the fundamental system $y_1 = \exp\left(\frac{1}{2}A\right)$, $y_2 = x\exp\left(\frac{1}{2}A\right)$. This result shows: A second-order lode with constant coefficients is always completely reducible.

1.8. According to Lemma 1.1, case (i), the coefficient a of a first-order factor $D + a$ has to satisfy

$$a'' - 3a'1 + \left(1 - \frac{1}{x}\right)a' + a^3 - \left(1 - \frac{1}{x}\right)a^2 + \left(x - \frac{2}{x}\right)a - \frac{2}{x^2} = 0.$$

This second-order Riccati equation does not have a rational solution; thus a first-order right factor does not exist. According to case (ii) of the same lemma, the coefficient b of a second-order factor $D^2 + bD + c$ follows from

$$b'' - 3b'b + \left(2 - \frac{2}{x}\right)b' + b^3 - \left(2 - \frac{2}{x}\right)b^2 + \left(x + 1 - \frac{4}{x} + \frac{2}{x^2}\right)b - x = \frac{2}{x} - \frac{2}{x^2} = 0.$$

Its single rational solution $b = 1$ leads to $c = x - \frac{1}{x}$ and yields the second-order factor given in Example 1.3.

1.9. Let y_1, y_2, and y_3 be a fundamental system for the homogeneous equation and W its Wronskian. The general solution may be written as

$$y = C_1 y_1 + C_2 y_2 + C_3 y_3 + y_1 \int \frac{r}{W}(y_2 y_3' - y_2' y_3)dx$$

$$-y_2 \int \frac{r}{W}(y_1 y_3' - y_1' y_3)dx + y_3 \int \frac{r}{W}(y_1 y_2' - y_1' y_2)dx$$

where C_1, C_2 and C_3 are constants. For $y''' = r$ with the fundamental system $y_1 = 1$, $y_2 = x$ and $y_3 = x^2$ there follows

$$y = C_1 + C_2 x + C_3 x^2 + \tfrac{1}{2} \int x^2 r dx - x \int x r dx + \frac{1}{2} x^2 \int r dx.$$

1.10. A simple calculation shows that the commutator between $l_1 = D + a_1$ and $l_2 = D + a_2$ vanishes if $a_1' - a_2' = 0$, i.e. if a_1 and a_2 differ by a constant. If this is true, the representation (1.17) simplifies to $L = D^2 + (a_1 + a_2)D + a_1 a_2 + a_1'$; furthermore $L = l_1 l_2 = l_2 l_1$. An example of this case is

$$D^2 - 4xD + 4x^2 - 3 = Lclm(D - 2x + 1, D - 2x - 1).$$

An example of non-commutative first-order factors is

$$D^2 - \left(1 + \frac{1}{x}\right)D + \frac{1}{x} = Lclm\left(D - 1, D - 1 + \frac{1}{x+1}\right).$$

Chapter 2

2.1. The product of $l_1 \equiv \partial_x + a_1 \partial_y + b_1$ and $l_2 \equiv \partial_x + a_1 \partial_y + b_2$ is

$$l_1 l_2 = \partial_{xx} + (a_1 + a_2)\partial_{xy} + a_1 a_2 \partial_{yy} + (b_1 + b_2)\partial_x$$
$$+ (a_{2,x} + a_1 a_{2,y} + a_1 b_2 + a_2 b_1)\partial_y + b_{2,x} + a_1 b_{2,y} + b_1 b_2.$$

The product $l_2 l_1$ is obtained from this expression by interchange of all indices 1 and 2. Comparing those coefficients that are not symmetrical under this permutation leads to

$$a_{2,x} + a_1 a_{2,y} = a_{1,x} + a_2 a_{1,y}, \quad b_{2,x} + a_1 b_{2,y} = b_{1,x} + a_2 b_{1,y}.$$

Upon rearrangement, the conditions for case (i) given in Lemma 2.1 are obtained. The calculation for case (ii) is similar.

2.2. By definition of the Hilbert-Kolchin polynomial, the lc of an ideal with differential dimension $(0, k)$ is k, the dimension of the solution space of the corresponding system of pde's. Hence (2.8) for this special case reduces to the well known relation

$$dim\, V_{I+J} + dim\, V_{I \cap J} = dim\, V_I + dim\, V_J.$$

2.3. The third-order terms of the intersection ideal are

$$\partial_{xxx} - (a_1^2 + a_1 a_2 + a_2^2)\partial_{xyy} - (a_1 + a_2)a_1 a_2 \partial_{yyy}, \quad \partial_{xxy} + (a_1 + a_2)\partial_{xxy} + a_1 a_2 \partial_{yyy}.$$

2.4. The generator of the intersection ideal is

$$\partial_{xy} + \partial_{yy} + b_2 \partial_x + (a_1 b_2 + b_1)\partial_y + b_{2,x} + (a_1 b_2 + a_1)_y + b_1 b_2.$$

2.5. The solutions of the three equations $l_i z_i = 0$ are $z_1 = f(y)\exp(-2x)$, $z_2 = g(y-x)\exp(-x)$ and $z_3 = h(y-2x)$; f, g and h are undetermined functions of the respective argument. The system corresponding to the $Gcrd$ is $z_x + 2z = 0$, $z_y - z = 0$ with the solution $z = C \exp(y - 2x)$, C a constant. The undetermined functions have to be specialized as follows in order to obtain this latter solution:

$$f(y) = C \exp(y), \quad g(y - x) = C \exp(y - x), \quad h(y - 2x) = C \exp(y - 2x).$$

2.6. For constant coefficients the constraints (2.41), (2.46), and (2.47) are always satisfied. Thus, for subcase (b) the generator for the principal intersection ideal is

$$\partial_{xxx} + \sum a_i \partial_{xxy} + \sum_{i<j} a_i a_j \partial_{xyy} + \prod a_i \partial_{yyy} + \sum b_i \partial_{xx} + \sum_{i \neq j} a_i b_j \partial_{xy}$$

$$+ \sum_{i \neq j} a_i a_j b_k \partial_{yy} + \sum_{i<j} b_i b_j \partial_x + \sum_{i \neq j <k} a_i b_j b_k \partial_y + \prod b_i.$$

All indices run from 1 to 3. In order to satisfy the condition for subcase (a), $b_2 = (a_2 - a_3)b_1 + \frac{a_1 - a_2}{a_1 - a_3} b_3$ may be chosen. The third-order terms are not changed. The remaining expression is

$$\left(b_1 + b_3 + \frac{a_2 - a_3}{a_1 - a_3} b_1 + \frac{a_1 - a_2}{a_1 - a_3} \right) \partial_{xx}$$

$$+ \frac{2}{a_1 - a_3} \left[(a_1 a_2 - a_3^2) b_1 + (a_1^2 - a_2 a_3) b_3 \right] \partial_{xy}$$

$$+ \left[a_2(a_1 b_3 + a_3 b_1) + \frac{a_1 a_3}{a_1 - a_3} [(a_2 - a_3)b_1 + (a_1 - a_2)b_3] \right] \partial_{yy}$$

$$+ \left[2b_1 b_3 + \frac{a_2 - a_3}{a_1 - a_3} b_1^2 + \frac{a_1 - a_2}{a_1 - a_3} b_3^2 \right] \partial_x$$

$$+ \left[2a_2 b_1 b_3 + \frac{a_2 - a_3}{a_1 - a_3} a_3 b_1^2 + \frac{a_1 - a_2}{a_1 - a_3} a_1 b_3^2 \right] \partial_y$$

$$+ \frac{b_1 b_3}{a_1 - a_3} \left[(a_2 - a_3)b_1 + (a_1 - a_2)b_3 \right].$$

2.7. The number of derivatives in the ideal generated by the leading derivatives of I is $\frac{1}{2}(n-4)(n-5) + 2n - 7 = \frac{1}{2}n^2 - \frac{5}{2}n + 3$. Thus $H_I = 6n - 9$ and $d_I = (1, 6)$.

2.8. There is a single coherence condition
$$a_{1,y} - a_{2,y}c - a_{3,y}d - (b_{1,x} - b_{2,x}c - b_{3,x}d) + c_x b_2 - c_y a_2 + d_x b_3 - d_y a_3 = 0.$$

2.9. In order to determine the coherence conditions for the ideal $\mathbb{J}_{xxx}$ open the interactive user interface on the ALLTYPES website and define

```
L1:=Df(z,x,3)+A_1*Df(z,x,y,2)+A_2*Df(z,y,3)
        +A_3*Df(z,x,2)+A_4*Df(z,x,y)+A_5*Df(z,y,2)
        +A_6*Df(z,x)+A_7*Df(z,y)+A_8*z;
L2:=Df(z,x,2,y)+B_1*Df(z,x,y,2)+B_2*Df(z,y,3)
        +B_3*Df(z,x,2)+B_4*Df(z,x,y)+B_5*Df(z,y,2)
        +B_6*Df(z,x)+B_7*Df(z,y)+B_8*z;
T==|LDFMOD(DFRATF(Q,A_1,A_2,A_3,A_4,A_5,A_6,A_7,A_8,
        B_1,B_2,B_3,B_4,B_5,B_6,B_7,B_8, {x,y},GRLEX),
        {z},{x,y},GRLEX)|;
```

Then the coherence conditions given in Lemma 2.2 are displayed by submitting
```
cs:=IntegrabilityConditions({L1,L2}|T|);
```
A Janet basis representation of this system of conditions in various term orderings may be obtained by defining for example
```
Tp==|DFPOLID(Q,{B_8,B_7,B_6,B_5,B_4,B_3,B_2,B_1,
        A_8,A_7,A_6,A_5,A_4,A_3,A_2,A_1},{x,y},LEX)|;
jb:=JanetBasis(cs|Tp|);
```
The result comprises two alternatives; in either case there are two constraints for the $A's$, and a system from which the $B's$ may be determined. You may experiment with varying term orders and try to obtain a system of constraints as simple as possible. The calculations for the ideal $\mathbb{J}_{xxy}$ are similar.

2.10. The two third-order operators must be divisible by $\partial_x + a_i \partial_y + b_i, i = 1, 2$. The conditions that this division be exact yields two linear algebraic systems for the p_i and the q_i. The result for two highest coefficients is

$$p_1 = -(a_1^2 + a_1 a_2 + a_2^2), \quad p_2 = -a_1 a_2 (a_1 + a_2), \quad q_1 = a_1 + a_2, \quad q_2 = a_1 a_2. \quad \text{(A.2)}$$

The expressions for the remaining coefficients are too large to be given here.

2.11. Continuing the calculation given in the proof of Theorem 2.3 the following generator is obtained.

$$\partial_{xy} + a_1 \partial_{yy} + b_2 \partial_x + (a_{1,y} + b_1 + a_1 b_2) \partial_y + b_{2,x} + (a_1 b_2)_y + b_1 b_2.$$

2.12. The calculation is similar as for Exercise 2.10; the result is $p_1 = -a_1$, $p_2 = b_2, q_1 = a_1$ and $q_2 = 0$.

Chapter 3

3.1. The coefficients a_1, a_2 and a_3 have to satisfy

$$\begin{aligned}
a_1 z_{1,x} + a_2 z_{1,y} + a_3 z_1 &= -z_{1,xx}, \\
a_1 z_{2,x} + a_2 z_{2,y} + a_3 z_2 &= -z_{2,xx}, \quad \text{(A.3)} \\
a_1 z_{3,x} + a_2 z_{3,y} + a_3 z_3 &= -z_{3,xx}.
\end{aligned}$$

The systems for the coefficients b_i and c_i are obtained by replacing the second-order derivatives $z_{i,xx}$ by $z_{i,xy}$ or $z_{i,yy}$ respectively. Solving system (A.3) yields for the a_i

$$a_1 = -\frac{1}{w} \begin{vmatrix} z_{1,xx} & z_{1,y} & z_1 \\ z_{2,xx} & z_{2,y} & z_2 \\ z_{3,xx} & z_{3,y} & z_3 \end{vmatrix}, \quad a_2 = \frac{1}{w} \begin{vmatrix} z_{1,x} & z_{1,xx} & z_1 \\ z_{2,x} & z_{2,xx} & z_2 \\ z_{3,x} & z_{3,xx} & z_3 \end{vmatrix},$$

$$a_3 = -\frac{1}{w} \begin{vmatrix} z_{1,x} & z_{1,y} & z_{1,xx} \\ z_{2,x} & z_{2,y} & z_{2,xx} \\ z_{3,x} & z_{3,y} & z_{3,xx} \end{vmatrix} \quad \text{where} \quad w = \begin{vmatrix} z_{1,x} & z_{1,y} & z_1 \\ z_{2,x} & z_{2,y} & z_2 \\ z_{3,x} & z_{3,y} & z_3 \end{vmatrix}.$$

For the b_i and c_i there follows

$$b_1 = -\frac{1}{w}\begin{vmatrix} z_{1,xy} & z_{1,y} & z_1 \\ z_{2,xy} & z_{2,y} & z_2 \\ z_{3,xy} & z_{3,y} & z_3 \end{vmatrix}, \quad b_2 = \frac{1}{w}\begin{vmatrix} z_{1,x} & z_{1,xy} & z_1 \\ z_{2,x} & z_{2,xy} & z_2 \\ z_{3,x} & z_{3,xy} & z_3 \end{vmatrix}, \quad b_3 = -\frac{1}{w}\begin{vmatrix} z_{1,x} & z_{1,y} & z_{1,xy} \\ z_{2,x} & z_{2,y} & z_{2,xy} \\ z_{3,x} & z_{3,y} & z_{3,xy} \end{vmatrix},$$

$$c_1 = -\frac{1}{w}\begin{vmatrix} z_{1,yy} & z_{1,y} & z_1 \\ z_{2,yy} & z_{2,y} & z_2 \\ z_{3,yy} & z_{3,y} & z_3 \end{vmatrix}, \quad c_2 = \frac{1}{w}\begin{vmatrix} z_{1,x} & z_{1,yy} & z_1 \\ z_{2,x} & z_{2,yy} & z_2 \\ z_{3,x} & z_{3,yy} & z_3 \end{vmatrix}, \quad c_3 = -\frac{1}{w}\begin{vmatrix} z_{1,x} & z_{1,y} & z_{1,yy} \\ z_{2,x} & z_{2,y} & z_{2,yy} \\ z_{3,x} & z_{3,y} & z_{3,yy} \end{vmatrix}.$$

3.2. System (3.4) has to satisfy the coherence conditions (B.11) given in Appendix B. Proper substitution of the variables $A_1, \ldots, B_3$ by the coefficients of (3.4) yields the following conditions.

$$B_{1,yy} - 2B_{2,y} - A_{1,x} - ((A_1 B_1)_y = 0,$$

$$B_{2,yy} + (B_{2,y} - A_2 B_1)A_1 - A_{2,x} - (B_{1,y} - A_1 B_1)A_2 = 0.$$

Upon reduction w.r.t. the integrability conditions for the ideal of type $\mathbb{J}_1^{(0,3)}$ in Proposition 2.1 they vanish. The proof for system (3.5) is similar. This example shows the importance of knowing the coherence conditions for any system of pde's; without knowing them such a system is meaningless.

3.3. Depending on the divisor I_1 the left factor M may be of type $\mathrm{M}_1^{(0,1)}$ or $\mathrm{M}_2^{(0,1)}$ respectively.

3.4. Define $z_1 \equiv z_x + az$ and $z_2 \equiv z_y + bz$. For the type $\mathscr{L}_{yy,x}^1$ decomposition the syzygy and the quotient yield the system

$$z_{1,y} - z_{2,x} = 0, \quad z_{2,y} + (A_1 - b)z_2 = 0, \quad z_1 + B_1 z_2 = 0.$$

In $grlex$, $z_1 \succ z_2$, $x \succ y$ term order the Janet basis for the exact quotient module is obtained in the form

$$z_{2,x} + (B_{1,y} - A_1 B_1 + B_1 b)z_2 = 0, \quad z_{2,y} + (A_1 - b)z_2 = 0, \quad z_1 + B_1 z_2 = 0.$$

With the notation of Proposition 3.2 the decomposition may be written as

$$I = \left\langle \begin{pmatrix} 0 & \partial_x + B_{2,y} + A_1 B_2 - B_2 \\ 0 & \partial_y + A_1 - r \\ 1 & B_1 \end{pmatrix} \begin{pmatrix} \partial_x + B_2 - B_1 r \\ \partial_y + r \end{pmatrix} \right\rangle.$$

For the type $\mathscr{L}_{xx,y}^1$ decomposition a similar calculation yields

$$z_{1,x} + (A_1 - a)z_1 = 0, \quad z_{1,y} = 0, \quad z_2 = 0$$

and the decomposition

$$I = \left\langle \begin{pmatrix} \partial_x + A_1 - r & 0 \\ \partial_y & 0 \\ 0 & 1 \end{pmatrix} \begin{pmatrix} \partial_x + r \\ \partial_y + B \end{pmatrix} \right\rangle.$$

3.5. The system (3.6) on page 70 for this ideal is

$$a_x - a^2 + \tfrac{4}{x}a - \tfrac{2}{x^2} = 0, \quad a_y - ab + \tfrac{1}{x}b = 0,$$
$$b_x - ab + \tfrac{1}{x}b = 0, \quad b_y - b^2 - \tfrac{x}{y^2}a + \tfrac{1}{y}b + \tfrac{2}{y^2} = 0.$$

The first equation is a Riccati ode for the x-dependence of a with the general solution $a = \tfrac{1}{x}\tfrac{Cx-2}{Cx-1}$. Two special solutions for $C \to \infty$ and $C = 0$ are $a = \tfrac{1}{x}$ and $a = \tfrac{2}{x}$ respectively. In the latter case the only choice for b is $b = 0$. In the former the equations $b_x = 0$ and $b_y - b^2 + \tfrac{1}{y}b + \tfrac{1}{y} = 0$ for b are obtained. Two special solutions are $b = \pm\tfrac{1}{y}$. Thus there are three divisors

$$l_1 = \langle\langle \partial_x + \tfrac{1}{x}, \partial_y + \tfrac{1}{y} \rangle\rangle, \quad l_2 = \langle\langle \partial_x + \tfrac{1}{x}, \partial_y - \tfrac{1}{y} \rangle\rangle, \quad l_3 = \langle\langle \partial_x + \tfrac{2}{x}, \partial_y \rangle\rangle;$$

the given ideal may be represented as $Lclm(l_1, l_2, l_3)$.

3.6. The system to be solved is $z_{yyy} + \tfrac{3}{x}z_{yy} = 0, z_x + \tfrac{y}{x}z_y = 0$. Two first-order right divisors determined in Example 3.9 are $\langle\langle \partial_x, \partial_y \rangle\rangle$ and $\langle\langle \partial_x + \tfrac{1}{x}, \partial_y - \tfrac{1}{y} \rangle\rangle$; they yield the basis elements 1 and $\tfrac{y}{x}$. The exact quotient module generates the system

$$z_{1,x} + \left(\tfrac{2}{x} - \tfrac{3y}{x^2}\right)z_1 = 0, \quad z_{1,y} + \tfrac{3}{x}z_1 = 0, \quad z_2 = 0.$$

The special solution $z_1 = \tfrac{1}{x^2}\exp\left(-\tfrac{3y}{x}\right)$, $z_2 = 0$ leads to the inhomogeneous system $z_{yy} = \tfrac{1}{x^2}\exp\left(-\tfrac{3y}{x}\right)$, $z_x + \tfrac{y}{x}z_y = 0$; its special solution $\exp\left(-\tfrac{3y}{x}\right)$ yields the third basis element, i.e. a fundamental system is $\{1, \tfrac{y}{x}, \exp\left(-\tfrac{3y}{x}\right)\}$.

Chapter 4

4.1. From (4.4) the two equations $a_i^2 - A_1 a_i + A_2 = 0$ for $i = 1, 2$ are obtained; they yield $A_1 = a_1 + a_2$ and $A_2 = a_1 a_2$. Equation 4.5 leads to

$$a_{i,x} + (A_1 - a_i)a_{i,y} + A_3 a_i + (A_1 - 2a_i)b_i = A_4 \quad \text{for} \quad i = 1, 2.$$

These equations may be solved for A_3 and A_4 with the result

$$A_3 = b_1 + b_2 - \frac{(a_1 - a_2)_x}{a_1 - a_2} - \frac{a_{1,y}a_2 - a_{2,y}a_1}{a_1 - a_2},$$

$$A_4 = a_1 b_2 - a_2 b_1 + a_1 a_2 \frac{b_1 - b_2}{a_1 - a_2} - \frac{a_{1,x}a_2 - a_{2,x}a_1}{a_1 - a_2} - \frac{a_{1,y}a_1^2 - a_{2,y}a_1^2}{a_1 - a_2}.$$

Substituting these values into (4.6), e.g. for $i = 1$, there follows

$$A_5 = b_1 b_2 - b_2 \frac{(a_1 - a_2)_x}{a_1 - a_2} - b_2 \frac{a_{1,y}a_2 - a_{2,y}a_1}{a_1 - a_2} + b_{2,x} + a_1 b_{2,y}.$$

In addition there is a constraint for solvability of the linear system for $A_1, \ldots, A_5$. It is most easily obtained by taking the difference of the two equations obtained from (4.6) for $i = 1, 2$ and substituting $A_1, \ldots, A_4$ obtained above. It leads to the constraint

$$\frac{(b_1 - b_2)_x}{a_1 - a_2} - \frac{b_1 - b_1}{a_1 - a_2}\frac{(a_1 - a_2)_x}{a_1 - a_2} = \frac{(a_1 b_2 - a_2 b_1)_y}{a_1 - a_2} - \frac{a_1 b_2 - a_2 b_1}{a_1 - a_2}\frac{(a_1 - a_2)_y}{a_1 - a_2}.$$

By means of a few simple manipulations it may be shown that it is identical to the condition of case (i) in Theorem 2.2, i.e. it proves Lemma 2.6 for this special case.

4.2. Substituting $z = f_0 F(y) + f_1 F'(y) + g_0 G(x) + g_1 G'(x)$ into the equation $z_{xy} + A_1 z_x + A_2 z_y + A_3 z = 0$ yields an expression that is linear and homogeneous in F, G and its derivatives. It vanishes if all coefficients vanish. This leads to the following system of equations involving f_0 and f_1.

$$f_{1,x} + A_2 f_1 = 0,$$

$$f_{0,x} + f_{1,xy} + A_1 f_{1,x} + A_2(f_0 + f_{1,y}) + A_3 f_1 = 0, \qquad (A.4)$$

$$f_{0,xy} + A_1 f_{0,x} + A_2 f_{0,y} + A_3 f_0 = 0.$$

There is a similar system involving g_0 and g_1.

$$g_{1,y} + A_1 g_1 = 0,$$

$$g_{0,y} + g_{1,xy} + A_1(g_0 + g_{1,x}) + A_2 g_{1,y} + A_3 g_1 = 0, \qquad (A.5)$$

$$g_{0,xy} + A_1 g_{0,x} + A_2 g_{0,y} + A_3 g_0 = 0.$$

System (A.4) has the solution $A_2 = -\dfrac{f_{1,x}}{f_1}$,

$$A_1 = \frac{f_0}{f_1} + \frac{1}{\begin{vmatrix} f_0 & f_1 \\ f_{0,x} & f_{1,x} \end{vmatrix}} \left(\frac{f_{1,x}}{f_1} \begin{vmatrix} f_0 & f_1 \\ f_{0,y} & f_{1,y} \end{vmatrix} - \begin{vmatrix} f_0 & f_1 \\ f_{0,xy} & f_{1,xy} \end{vmatrix} \right),$$

$$A_3 = -\frac{f_{0,x}}{f_1} + \frac{1}{\begin{vmatrix} f_0 & f_1 \\ f_{0,x} & f_{1,x} \end{vmatrix}} \left(\frac{f_{1,x}}{f_1} \begin{vmatrix} f_{0,x} & f_{1,x} \\ f_{0,y} & f_{1,y} \end{vmatrix} - \begin{vmatrix} f_{0,x} & f_{1,x} \\ f_{0,xy} & f_{1,xy} \end{vmatrix} \right).$$

If these values for A_1, A_2 and A_3 are substituted into (A.5), constraints between the f_0, f_1, g_0 and g_1 are obtained.

4.3. If the single integrability condition $\partial_x L = \partial_y l_{2,xx}$ is reduced w.r.t. L and $l_{2,xx}$, the following system of equations is obtained.

$$a_1 A_1 A_2 + a_1 A_{2,y} - a_1 A_3 + A_{1,x} A_2 + A_1 A_{2,x} - 2 A_1 A_2^2$$
$$+ A_{2,xy} - 2 A_{2,y} A_2 + 2 A_2 A_3 - A_{3,x} = 0,$$
$$a_{1,y} - A_{1,x} + A_1 A_2 - A_3 = 0, \quad a_2 - a_1 A_2 - A_{2,x} + A_2^2 = 0.$$

A few elimination steps yield

$$a_1 = 2 A_2 - \log{(A_{2,y} + A_1 A_2 - A_3)}_x,$$
$$a_2 = A_{2,x} + A_2^2 - A_2 \log{(A_{2,y} + A_1 A_2 - A_3)}_x$$

and the condition

$$2 A_{2,y} - \log{(A_{2,y} + A_1 A_2 - A_3)}_{xy} - A_{1,x} + A_1 A_2 - A_3 = 0.$$

Performing the same reduction steps to $L - z_1$ and $l_{2,xx} - z_2$, the relation

$$z_{1,x} + (a_1 - A_2)z_1 - z_{2,y} - A_1 z_2 = 0$$

follows, i.e. there is a single syzygy $(\partial_x + a_1 - A_2, -\partial_y - A_1)$. The exact quotient module is generated by $((1,0), (\partial_x + a_1 - A_2, -\partial_y - A_1))$; the corresponding Janet basis is $((1,0), (0, \partial_y + A_1))$.

4.4. Define $\Delta \equiv \sqrt{A_1^2 - 4 A_2}$. The two roots of (4.4) are $a_{1,2} = \frac{1}{2}A_1 \pm \frac{1}{2}\Delta$. If $\Delta \neq 0$ there follows $b_{1,2} = \frac{1}{2}A_3 \pm \frac{1}{2\Delta}(A_1 A_3 - 2 A_4)$. The constraint (4.6) is $A_5(A_1^2 - 4 A_2) + (A_2 A_3 - A_1 A_4)A_3 + A_4^2 = 0$. If it is satisfied there are two factors

$$l_{1,2} = \partial_x + \frac{1}{2}(A_1 \pm \Delta)\partial_y + \frac{1}{2}A_3 \pm \frac{1}{2\Delta}(A_1 A_3 - 2 A_4).$$

If $\Delta = 0$ there follows $a_1 = a_2 = \frac{1}{2}A_1$; condition (4.2) is $A_4 - \frac{1}{2}A_1 A_3 = 0$. If it is satisfied, b has to be determined from $b_x + \frac{1}{2}A_1 b_y - b^2 + A_3 b - A_5 = 0$. Applying the results of Exercise B.2, the following factorizations are obtained. If $A_3^2 - 4 A_5 \neq 0$ there are the two factors

$$l_{1,2} = \partial_x + \frac{1}{2}A_1 \partial_y + \frac{1}{2}A_3 \pm \frac{1}{2}(4 A_5 - A_3^2).$$

Finally, if $A_3^2 - 4A_5 = 0$ there is the right factor

$$l = \partial_x + \tfrac{1}{2}A_1\partial_y + \tfrac{1}{2}A_3 - \frac{1}{F(y - \tfrac{1}{2}xA_1) + x}.$$

4.5. By Proposition 4.2 and Corollary 4.2 the type $\mathcal{L}_{xy}^3$ decomposition $L = Lclm(\partial_x + y, \partial_y + x + \tfrac{1}{y})$ is obtained.

Chapter 5

5.1. The given equation factorizes according to

$$(\partial_x - x\partial_y - 1)(\partial_x + (y^2 - x)\partial_y + 1)z = 0,$$

i.e. the order of the factors of the equation in Example 5.2 is reversed. Because the right factor equation now does not allow a Liouvillian solution, the full second order equation does not have one either. The left-factor equation $w_x + xw_y - w = 0$ has the solution $w = e^x F(y + \tfrac{1}{2}x^2)$. Therefore a second element of a fundamental system is determined by

$$z_x + (y^2 - x)z_y + z = e^x F(y + \tfrac{1}{2}x^2).$$

5.2. The same steps as in Example 5.18 are performed. The inhomogeneity has to satisfy $r_y + \frac{2}{x-y}r = y$; its general solution is

$$r = C(x - y)^2 + (x - y)^2 \log(x - y) + (x - y)y.$$

Choosing $C = 0$ the equation

$$z_{xx} - \frac{2}{x-y}z_x + \frac{2}{(x-y)^2}z = (x-y)^2 \log(x-y) + (x-y)y$$

has to be solved. Substituting its general solution

$$z = C_1(x-y) + C_2 x(x-y) + \tfrac{1}{6}(x-y)^2 \log(x-y) - \tfrac{1}{36}x(x-y)(5x^2 - 33xy + 6y^2)$$

into the given equation leads to the constraint $C_{1,y} + yC_{2,y} - C_2 + \tfrac{1}{3}y^2 = 0$. A special solution is $C_1 = 0$, $C_2 = -\tfrac{1}{3}y^2$, it yields

$$z_0(x, y) = \tfrac{1}{6}(x-y)^4 \log(x-y) - \tfrac{1}{36}x(x-y)(x-6y)(5x-3y).$$

5.3. The difference of any two special solutions is a solution of the corresponding homogeneous equation. For the three alternatives in Example 5.19 these differences are

$$x^2, \quad \frac{x^2 y^4 + 2xy^2 - 1}{y^3}, \quad \frac{x^2 y^3(y-1) + 2xy^2 - 1}{y^3}.$$

They satisfy the homogeneous equation $z_{xy} + xyz_x - 2yz = 0$. The general solution of this equation corresponding to the Laplace divisor has been determined in Example 5.10, Eq. (5.3). It must be possible to obtain the above differences by suitable choices of the undetermined function $F(y)$. To this end, they may be considered as inhomogeneities of a linear ordinary differential equation for F. For the first alternative this leads to

$$F'' + \frac{2xy^2 - 1}{y}F' + x^2y^2F = x^2y^2;$$

its special solution $F_0 = 1$ is the desired result. For the remaining two cases the result is $F_0 = y$ and $F_0 = y - 1$ respectively.

5.4. The equation $z_{xy} + xyz_x - 2yz = \frac{x^2 + 1}{y}$ has the special solution

$$z_0(x, y) = (y^4 - 4xy^2 + 8)\frac{1}{y^6}\log x - \frac{1}{y^6}\left(\tfrac{3}{2}x^2y^4 - 12\right)$$
$$- \frac{16}{y^6}\exp\left(-\tfrac{1}{2}xy^2\right)\int \exp\left(\tfrac{1}{2}xy^2\right)\frac{dy}{y}.$$

The equation $z_{xy} + xyz_x - 2yz = \frac{1}{xy}$ has the special solution

$$z_0(x, y) = \left(\tfrac{1}{8}x^2y^4 + xy^2 + 1\right)\exp\left(\tfrac{1}{2}xy^2\right)\int \exp\left(-\tfrac{1}{2}xy^2\right)\int \exp\left(\tfrac{1}{2}xy^2\right)\frac{dy}{y}\frac{dx}{x}$$
$$+\left(\tfrac{1}{4}xy^2 + \tfrac{3}{2}\right)\int \exp\left(\tfrac{1}{2}xy^2\right)\frac{dy}{y} + \tfrac{1}{8}(xy^2 + 7)\exp\left(\tfrac{1}{2}xy^2\right).$$

5.5. In general, the coefficients $A_1, \ldots, B_5$ of

$$I \equiv \langle \partial_{xx} + A_1\partial_{yy} + A_2\partial_x + A_3\partial_y + A_4, \partial_{xy} + B_1\partial_{yy} + B_2\partial_x + B_3\partial_y + B_4 \rangle$$

generated by a Janet basis in $grlex$, $x \succ y$ term order are related by $A_1 + B_1^2 = 0$ and

$$A_{4,y} - B_{4,x} + B_{4,y}B_1 - A_2B_4 + A_4B_2 - B_1B_2B_4 + B_3B_4 = 0,$$
$$A_{3,y} - B_{3,x} + B_{3,y}B_1 - A_2B_3 + A_3B_2 + A_4 - B_1B_2B_3 + B_1B_4 + B_3^2 = 0,$$
$$A_{2,y} - B_{2,x} + B_{2,y}B_1 - B_1B_2^2 + B_2B_3 - B_4 = 0,$$
$$A_{1,y} - B_{1,x} + B_{1,y}B_1 + A_1B_2 - A_2B_1 + A_3 - B_1^2B_2 + 2B_1B_3 = 0.$$

If $A_1, \ldots, A_4$ are given, this system determines $B_1, \ldots, B_4$ such that the integrability conditions are satisfied. For the special values $A_1 = 0$, $A_2 = -1$, $A_3 = -\frac{2}{x}$ and $A_4 = A_5 = 0$ the solutions are $B_1 \pm 1$, $B_2 = \pm\frac{1}{x}$ and $B_3 = B_4 = 0$. They yield the two ideals

$$I_1 \equiv \langle\!\langle \partial_{xx} - \partial_{xy} - \tfrac{2}{x}\partial_x, \partial_{xy} + \partial_{yy} + \tfrac{1}{x}\partial_x \rangle\!\rangle \quad \text{and}$$

$$I_2 \equiv \langle\!\langle \partial_{xx} - \partial_{xy} - \tfrac{2}{x}\partial_x, \partial_{xy} - \partial_{yy} - \tfrac{1}{x}\partial_x \rangle\!\rangle$$

such that $I = I_1 \cap I_2$; I_1 and I_2 correspond to the components involving F and G respectively in the solution $Lz = 0$ (compare Example 5.23).

5.6. The equation may be represented as $Lclm(\partial_x - \tfrac{1}{x}\partial_y, \partial_x - y\partial_y)z = 0$. This yields $z_1(x, y) = F(xe^y)$ and $z_2(x, y) = G(ye^x)$; F and G are undetermined functions.

5.7. The homogeneous part of the equation allows a type $\mathcal{L}_{xx}^1$ factorization, it yields the representation

$$\left(\partial_x - \frac{y}{x}\partial_y - \frac{1}{x}\right)(\partial_x + \partial_y)z = -xy.$$

Applying Proposition 5.1, case (i), the differential fundamental system

$$z_1(x, y) = F(y - x), \quad z_2(x, y) = \int xG(xy)\big|_{y=\bar{y}+x}dx\big|_{\bar{y}=y-x}$$

is obtained; F and G are undetermined functions. Proposition 5.4, case (i), yields the special solution

$$z_0(x, y) = \tfrac{1}{12}x^3\left(x \log x - 4y \log x - \tfrac{7}{12}x + \tfrac{4}{3}y\right).$$

5.8. The equation does not have any first-order factor. Introducing new variables by $u = x + iy$ and $v = x - iy$ the equation $w_{uv} + \dfrac{2w}{(1 + uv)^2} = 0$ for $w(u, v) \equiv f(x, y)$ is obtained. The corresponding operator has the two Laplace divisors

$$\left\langle\!\left\langle \partial_{uv} + \frac{2}{(1 + uv)^2}, \partial_{uu} + \frac{2v}{1 + uv}\partial_u \right\rangle\!\right\rangle \quad \text{and} \quad \left\langle\!\left\langle \partial_{uv} + \frac{2}{(1 + uv)^2}, \partial_{vv} + \frac{2u}{1 + uv}\partial_v \right\rangle\!\right\rangle.$$

This type $\mathcal{L}_{uv}^4$ decomposition yields the solutions

$$w_1(u, v) = \frac{2v}{1 + uv}f(u) - f'(u) \quad \text{and} \quad w_2(u, v) = \frac{2u}{1 + uv}g(v) - g'(v);$$

f and g are undetermined functions. Upon substitution of the variables u and v the corresponding expressions for $z_1(x, y)$ and $z_2(x, y)$ are obtained.

5.9. Dividing out the factor ∂_y leads to the representation $L = \left(\partial_x - \frac{1}{x + y}\right)\partial_y$. The left-factor equation $w_x - \frac{1}{x + y}w = 0$ has the solution $w(x, y) = (x + y)G(y)$ where $G(y)$ is an undetermined function of y. The inhomogeneous right-factor equation $z_y = (x + y)G(y)$ yields the second element of a fundamental system $z_2(x, y) = \int yG(y)dy + x\int G(y)dy$. The identifications $C_1(y) = \int yG(y)dy$ and $C_2(y) = \int G(y)dy$ show that it is identical to $z_2(x, y)$ obtained in Example 5.11 as it was to be expected. However, the representation of $z_2(x, y)$ obtained from the single factor ∂_y is unnecessarily complicated due to the occurrence of integrals. In general, it is not obvious how to simplify a solution obtained from an incomplete factorization.

5.10. The Loewy decomposition is

$$Lz = \begin{pmatrix} 1 & 0 \\ 0 & \partial_y - \frac{2}{y} \end{pmatrix} \begin{pmatrix} w_1 \equiv z_{xy} - \frac{2}{y}z_x - yz_y \\ w_2 \equiv z_{xx} - 2yz_x + y^2z \end{pmatrix}.$$

The equation $z_{xx}-2yz_x+y^2z = 0$ has the solution $z = C_1\exp(xy)+C_2x\exp(xy)$; C_1 and C_2 are undetermined functions of y. Substitution into $Lz = 0$ leads to $yC_2' - yC_1 - 2C_2 = 0$, i.e. $C_1 = C_2' - \frac{2}{y}C_2$. This yields

$$z_1(x, y) = \exp(xy)\big(F'(y) + (x - \frac{2}{y})F(y)\big)$$

where $F \equiv C_2$. The equation $w_{2,y} - \frac{2}{y}w_2 = 0$ has the solution $w_2 = G(x)y^2$; G is an undetermined function of x. A special solution of the inhomogeneous equation $z_{xx} - 2yz_x + y^2z = Gy^2$ yields the second member of a fundamental system

$$z_2(x, y) = xy^2\exp(xy)\int\exp(-xy)G(x)dx-y^2\exp(xy)\int x\exp(-xy)G(x)dx.$$

5.11. The Loewy decomposition is

$$Lz = \begin{pmatrix} 1 & 0 \\ 0 & \partial_y \end{pmatrix} \begin{pmatrix} w_1 \equiv z_{xy} + (xy - 1)z_y + 2xz \\ w_2 \equiv z_{xx} + (2xy - 2 - \frac{1}{x})z_x + (x^2y^2 - 2xy + 1 + \frac{1}{x})z \end{pmatrix}.$$

Proceeding similar as in the preceding problem the fundamental system

$$z_1(x, y) = \exp\left(x - \tfrac{1}{2}x^2y\right)\big(2F'(y) - x^2F(y)\big),$$

$$z_2(x, y) = \exp\left(x - \tfrac{1}{2}x^2y\right)\bigg(x^2\int\exp\left(\tfrac{1}{2}x^2y - x\right)G(x)\frac{dx}{x} - \int\exp\left(\tfrac{1}{2}x^2y - x\right)G(x)xdx\bigg)$$

is obtained. F and G are undetermined functions.

5.12. The operator L is completely reducible, its Loewy decomposition of type $\mathscr{L}^4_{xy}$ is

$$L = Lclm\big(\langle\!\langle L, \partial_{xx}\rangle\!\rangle, \langle\!\langle L, \partial_{yy} + \tfrac{2}{y}\partial_y\rangle\!\rangle\big).$$

It yields the fundamental system

$$z_1(x, y) = \frac{xy + 1}{y}F'(y) - \frac{2x}{y}F(y), \quad z_2(x, y) = \frac{xy + 1}{xy}G'(x) - \frac{3xy + 1}{x^2 y}G(x);$$

F and G are undetermined functions.

5.13. Substituting $z = \varphi(x, y)w$ into the equation $z_{xx} + Az_x + Bz_y + Cz = 0$ yields

$$w_{xx} + \left(2\frac{\varphi_x}{\varphi} + A\right)w_x + Bw_y + \left(\frac{\varphi_{xx}}{\varphi} + A\frac{\varphi_x}{\varphi}\right) + B\frac{\varphi_y}{\varphi} + Cw = 0.$$

The coefficient of w_x vanishes if $\frac{\varphi_x}{\varphi} = \frac{1}{2}A$, i.e. if $\varphi = \exp\left(-\frac{1}{2}\int A\,dx\right)$. Substituting it into the coefficient of w yields the result of case (i).

In case (ii), the same substitution into the equation $z_{xy} + Az_x + Bz_y + Cz = 0$ yields

$$w_{xy} + \left(A + \frac{\varphi_y}{\varphi}\right)w_x + \left(B + \frac{\varphi_x}{\varphi}\right)w_y + \left(C + B\frac{\varphi_y}{\varphi} + A\frac{\varphi_x}{\varphi} + \frac{\varphi_{xy}}{\varphi}\right)w = 0.$$

The derivative w_x does not occur in this equation if φ is a solution of the equation $\varphi_y + A\varphi = 0$; a special solution is $\varphi = \exp\left(-\int A\,dy\right)$. Substituting this expression into the above equation for w yields the second case of Corollary 5.4.

In order that the derivative w_y vanishes as well, φ has to obey the system $\varphi_x + B\varphi = 0$, $\varphi_y + A\varphi = 0$. In order that a nontrivial solution exists, the condition $A_x = B_y$ must be valid. Then $w_{xy} + (C - B_y - AB)w = 0$.

5.14. Substituting $z = \varphi w$ into the given equation yields

$$w_{xx} + A_1 w_{xy} + A_2 w_{yy}$$

$$+\left(2\frac{\varphi_x}{\varphi} + A_1\frac{\varphi_y}{\varphi} + A_3\right)w_x + \left(A_1\frac{\varphi_x}{\varphi} + 2A_2\frac{\varphi_y}{\varphi} + A_4\right)w_y$$

$$+\left(\frac{\varphi_{xx}}{\varphi} + A_1\frac{\varphi_{xy}}{\varphi} + A_2\frac{\varphi_{yy}}{\varphi} + A_3\frac{\varphi_x}{\varphi} + A_4\frac{\varphi_y}{\varphi} + A_5\right)w = 0.$$

The two first-order derivatives disappear if

$$2\frac{\varphi_x}{\varphi} + A_1\frac{\varphi_y}{\varphi} + A_3 = 0, \quad A_1\frac{\varphi_x}{\varphi} + 2A_2\frac{\varphi_y}{\varphi} + A_4 = 0;$$

if $A_1^2 - 4A_2 \neq 0$ this algebraic system has the solution

$$\frac{\varphi_x}{\varphi} = \frac{A_1 A_4 - 2A_2 A_3}{A_1^2 - 2A_2} \equiv p \quad \text{and} \quad \frac{\varphi_y}{\varphi} = \frac{A_1 A_3 - 4A_2}{A_1^2 - 4A_2} \equiv q;$$

its integrability condition is $p_y = q_x$. If it is satisfied, integration yields the solution given in case (i).

If $A_1^2 - 4A_2 = 0$, the above algebraic system is consistent if $2A_4 - A_1 A_3 = 0$ and the equation

$$\varphi_x + \tfrac{1}{2}A_2\varphi_y + \tfrac{1}{2}A_3\varphi = 0$$

for φ follows. According to Corollary B.1 its solution is

$$\varphi(x, y) = \Phi(\bar{y}) \exp\left(-\tfrac{1}{2} \int A_3(x, \bar{\psi}(x, \bar{y}))dx \Big|_{\bar{y}=\psi(x,y)} \right);$$

$\psi(x, y)$ is a first integral of $\frac{dy}{dx} = \tfrac{1}{2}A_1$; $\bar{y} \equiv \psi(x, y)$ and $y = \bar{\psi}(x, \bar{y})$. This is case (ii).

5.15. Substitution of $z = \varphi(x, y)w$ into the given equation yields

$$w_{xy} + \left(A_1\frac{\varphi_y}{\varphi}\right)w_x + \left(A_2\frac{\varphi_y}{\varphi}\right)w_y + \left(A_3 + A_2\frac{\varphi_y}{\varphi} + A_1\frac{\varphi_x}{\varphi} + \frac{\varphi_{xy}}{\varphi}\right)w = 0.$$

Solvability of the conditions $\frac{\varphi_x}{\varphi} + A_2 = 0$ and $\frac{\varphi_y}{\varphi} + A_1 = 0$ requires $A_{1,x} = A_{2,y}$. If it is satisfied, the transformed equation is

$$w_{xy} + (A_3 - A_1 A_2 - A_{2,y})w = 0.$$

Obviously its coefficients are in the base field of the originally given equation.

Chapter 6

6.1. The derivatives up to order three are

$$z = F + Ge^{-x}, \quad z_x = -xF' - Ge^{-x}, \quad z_y = F' + G'e^{-x},$$

$$z_{xx} = -F' + x^2 F'' + Ge^{-x}, \quad z_{xy} = -xF'' - G'e^{-x}, \quad z_{yy} = F'' + G''e^{-x},$$

$$z_{xxx} = 3xF'' - x^3 F''' - Ge^{-x}, \quad z_{xxy} = -F'' + x^2 F''' + G'e^{-x},$$

$$z_{xyy} = -xF''' - G''e^{-x}, \quad z_{yyy} = F''' + G'''e^{-x}.$$

The $'$ means the derivative w.r.t. the argument of F or G. These ten equations may be considered as an inhomogeneous linear system for the eight indeterminates $F, G, F', \ldots G'''$; it may be solved by Gauss elimination. The two consistency conditions yield the desired generators of the $Lclm$.

6.2. The answer is obvious from the representations

$$L_1 = \left(\partial_{xx} - x^2\partial_{yy} + 2\partial_x + (2x+3)\partial_y\right)(\partial_x + 1)$$
$$= \left(\partial_{xx} - x\partial_{xy} + 3\partial_x - (x-1)\partial_y + 2\right)(\partial_x + x\partial_y)$$

and

$$L_2 = \left(\partial_{xy} + x\partial_{yy} - \tfrac{1}{x}\partial_x + \left(1 + \tfrac{1}{x}\right)\partial_y\right)(\partial_x + 1)$$
$$= \left(\partial_{xy} - \tfrac{1}{x}\partial_x + \left(1 - \tfrac{1}{x}\right)\partial_y - \tfrac{1}{x}\right)(\partial_x + x\partial_y)$$

6.3. The Loewy decomposition is

$$L = \begin{pmatrix} 1 & 1 \\ 0 & \partial_x - (x-1)\partial_y + 2 \end{pmatrix} \begin{pmatrix} L_1 \\ L_2 \end{pmatrix};$$

L_1 and L_2 are defined in (6.20), i.e. the operator L is in the same ideal as Blumberg's operator (6.17). L may be represented as $L = L_1 + L_2$.

6.4. By Proposition 6.3 there is no first-order right factor. Up to order $m = 10$ a Laplace divisor $\mathbb{L}_{y^n}(L)$ does not exist. However, there is a Laplace divisor $\mathbb{L}_{x^2}(L)$ corresponding to $l_2 \equiv \partial_{xx} + \tfrac{2}{x}\partial_x$; it yields the type $\mathscr{L}_{xyy}^{14}$ Loewy decomposition

$$L = \begin{pmatrix} 1 & 1 \\ 0 & \partial_{xy} - (xy - y)\partial_y - \tfrac{2}{y} \end{pmatrix}$$
$$\begin{pmatrix} \partial_{xyy} + (xy - y)\partial_{xy} + \tfrac{1}{x}\partial_{yy} - \tfrac{2}{y}\partial_x - \tfrac{y}{x}\partial_y - \tfrac{2}{xy} \\ \partial_{xx} + \tfrac{2}{x}\partial_x \end{pmatrix}.$$

Chapter 7

7.1. The solutions $z_1(x, y)$ and $z_2(x, y)$ are easily shown to annihilate the left-hand side of (7.2) by substitution. Let $\varphi = \log(y+1) - x$ and $\psi = \exp(\bar{y} + x) - 1$. Define

$$\bar{H}(x, \bar{y}) \equiv H(y)\big|_{y=\psi(x,\bar{y})} \quad \text{and} \quad z \equiv \int (x-1)\bar{H}(x, \bar{y})dx\big|_{\bar{y}=\varphi(x,y)}.$$

Applying the chain rule, the following expressions for the derivatives of z are obtained.

$$z_x = (x - 1)H(y) + \varphi_x \int (x - 1)\bar{H}_{\bar{y}}(x, \bar{y})dx\Big|_{\bar{y}=\varphi(x,y)},$$

$$z_y = \varphi_y \int (x - 1)\bar{H}_{\bar{y}}(x, \bar{y})dx\Big|_{\bar{y}=\varphi(x,y)},$$

$$z_{xx} = H(y) + \varphi_{xx} \int (x - 1)\bar{H}_{\bar{y}}(x, \bar{y})dx\Big|_{\bar{y}=\varphi(x,y)}$$

$$+ (x - 1)\varphi_x \bar{H}_{\bar{y}}(x, \bar{y})\Big|_{\bar{y}=\varphi(x,y)} + \varphi_x^2 \int (x - 1)\bar{H}_{\bar{y}\bar{y}}(x, \bar{y})dx\Big|_{\bar{y}=\varphi(x,y)},$$

$$z_{xy} = (x - 1)H_y(y) + \varphi_{xy} \int (x - 1)\bar{H}_{\bar{y}}(x, \bar{y})dx\Big|_{\bar{y}=\varphi(x,y)}$$

$$+ \varphi_x \varphi_y \int (x - 1)\bar{H}_{\bar{y}\bar{y}}(x, \bar{y})dx\Big|_{\bar{y}=\varphi(x,y)}.$$

Substitution into (7.2) yields

$$z = xH(y) + (x - 1)\big(\varphi_x \psi_{\bar{y}}|_{\bar{y}=\varphi(x,y)} + y + 1\big)H_y(y)$$

$$+ \big(\varphi_x^2 + (y + 1)\varphi_x \varphi_y\big) \int (x - 1)\bar{H}_{\bar{y}\bar{y}}(x, \bar{y})dx\Big|_{\bar{y}=\varphi(x,y)}$$

$$+ \big(\varphi_{xx} + (y + 1)\varphi_{xy} + \varphi_x + (y + 1)\varphi_y\big) \int (x - 1)\bar{H}_{\bar{y}}(x, \bar{y})dx\Big|_{\bar{y}=\varphi(x,y)}.$$

If the above expressions for φ and ψ are substituted, the coefficients of H_y and the integrals vanish, i.e. only the first term $xH(y)$ remains; this is the right hand side of (7.2).

7.2. The first-order right factor in the second line of (6.19) yields $z_1 = F\big(y - \frac{1}{2}x^2\big)$. The two first-order factors in the argument of the $Lclm$ yield

$$w_1 = G(y)e^{-x} \quad \text{and} \quad w_2 = H(y)xe^{-x}.$$

They lead to the inhomogeneous equations $z_x + xz_y = w_i$ with special solutions

$$z_2(x, y) = \int G\big(\bar{y} + \tfrac{1}{2}x^2\big)e^{-x}dx\Big|_{\bar{y}=y-\frac{1}{2}x^2} \quad \text{and}$$

$$z_3(x, y) = \int G\big(\bar{y} + \tfrac{1}{2}x^2\big)xe^{-x}dx\Big|_{\bar{y}=y-\frac{1}{2}x^2}$$

respectively. The differential fundamental system obtained in this way is more complicated than the one obtained in Example 7.8.

7.3. Introducing new variables $u \equiv \varphi(x, y) = y$ and $v \equiv \psi(x, y) = y - \frac{1}{2}x^2$ into the left factor $\partial_{xx} + x\partial_{xy} + \partial_x + (x + 2)\partial_y$ yields, according to case (i) of Lemma 5.4,

$$D_{uv} \equiv \partial_{uv} - \frac{\sqrt{2(u - v)} + 2}{2(u - v)}\partial_u - \frac{1}{2(u - v)}\partial_v.$$

According to Proposition 4.2 a first-order right factor does not exist. However, there is the operator $D_{uu} \equiv \partial_{uu} + \frac{1}{2(u-v)}\partial_u$ such that $\langle D_{uv}, D_{uu} \rangle$ is a Laplace divisor generated by a Janet basis; it yields the solution $w_1 = F(v) - \sqrt{2(u-v)}\,F'(v)$.

Up to order 5 there is no Laplace divisor involving the variable v. Therefore the exact quotient module

$$\langle\!\langle (1,0), \left(0, \partial_v - \frac{\sqrt{2(u-v)}+2}{2(u-v)}\right)\rangle\!\rangle$$

is set up. The equation corresponding to the latter operator has the solution $\frac{H(u)}{u-v}\exp\left(-\sqrt{2(u-v)}\right)$. It leads to the inhomogeneous equation

$$w_{2,uu} + \frac{1}{2(u-v)}w_{2,u} = \frac{H(u)}{u-v}\exp\left(-\sqrt{2(u-v)}\right)$$

with the solution

$$w_2 = \sqrt{u-v}\int \frac{H(u)}{\sqrt{u-v}}\exp\left(-\sqrt{2(u-v)}\right)du - \int H(u)\exp\left(-\sqrt{2(u-v)}\right)du.$$

In the original variables the solutions are

$$z_1 = w_1\big|_{u=y,v=y-\frac{1}{2}x^2} = F(y-\tfrac{1}{2}x^2) - xF'(y-\tfrac{1}{2}x^2) \quad\text{and}\quad z_2 = w_2\big|_{u=y,v=y-\frac{1}{2}x^2}.$$

Again, this result is more complicated than that given in Example 7.8.

7.4. The operator

$$L \equiv \partial_{xxx} - x^2\partial_{xyy} + 3\partial_{xx} + (2x+3)\partial_{xy} - x^2\partial_{yy} + 2\partial_x + (2x+3)\partial_y$$

defined by the given equation has the right factors $\partial_x + 1$ and $\partial_x + x\partial_y$, i.e. it is contained in the ideal (6.20) generated by L_1 and L_2. With the notation of Example 6.9 there follows $L = L_1$, i.e. it is contained in the same ideal as Blumberg's example but is a different combination of the two generators L_1 and L_2. Its Loewy decomposition is

$$L = \begin{pmatrix} 1 & 0 \\ 0 & \partial_x - x\partial_y + 2 + \frac{1}{x} \end{pmatrix}\begin{pmatrix} L_1 \\ L_2 \end{pmatrix}.$$

Equations 7.5 yield $w_1 = 0$ and $w_2 = \frac{1}{x}\exp(-2x)F\left(y+\frac{1}{2}x^2\right)$; F is an undetermined function. The system $L_1 z = w_1$ and $L_2 z = w_2$ is solved by Theorem 5.1. The constraint (5.24) is satisfied. The system (5.26) is

$$r_{xy} - \frac{1}{x}r_x + \left(1 + \frac{2}{x}\right)r_y - \frac{1}{x}r = \frac{1}{x}\exp\left(-2x\right)F\left(y + \frac{1}{2}x^2\right),$$

$$r_{yy} - \frac{2}{x}r_y + \frac{1}{x^2}r = \frac{1}{x^2}\exp\left(-2x\right)F\left(y + \frac{1}{2}x^2\right)$$

with the special solution

$$r_0(x, y) = \frac{1}{x^2}\exp\left(\frac{y}{x} - 2x\right)\left(y\int \exp\left(-\frac{y}{x}\right)F\left(y + \frac{1}{2}x^2\right)dy\right.$$

$$\left. - \int y\exp\left(-\frac{y}{x}\right)F\left(y + \frac{1}{2}x^2\right)dy\right).$$

Substitution into (5.25) finally yields

$$z_3(x, y) = \exp\left(x\right)\int r_0(x, y)\exp\left(-x\right)dx - \int r_0(x, y)dx.$$

7.5. In general, a differential type 0 solution may be obtained from a system $z_x + az = 0$, $z_y + bz = 0$, $a_y = b_x$, that reduces the given third order equation to zero. For the problem at hand, this leads to the following constraint for the coefficients a and b.

$$b_{xx} + xb_{xy} - \frac{1}{y+1}a_x + b_x + b_y - \frac{1}{y+1}(a + b) - a_xb - 2b_xa - 2xb_xb$$

$$+ \frac{1}{y+1}a^2 - ab - b^2 - xab_y + xab^2 + a^2b = 0.$$

A solution scheme for this nonlinear second-order equation does not seem to exist. A special solution may be obtained from the requirement that a solution with no dependence on x is searched. It yields the result $z = (y+1)^2$ that cannot be obtained by specialization of F in the solution given in Example 7.19. The corresponding values of a and b are $a = 0$ and $b = -\frac{2}{y+1}$; they obey the above equation.

7.6. The three alternatives may be described as follows.

$$L_1 \equiv (\partial_x + y)(\partial_y - x)(\partial_y - y)$$

$$= \partial_{xyy} - (x + y)\partial_{xy} + y\partial_{yy} + (xy - 1)\partial_x + (xy + y^2 + 1)\partial_y + xy^2;$$

$L_1z = 0$ has fundamental system

$$z_1(x, y) = F(x)\exp\left(\frac{1}{2}y^2\right), \quad z_2(x, y) = G(x)\exp\left(\frac{1}{2}y^2\right)\int \exp\left(xy - \frac{1}{2}y^2\right)dy,$$

$$z_3(x, y) = \exp\left(\frac{1}{2}y^2\right)\left(\int \exp\left(xy - \frac{1}{2}y^2\right)dy \int \exp\left(-2xy\right)H(y)dy\right.$$

$$\left. - \int \exp\left(-2xy\right)H(y)\int \exp\left(xy - \frac{1}{2}y^2\right)dy dy\right).$$

$$L_2 \equiv (\partial_y - x)(\partial_y - y)(\partial_x + y)$$
$$= \partial_{xyy} - (x + y)\partial_{xy} + y\partial_{yy} + (xy - 1)\partial_x + (xy + y^2 - 2)\partial_y + xy^2 - x - 2y;$$

$L_2 z = 0$ has fundamental system

$$z_1(x, y) = F(y)\exp(-xy), \quad z_2(x, y) = \exp\left(\tfrac{1}{2}y^2 - xy\right)\int \exp(xy)G(x)dx,$$

$$z_3(x, y) = \exp(-xy)\int \exp\left(xy + \tfrac{1}{2}y^2\right)H(x)\int \exp\left(xy - \tfrac{1}{2}y^2\right)dydx.$$

$$L_3 \equiv (\partial_y - x)(\partial_x + y)(\partial_y - y)$$
$$= \partial_{xyy} - (x + y)\partial_{xy} + y\partial_{yy} + (xy - 1)\partial_x + (xy + y^2 - 1)\partial_y + xy^2 - 2y;$$

$L_3 z = 0$ has fundamental system

$$z_1(x, y) = F(x)\exp\left(\tfrac{1}{2}y^2\right), \quad z_2(x, y) = \exp\left(\tfrac{1}{2}y^2\right)\int \exp\left(-xy - \tfrac{1}{2}y^2\right)G(y)dy,$$

$$z_3(x, y) = \exp\left(\tfrac{1}{2}y^2\right)\int \exp\left(-xy - \tfrac{1}{2}y^2\right)\int \exp(2xy)H(x)dxdy.$$

Despite the fact that all six alternatives considered in Examples 7.21–7.23 have identical symbols, the structure of the solutions differs significantly. There are three different elements $z_1(x, y)$ corresponding to the three possible rightmost factors; hence, there are three pairs of permutations with identical elements $z_1(x, y)$. There is not a single pair with identical elements $z_2(x, y)$; the most significant distinguishing feature is the occurrence of the undetermined functions under the integral sign or not. In the element $z_3(x, y)$ there occurs always a repeated integral over x and y or a twofold integral over y. The undetermined function occurs always under the integral sign, either under the innermost integration or the outermost one.

   7.7. The operator $L \equiv \partial_{xyy} + (x + 1)\partial_{xy} - \partial_y$ allows the factor ∂_y and in addition the Laplace divisor $\langle\!\langle \partial_{xyy} + (x + 1)\partial_{xy} - \partial_y, \partial_{xx}\rangle\!\rangle$; its type $\mathcal{L}^{15}_{xyy}$ Loewy decomposition is

$$L = \begin{pmatrix} 0, & 1 \\ \partial_y + x + 1, & 0 \end{pmatrix}\begin{pmatrix} K_1 \equiv \partial_{xxy} \\ K_2 \equiv \partial_{xyy} + (x + 1)\partial_{xy} - \partial_y \end{pmatrix}. \qquad \Box$$

It leads to the fundamental system is $z_1(x, y) = F(x)$, $z_2(x, y) = (x + 1)G(y) + G'(y)$, $z_3(x, y) = \int\int H(x)\exp(-(x+1)y)dxdx$; F, G and H are undetermined functions.

Appendix B

B.1. If $\Delta \equiv \sqrt{a^2 - 4b} \neq 0$ there are two special solutions $z_{1,2} = -\frac{1}{2}a \pm \frac{1}{2}\Delta$. They yield the two representations of the general solution

$$z(x) = \frac{\pm \Delta \exp\left(\mp \Delta x\right)}{C - \exp\left(\mp \Delta x\right)} - \frac{1}{2}a \pm \frac{1}{2}\Delta;$$

C is a constant. For $C = 0$ and $C \to \infty$ the special solutions z_1 and z_2 respectively are obtained; they are the only rational solutions in this case.

If $\Delta = 0$ the single solution $z_1 = z_2 = -\frac{1}{2}a$ follows. It yields the general rational solution $z(x) = -\frac{1}{2}a + \frac{1}{C + x}$.

B.2. Define $\Delta \equiv \sqrt{c^2 - 4bd}$ now. If $\Delta \neq 0$ there are two special rational solutions $z_{1,2} = -\frac{c}{2b} \pm \frac{1}{2b}\Delta$. The general solution is

$$z(x, y) = \frac{1}{b}\frac{\pm \Delta \exp\left(\mp \Delta x\right)}{F(y - ax) - \exp\left(\mp \Delta x\right)} - \frac{c}{2b} \pm \frac{1}{2b}\Delta.$$

F is an undetermined function. If $\Delta = 0$ the general solution is

$$z(x, y) = -\frac{c}{2b} + \frac{1}{b}\frac{1}{F(y - ax) + x}.$$

Appendix C

C.1. Applying $z = \lambda(x, y)w$ to the expression E defined at the beginning of Appendix C yields an equation of the form $w_{xy} + \bar{a}w_x + \bar{b}w_y + \bar{c}w = 0$ where

$$\bar{a} = a + \frac{\partial \log \lambda}{\partial y}, \quad \bar{b} = b + \frac{\partial \log \lambda}{\partial x}, \quad \bar{c} = c + a\frac{\partial \log \lambda}{\partial x} b\frac{\partial \log \lambda}{\partial y} + \frac{1}{\lambda}\frac{\partial^2 \lambda}{\partial x \partial y}.$$

Substituting these expressions into $\bar{a}_x + \bar{a}\bar{b} - \bar{c}$ and $\bar{b}_y + \bar{a}\bar{b} - \bar{c}$ makes the invariance explicit.

C.2. Forsyth's example. The coefficients are $a = \frac{2}{x - y}$, $b = -\frac{2}{x - y}$, and $c = -\frac{4}{(x - y)^2}$. The invariants $h = k = -\frac{2}{(x - y)^2}$ and $h_1 = k_{-1} = 0$ explain the solution given in Exercise 4.9.

Imschenetzky's example $z_{xy} + xyz_x - 2yz = 0$. Here $a = xy$, $b = 0$ and $c = -2y$. The transformed equations are

$$z_{1,xy} + \left(xy - \frac{1}{y}\right)z_{1,x} - 3yz_1 = 0, \quad z_{2,xy} + \left(xy - \frac{2}{y}\right)z_{2,x} - 4yz_2 = 0$$

and in general

$$z_{n,xy} + \left(xy - \frac{n}{y}\right)z_{n,x} - (n + 2)yz_n = 0$$

for any positive integer n. For negative indices the corresponding equations are

$$z_{-1,xy} + xyz_{-1,x} - yz_{-1} = 0, \; z_{-2,xy} + xyz_{-2,x} = 0, \; z_{-3,xy} + xyz_{-3,x} + yz_{-3} = 0,$$

$$z_{-n,xy} + xyz_{-n,x} - (n+2)yz_{-n} = 0 \; \text{ for } n \geq 4.$$

The invariants $k_0 = 2y$, $k_{-1} = y$, and $k_{-2} = 0$ explain the existence of solution z_1 in Example 5.10. The two lowest h–invariants are $h_0 = 3y$ and $h_1 = 4y$. Due to the fact that they depend only on y the recurrence (C.4) simplifies to $h_{i+1} = 2h_i - h_{i-1}$. The solution satisfying the above initial conditions for $i = 0$ and $i = 1$ is $h_i = (i+1)y$; it does not vanish for $i \geq 2$. Hence, for this particular case it is proved that a Laplace divisor $\langle\langle L, \mathfrak{k}_n \rangle\rangle$ does not exist for any n.

Appendix D

D.1. By Proposition 4.2, the generic equation $z_{xy} + A(x, y)z_y + B(x, y)z = 0$ has a right factor with leading derivative ∂_x if $B(x, y) = A_y(x, y)$; then $\partial_y(z_x + A(x, y)z) = 0$, i.e. the left hand side is a total derivative. There is a right factor with leading derivative ∂_y if $B(x, y) = 0$, i.e. $(\partial_x + A(x, y))z_y = 0$. There are both right factors if $B(x, y) = 0$ and $A_y(x, y) = 0$, i.e. if A does not depend on y. In this case $Lclm(\partial_x + A(x), \partial_y)z = 0$, i.e. the equation is completely reducible.

In special cases one may get even further. In cases (i) and (iii) the above condition for a right factor requires $A = y + C$, C a constant; it yields the factorization $\partial_{xy} + (y + C)\partial_y + 1 = \partial_y(\partial_x + y + C)$. If $B = A_y$ does not hold, there may be a Laplace divisor, e.g. $\langle\langle \partial_{xy} - (y + C)\partial_y + 1, \partial_{yy} \rangle\rangle$; it may be found with the help of the function `LaplaceDivisor` provided by the ALLTYPES system.

In case (iv), for $A = B$ the factorization $\partial_y(\partial_x + \frac{A}{x-y})$ is obtained. A nontrivial Loewy-decomposition allows finding the general solution of the corresponding differential equation involving two undetermined functions. A symmetry yields only so-called similarity solutions of less generality.

Appendix B
Solving Riccati Equations

Abstract Historically Riccati equations were the first non-linear ordinary differential equations that have been systematically studied. A good account of these efforts may be found in the book by Ince [29]. Originally they were of first order, linear in the first derivative, and quadratic in the dependent variable. Its importance arises from the fact that they occur as subproblems in many more advanced applications. Later on, ordinary Riccati equations of higher order have been considered. *Partial Riccati equations* are introduced as a straightforward generalization of the ordinary ones. All derivatives are of first order and occur only linearly, whereas the dependent variables may occur quadratically.

In the first section a few results for ordinary Riccati equations of first and second order are given. Thereafter various generalizations to partial Riccati-equations or Riccati-like systems of equations are considered.

B.1 Ordinary Riccati Equations

In the subsequent lemma the following terminology is applied. Two rational functions $p, q \in \mathbb{Q}(x)$ are called equivalent if there exists another function $r \in \mathbb{Q}(x)$ such that $p - q = \frac{r'}{r}$ holds, i.e. if p and q differ only by a logarithmic derivative of another rational function. This defines an equivalence relation on $\mathbb{Q}(x)$. A *special rational solution* does not contain a constant.

Lemma B.1. *If a first order Riccati equation $z' + z^2 + az + b = 0$ with $a, b \in \mathbb{Q}(x)$ has rational solutions, one of the following cases applies.*

(i) *The general solution is rational and has the form*

$$z = \frac{r'}{r + C} + p \tag{B.1}$$

where $p, r \in \bar{\mathbb{Q}}(x)$; $\bar{\mathbb{Q}}$ is a suitable algebraic extension of $\mathbb{Q}$, and C is a constant.

(ii) *There is only one, or there are two inequivalent special rational solutions.*

F. Schwarz, *Loewy Decomposition of Linear Differential Equations*, Texts & Monographs in Symbolic Computation, DOI 10.1007/978-3-7091-1286-1,
© Springer-Verlag/Wien 2012

Analogous results for Riccati equations of second order are given next.

Lemma B.2. *If a second order Riccati equation*

$$z'' + 3zz' + z^3 + a(z' + z^2) + bz + c = 0$$

with $a, b, c \in \mathbb{Q}(x)$ has rational solutions, one of the following cases applies.

(i) *The general solution is rational and has the form*

$$z = \frac{C_2 u' + v'}{C_1 + C_2 u + v} + p \tag{B.2}$$

   where $p, u, v \in \bar{\mathbb{Q}}(x)$; $\bar{\mathbb{Q}}$ is a suitable algebraic extension of $\mathbb{Q}$, C_1 and C_2 are constants.

(ii) *There is a single rational solution containing a constant, it has the form shown in equation (B.1).*

(iii) *There is a rational solution containing a single constant as in the preceding case, and in addition a single special rational solution.*

(iv) *There is only a single one, or there are two or three special rational solutions that are pairwise inequivalent.*

The proofs of Lemma B.1 and B.2 may be found in Chap. 2 of the book by Schwarz [61].

B.2   Partial Riccati Equations

At first general first-order linear pde's in x and y are considered; they may be obtained as specializations of a Riccati pde if the quadratic term in the unknown function is missing.

Lemma B.3. *Let the first-order linear pde $z_x + az_y + bz = c$ for $z(x, y)$ be given where $a, b, c \in \mathbb{Q}(x, y)$. Define $\varphi(x, y) = const$ to be a rational first integral of $\frac{dy}{dx} = a(x, y)$; assign $\bar{y} = \varphi(x, y)$ and the inverse $y = \psi(x, \bar{y})$ which is assumed to exist. Define*

$$\mathscr{E}(x, y) \equiv \exp\left(-\int b(x, y)|_{y=\psi(x,\bar{y})} dx\right)\Big|_{\bar{y}=\varphi(x,y)}. \tag{B.3}$$

The general solution $z = z_1 + z_0$ of the given first-order pde is

$$z_1(x, y) = \mathscr{E}(x, y)\Phi(\varphi), \quad z_0 = \mathscr{E}(x, y)\int \frac{c(x, y)}{\mathscr{E}(x, y)}\Big|_{y=\psi(x,\bar{y})} dx\Big|_{\bar{y}=\varphi(x,y)}; \tag{B.4}$$

Φ is an undetermined function.

Proof. Introducing a new variable $\bar{y} = \varphi(x, y)$ as defined above leads to the first-order ode $\bar{z}_x + \bar{b}(x, \bar{y})\bar{z} = \bar{c}(x, \bar{y})$. Upon substitution of $\bar{y}$ into its general solution, the solution (B.4) in the original variables is obtained.   □

For $b = 0$ or $c = 0$ the expressions (B.4) simplify considerably as shown next.

Corollary B.1. *With the same notations as in the preceding lemma, the homogeneous equation $z_x + az_y + bz = 0$ has the solution*

$$z_1(x, y) = \mathscr{E}(x, y)\Phi(\varphi). \tag{B.5}$$

The equation $z_x + az_y = c$ has the solution $z = z_1 + z_0$ where

$$z_1(x, y) = \Phi(\varphi), \quad z_0(x, y) = \int c(x, y)|_{y=\psi(x,\bar{y})}dx \,\Big|_{\bar{y}=\varphi(x,y)} \tag{B.6}$$

with $\bar{y}$ and Φ as defined in Lemma B.3.

It should be noticed that Lemma B.3 in general does not allow solving a linear pde algorithmically. To this end, a rational first integral of $\frac{dy}{dx} = a(x, y)$ is required. This problem is discussed below. The subject of the next lemma is the general first-order Riccati pde in x and y.

Lemma B.4. *If the partial Riccati equation*

$$z_x + az_y + bz^2 + cz + d = 0 \tag{B.7}$$

where $a, b, c, d \in \mathbb{Q}(x, y)$ has rational solutions, two cases may occur.

(i) The general solution is rational and has the form

$$z = \frac{1}{a} \left(\frac{r_x(x, \bar{y})}{r(x, \bar{y}) + \Phi(\bar{y})} + p(x, \bar{y}) \right)\Big|_{\bar{y}=\varphi(x,y)} \tag{B.8}$$

where $\varphi(x, y)$ is a rational first integral of $\frac{dy}{dx} = a(x, y)$, r and p are rational functions of its arguments and Φ is an undetermined function.

(ii) There is a single rational solution, or there are two inequivalent rational solutions which do not contain undetermined elements.

Proof. Introducing the new dependent variable w by $z = \frac{w}{b}$, (B.7) is transformed into

$$w_x + aw_y + w^2 + \left(c - \frac{b_x}{b} - a\frac{b_y}{b}\right)w + bd = 0. \tag{B.9}$$

Assume that the first integral $\varphi(x, y) \equiv \bar{y}$ of $\frac{dy}{dx} = a(x, y)$ is rational and the inverse $y = \bar{\varphi}(x, \bar{y})$ exists. Replacing y by $\bar{y}$ leads to

$$\bar{w}_x + \bar{w}^2 + \left(\bar{c} - \frac{\bar{b}_x}{\bar{b}} - \bar{a}\frac{\bar{b}_y}{\bar{b}}\right)\bar{w} + \bar{b}\bar{d} = 0$$

where $\bar{w}(x, \bar{y}) \equiv w(x, y)|_{y=\bar{y}}$, $\bar{a}(x, \bar{y}) \equiv a(x, y)|_{y=\bar{y}}$ and similar for the other coefficients. This is an ordinary Riccati equation for $\bar{w}$ in x with parameter $\bar{y}$. If its general solution is rational, it has the form $\dfrac{r_x}{r + \Phi(\bar{y})} + p$ where r and p are rational functions of x and $\bar{y}$. Back substitution of the original variables yields (B.8). If the general solution is not rational, one or two special rational solutions may exist leading to case (ii). $\square$

B.3   Partial Riccati-Like Systems

In several problems discussed in the main part of this monograph there occur special systems of first-order pde's in two independent, and one or two dependent variables, involving quadratic terms of the latter. They have been baptized *partial Riccati-like systems* by Li et al. [43]; see also Chap. 2 of the book by Schwarz [61]. At first a system for a single function is considered.

Theorem B.1. *The first order Riccati-like system of pde's*

$$e_1 \equiv z_x + A_1 z^2 + A_2 z + A_3 = 0, \quad e_2 \equiv z_y + B_1 z^2 + B_2 z + B_3 = 0 \quad (B.10)$$

is coherent and its general solution depends on a single constant if its coefficients satisfy the constraints

$$A_{1,y} - B_{1,x} - A_1 B_2 + A_2 B_1 = 0,$$
$$A_{2,y} - B_{2,x} - 2A_1 B_3 + 2B_1 A_3 = 0, \quad (B.11)$$
$$A_{3,y} - B_{3,x} - A_2 B_3 + A_3 B_2 = 0.$$

Let A_k, $B_k \in \mathbb{Q}(x, y)$, $A_1 \neq 0$, $B_1 \neq 0$ and system (B.11) be satisfied. If (B.10) has a rational solution, one of the following alternatives applies.

(i) *The general solution is rational and has the form*

$$z = \frac{1}{A_1} \frac{r_x}{r + C} + p = \frac{1}{B_1} \frac{r_y}{r + C} + p \quad (B.12)$$

   where p, $r \in \mathbb{Q}(x, y)$ and C is the integration constant.
(ii) *There is a single one, or there are two inequivalent special rational solutions.*

The proof may be found in the above quoted literature. A system with two dependent variables is

$$e_1 \equiv z_{1,x} + z_1^2 + A_1 z_2 + A_2 z_1 + A_3 = 0,$$
$$e_2 \equiv z_{1,y} + z_1 z_2 + B_1 z_2 + B_2 z_1 + B_3 = 0,$$
$$e_3 \equiv z_{2,x} + z_1 z_2 + B_1 z_2 + B_2 z_1 + B_3 = 0, \quad (B.13)$$
$$e_4 \equiv z_{2,y} + z_2^2 + D_1 z_2 + D_2 z_1 + D_3 = 0.$$

It is coherent if the coefficients $A_1, \ldots, D_3$ satisfy

$$
\begin{aligned}
A_{1,y} - B_{1,x} - A_2 B_1 + A_1 B_2 - A_1 D_1 + A_3 + B_1^2 &= 0, \\
A_{2,y} - B_{2,x} - A_1 D_2 + B_1 B_2 - B_3 &= 0, \\
A_{3,y} - B_{3,x} - A_2 B_3 - A_1 D_3 + A_3 B_2 + B_1 B_3 &= 0 \\
B_{1,y} - D_{1,x} + A_1 D_2 - B_1 B_2 + B_3 &= 0, \\
B_{2,y} - D_{2,x} + A_2 D_2 - B_2^2 + B_2 D_1 - B_1 D_2 - D_3 &= 0, \\
B_{3,y} - D_{3,x} + A_3 D_2 - B_2 B_3 - B_1 D_3 + B_3 D_1 &= 0.
\end{aligned}
\tag{B.14}
$$

The rational solutions of the system (B.13) are described next.

Theorem B.2. *If the coherent system (B.13) has a rational solution, one of the following alternatives applies with $r, s, p, q \in \mathbb{Q}(x, y)$ and $p_y = q_x$.*

(*i*) *The general solution is rational and contains two constants. It may be written in the form*

$$
z_1 = \frac{C_2 r_x + s_x}{C_1 + C_2 r + s} + p, \quad z_2 = \frac{C_2 r_y + s_y}{C_1 + C_2 r + s} + q.
$$

(*ii*) *There is a rational solution containing a single constant, it may be written in the form*

$$
z_1 = \frac{r_x}{C + r} + p, \quad z_2 = \frac{r_y}{C + r} + q.
$$

(*iii*) *There is a solution as described in the preceding case, and in addition there is a special rational solution not equivalent to it.*

(*iv*) *There is a single one, or there are two or three special rational solutions which are pairwise inequivalent.*

The proof may be found in the above references [43] and [61].

B.4 First Integrals of Differential Equations

In several places above solving first-order ode's occurred as a subproblem. Usually however the solution is not required explicitly but in terms of a first integral of the form $\varphi(x, y) = C$ where x and y are the independent and dependent variable, and C is a constant. The relevant features of this problem are discussed in the remaining part of this appendix. It is based on articles by Prelle and Singer [56] and Man and MacCallum [73, 74].

Let a first-order differential equation

$$
\frac{dy}{dx} = \frac{Q}{P} \quad \text{or equivalently} \quad Q\,dx - P\,dy = 0
\tag{B.15}
$$

be given with polynomials $P, Q \in \mathbb{Q}[x, y]$. A first integral of (B.15) is a function $F(x, y)$ that is constant on its solutions, i.e. for which

$$dF = \partial_x F + y' \partial_y F = (P \partial_x + Q \partial_y) F = 0.$$

Prelle and Singer [56] prove that if a differential equation (B.15) has an elementary first integral then it must have a very special form.

Theorem B.3. *(Prelle and Singer [56]) If a differential equation (B.15) has an elementary first integral, it has the form*

$$F(x, y) = w_0(x, y) + \sum c_i \log w_i(x, y) \tag{B.16}$$

where the c_i are constants and the w_i are algebraic functions of x and y.

This result provides strong restrictions on the form of an elementary first integral, if there is any. There remains the question how to determine a first integral explicitly for any given equation (B.15). The following procedure has been given by Man [73], based on the results of Prelle and Singer. In addition to the operator $D \equiv P \partial_x + Q \partial_y$ an upper bound for the degree d of the polynomials occurring in the numerators and denominators of the $w_i(x, y)$ in (B.16) must be provided as input. It is a semi-decision procedure for the existence of an elementary first integral; up to degree d the correct answer is guaranteed if a first integral is known to exist; beyond it there is no conclusion. To make it into an algorithm a bound for the polynomials involved is required. Despite considerable efforts [10, 16], an answer seems not to be known at present (Singer, 2006, private communication).

Algorithm PrelleSinger(D,d) Given an operator $D = P \partial_x + Q \partial_y$ where $P, Q \in \mathbb{Q}[x, y]$, and $d \in \mathbb{N}, d \geq 1$, an elementary first integral of order not higher than d is returned if it exists, or "*failed*" otherwise. Assign $N = 0$.

$S1$: Set $N = N + 1$. If $N > d$ return "failed";

$S2$: Find all monic irreducible polynomials $f_i \in \mathbb{Q}[x, y]$ such that $deg\ f_i \leq N$ and $f_i | Df_i$; define $g_i = \dfrac{Df_i}{f_i}$.

$S3$: Decide if there are constants n_i, not all zero, such that $\sum n_i g_i = 0$. If such n_i exist return $\prod f_i^{n_i}$.

$S4$: Decide if there are constants n_i, not all zero, such that $\sum n_i g_i = -(P_x + Q_y)$. If such n_i exist, set $R = \prod f_i^{n_i}$ and return $\oint RP\, dx + RQ\, dy$, otherwise goto S1.

The polynomials f_i determined in step $S2$ with the property that f_i divides Df_i are called *Darboux polynomials*.

If this procedure is left in step $S3$ there are no logarithmic terms, i.e. a rational first integral has been found. This case is most interesting for the applications in this monograph; two examples are given next.

Example B.1. Consider the equation $(x^2 + x + 2y)y' + 3x^2 + 2xy + y = 0$. It defines the operator $D = (x^2 + x + 2y)\partial_x - (3x^2 + 2xy + y)\partial_y$. For $n = 2$ the following Darboux polynomials are obtained.

Table B.1 The polynomials f and g for the differential equation in Example B.3 for degree of f not higher than 5

f_i	g_i	Range
x^k	k	$1 \le k \le 5$
y^k	$-k(3x+y)y$	$1 \le k \le 5$
$x^k y$	$-(3x+y)y+k$	$1 \le k \le 4$
xy^k	$ky(3x+y)+1$	$2 \le k \le 4$
$x^2 y^2$	$-2y(3x+y)+k$	$2 \le k \le 3$
$x^2 y^2$	$-3y(3x+y)+1$	

$$f_1 = (x+y)^2, \quad g_1 = -4x+2, \quad f_2 = x^2+y,$$
$$g_2 = 2x-1, \quad f_3 = x+y, \quad g_3 = -2x+1.$$

The constraint $n_1 g_1 + n_2 g_2 + n_3 g_3 = 0$ yields $n_2 = 2n_1 + n_3$; it leads to the first integral $F(x,y) = (x+y)^{2n_1}(x^2+y)^{n_2}(x+y)^{n_3} = (x+y)(x^2+y)$ where in the last step the substitution $n_2 = 1$ has been made. □

Example B.2. Equation $x^2 y' + x^2 y^2 + 4xy + 2 = 0$, no. 1.140 of Chap. C.1 in the collection by Kamke [32], defines the operator $D = x^2 \partial_x - (x^2 y^2 + 4xy + 2)\partial_y$. For $n = 2$ the Darboux polynomials

$$f_1 = xy+2, \quad g_1 = -x^2 y - x, \quad f_2 = xy+1, \quad g_2 = -x^2 y - 2x, \quad f_3 = x, \quad g_3 = x$$

are obtained. The constraint $n_1 g_1 + n_2 g_2 + n_3 g_3 = 0$ yields $n_2 = n_3 = -n_1$; it leads to the first integral

$$F(x,y) = (xy+2)^{n_1}(xy+1)^{n_2} x^{n_3} = \frac{xy+2}{x(xy+1)};$$

in the last step the substitution $n_2 = n_3 = -1$ has been made. □

The answers in the preceding two examples are easily obtained because there does exist a rational first integral of low order. This is different if a rational first integral does not seem to exist as shown in the next example.

Example B.3. The Abel equation $xy' + y^3 + 3xy^2 = 0$, no. 1.111 of Chap. C.1 in the collection by Kamke [32], defines the operator $D \equiv x\partial_x - (3xy^2 + y^3)\partial_y$. Up to order $n \le 5$, the Darboux polynomials listed in Table B.1 are obtained. The f's cover all monomials of total degree up to order 5. Any system that may be constructed from these polynomials $\sum n_i g_i = 0$ has only the trivial solution $n_i = 0$ for all i; consequently, a nontrivial rational first integral up to order 5 does not exist. In order to prove non-existence for any order, it has to be shown that the structure of these system remains the same as shown in Table B.1. □

In the preceding example it may be possible to show that a rational first integral does not exist for any given order n. However, for a generic first order equation there is only a partial answer. If it is known that a rational first integral exists, the above procedure of Prelle and Singer returns the result; otherwise it does not provide a conclusion.

The situation is similar as for solving diophantine equations. These problems are semi-decidable, i.e if an equation is known to have any integer solution it can always be found, otherwise in general there is no answer. Even more: It has been shown by Matyasevich [48] that in general diophantine problems are undecidable. It remains an open problem to obtain a conclusive answer for the existence of a rational first integral of a generic first order differential equation, either by designing an algorithm for its solution or proving that it is undecidable.

B.5   Exercises

Exercise B.1. Discuss the solutions of the first-order ordinary Riccati equation $z' + z^2 + az + b = 0$ if a and b are constant.

Exercise B.2. The same problem for the first-order partial Riccati equation $z_x + az_y + bz^2 + cz + d = 0$ if a, b, c and d are constant.

Appendix C
The Method of Laplace

This appendix follows closely Chap. 2 of the book by Darboux [14] and Chap. 5 of the book by Goursat [18]; both are based on the original work of Laplace [40]. The method known under his name deals with equations of the form

$$E \equiv z_{xy} + az_x + bz_y + cz = 0;$$

a, b and c are functions of x and y without further specification. This equation may be written as

$$(\partial_x + b)(\partial_y + a)z = 0 \quad \text{if} \quad a_x + ab = c.$$

In this case the general solution is

$$z = \exp\left(-\int ady\right)\left(F(x) + \int G(y)\exp\left(\int ady - \int bdx\right)dy\right)$$

where F and G are undetermined functions of its argument. Similarly, if

$$(\partial_y + a)(\partial_x + b)z = 0 \quad \text{if} \quad b_y + ab = c$$

the general solution is

$$z = \exp\left(-\int bdx\right)\left[G(y) + \int F(x)\exp\left(\int bdx - \int ady\right)dx\right].$$

If both conditions $a_x + ab = c$ and $b_y + ab = c$ hold, the general solution is

$$z = F(x)\exp\left(-\int ady\right) + G(y)\exp\left(-\int bdx\right).$$

For the further discussion it is advantageous to define the two quantities

$$h_0 \equiv h \equiv a_x + ab - c \quad \text{and} \quad k_0 \equiv k \equiv b_y + ab - c. \tag{C.1}$$

F. Schwarz, *Loewy Decomposition of Linear Differential Equations*, Texts & Monographs in Symbolic Computation, DOI 10.1007/978-3-7091-1286-1,
© Springer-Verlag/Wien 2012

In Exercise C.1 it will be shown that h and k are invariants w.r.t. to transformations $z = \lambda(x, y)w$ of the Laplace equation E.

If both $h \neq 0$ and $k \neq 0$, the above factorizations do not apply. According to Laplace one may proceed as follows. A new dependent variable z_1 is introduced by $z_1 \equiv z_y + az$. Using it, the original equation may be written as $z_{1,x} + bz_1 = hz$. Eliminating z from the latter and substituting it into the definition of z_1, the equation

$$E_1 \equiv z_{1,xy} + a_1 z_{1,x} + b_1 z_{1,y} + c_1 z_1 = 0$$

of the same type as equation E is obtained where

$$a_1 = a - \frac{h_y}{h}, \quad b_1 = b, \quad c_1 = c - a_x + b_y - b\frac{h_y}{h}. \tag{C.2}$$

The corresponding invariants are

$$h_1 \equiv a_{1,x} + a_1 b_1 - c_1 = 2h - k - (\log h)_{xy}, \quad k_1 = h.$$

On the other hand, a new dependent variable z_{-1} may be introduced by $z_{-1} \equiv z_x + bz$. Using it, the original equation may be written as $z_{-1,y} + az_{-1} = kz$. Eliminating z from the latter and substituting it into the definition of z_{-1}, the equation

$$E_{-1} \equiv z_{-1,xy} + a_{-1} z_{-1,x} + b_{-1} z_{-1,y} + c_{-1} z_{-1} = 0$$

follows where

$$a_{-1} = a, \quad b_{-1} = b - \frac{k_x}{k}, \quad c_{-1} = c - b_y + a_x - a\frac{k_x}{k}. \tag{C.3}$$

Its invariants are

$$h_{-1} \equiv a_{1,x} + a_1 b_1 - c_1 = k, \quad k_{-1} = 2k - h - (\log k)_{xy}.$$

This proceeding may be repeated, generating a sequence of equations

$$\ldots E_{-2}, E_{-1}, E_0 \equiv E, E_1, E_2 \ldots$$

as long as the corresponding invariants are different from zero. They are related by

$$h_{i+1} = 2h_i - h_{i-1} - (\log h_i)_{xy}, \quad k_{i+1} = h_i. \tag{C.4}$$

Solving for h_i and k_i yields

$$h_i = k_{i+1}, \quad k_i = 2k_{i+1} - h_{i+1} - (\log k_{i+1})_{xy}.$$

This sequence of equations terminates for a positive integer i if $h_i = 0$; if this is true the corresponding equation E_i may be solved as described above. By back substitution a special solution of the original equation E may be obtained. It has the form

$$z = F(x)r_0(x, y) + F'(x)r_1(x, y) + \ldots + F^{(i)}(x)r_i(x, y);$$

F is an undetermined function of x, the $r_k(x, y)$ are functions of x and y which are determined by the problem.

Similarly, if the sequence of equations terminates for a negative value j because $k_j = 0$, a special solution of the form

$$z = G(y)s_0(x, y) + G'(y)s_1(x, y) + \ldots + G^{(j)}(y)s_j(x, y)$$

is obtained. Now G is an undetermined function of y; the $s_k(x, y)$ are functions of x and y which are determined by the originally given equation.

Obviously the success of the method hinges on the question whether any invariant vanishes for sufficiently large value of its index i or j. Right now, an upper bound for these values does not seem to exist. Consequently, the existence of a Laplace divisor is semi-decidable; if for a particular equation it is known to exist it may always be found, otherwise in general it remains an open question and may even turn out to be undecidable.

Laplace' method has been generalized to certain equations with leading derivative $\dfrac{\partial^{n+1}}{\partial x \partial y^n}$ for $n \geq 2$ by [53]. Another generalization has been given by [70].

C.1 Exercises

Exercise C.1. Show that the expressions h and k defined by (C.1) are invariants w.r.t. the transformations $z = \lambda(x, y)w$.

Exercise C.2. Determine the transformed equations $E_{\pm i}$ and the invariants h_i and k_i for the equations considered in Examples 5.9 and 5.10. To this end, apply the user-functions LaplaceTransformation and LaplaceInvariant provided on the website www.alltypes.de. Use the result to explain the solution behavior.

Appendix D
Equations with Lie Symmetries

Symmetries are special transformations of a differential equation leaving its form invariant; the totality of symmetries forms a group, the *symmetry group* of the respective equation. Good introductions into the subject may be found in the books by Olver [51] or Bluman and Kumei [4]; for symmetries of ode's see also the book by Schwarz [61].

Sophus Lie [44] discussed his symmetry methods for second-order linear pde's in two independent variables in full detail. These results are given in this appendix without proofs. He considers the general homogeneous equation for a function $z(x, y)$ in the form

$$R(x, y)z_{xx} + S(x, y)z_{xy} + T(x, y)z_{yy} + P(x, y)z_x + Q(x, y)z_y + Z(x, y)z = 0.$$

The coefficients $R, S, \ldots, Z$ are functions of the independent variables x and y without specifying its function field. According to Lemma 5.4 two cases with different leading derivatives are distinguished. If $R(x, y) = T(x, y) = 0$ there is only the mixed leading derivative z_{xy}; this case is considered first.

Theorem D.1. *Any equation* $z_{xy} + A(x, y)z_y + B(x, y)z = 0$ *has symmetry generators* $U_1 = z\partial_z$ *and* $U_\infty = \varphi(x, y)\partial_z$ *where* φ *is a solution of the given differential equation; its commutator is* $[U_1, U_\infty] = -U_\infty$. *There are four special cases with larger symmetry groups and corresponding canonical forms.*

(i) *An equation with canonical form* $z_{xy} + A(y)z_y + z = 0$ *has symmetry generators* $U_1 = z\partial_z$, $U_2 = \partial_x$, *and* U_∞ *with non-vanishing commutator* $[U_2, U_\infty] = (\log \varphi)_x U_\infty$.

(ii) *An equation with canonical form* $z_{xy} + P(x - y)z_y + Q(x - y)z = 0$, P *and* Q *undetermined functions of its single argument* $x - y$, *has symmetry generators* $U_1 = z\partial_z$, $U_2 = \partial_x + \partial_y$, *and* U_∞ *with non-vanishing commutator* $[U_2, U_\infty] = \big((log\varphi)_x + (\log \varphi)_y\big)U_\infty$.

(iii) *An equation with canonical form* $z_{xy} + Cyz_y + z = 0$, C *a constant, has symmetry generators* $U_1 = z\partial_z$, $U_2 = \partial_x$, $U_3 = x\partial_x - y\partial_y$, $U_4 = \partial_y - Cxz\partial_z$,

F. Schwarz, *Loewy Decomposition of Linear Differential Equations*, Texts & Monographs in Symbolic Computation, DOI 10.1007/978-3-7091-1286-1,
© Springer-Verlag/Wien 2012

and U_∞ with non-vanishing commutators

$$[U_2, U_3] = U_2, \quad [U_2, U_4] = -C U_1, \quad [U_3, U_4] = -U_4.$$

(iv) An equation with canonical form $z_{xy} + \dfrac{A}{x-y} z_y + \dfrac{B}{(x-y)^2} z = 0$, A and B constants, has symmetry generators

$$U_1 = z\partial_z, \quad U_2 = \partial_x + \partial_y, \quad U_3 = x\partial_x + y\partial_y, \quad U_4 = x^2\partial_x + y^2\partial_y - Axz\partial_z$$

and U_∞ with non-vanishing commutators

$$[U_2, U_3] = U_2, \quad [U_2, U_4] = 2U_3 - AU_1, \quad [U_3, U_4] = U_4.$$

If $S(x, y) = T(x, y) = 0$ the only second-order derivative is z_{xx}; the corresponding equations are considered next.

Theorem D.2. *Any equation* $z_{xx} + A(x, y)z_y + B(x, y)z = 0$ *has symmetry generators* $U_1 = z\partial_z$ *and* $U_\infty = \varphi(x, y)\partial_z$ *where* φ *is a solution of the differential equation; its commutator is* $[U_1, U_\infty] = -U_\infty$. *There are three special cases with larger symmetry groups and corresponding canonical forms.*

(i) *An equation with canonical form* $z_{xx} + z_y + B(x)z = 0$ *has symmetry generators* $U_1 = z\partial_z$, $U_2 = \partial_y$, *and* U_∞; *there are no non-vanishing commutators.*

(ii) *An equation with canonical form* $z_{xx} + z_y = 0$ *has symmetry generators*

$$U_1 = z\partial_z, \quad U_2 = \partial_x, \quad U_3 = \partial_y, \quad U_4 = x\partial_x + 2y\partial_y,$$

$$U_5 = 2y\partial_x + xz\partial_z, \quad U_6 = xy\partial_x + y^2\partial_y + (\tfrac{1}{4}x^2 - \tfrac{1}{2}y)z\partial_z, \quad \text{and } U_\infty;$$

its non-vanishing commutators are

$$[U_2, U_4] = U_2, \quad [U_2, U_5] = U_1, \quad [U_2, U_6] = \tfrac{1}{2}U_5, \quad [U_3, U_4] = 2U_3,$$

$$[U_3, U_5] = 2U_2, \quad [U_3, U_6] = U_4 - \tfrac{1}{2}U_1, \quad [U_4, U_5] = U_5, \quad [U_4, U_6] = 2U_6.$$

(iii) *An equation with canonical form* $z_{xx} + z_y + \dfrac{C}{x^2}z = 0$, C *a constant, has symmetry generators*

$$U_1 = z\partial_z, \quad U_2 = \partial_y, \quad U_3 = x\partial_x + 2y\partial_y,$$

$$U_4 = xy\partial_x + y^2\partial_y + (\tfrac{1}{4}x^2 - \tfrac{1}{2}y)z\partial_z, \quad \text{and } U_\infty;$$

its non-vanishing commutators are

$$[U_2, U_3] = 2U_2, \quad [U_2, U_4] = U_3 - \tfrac{1}{2}U_1, \quad [U_3, U_4] = 2U_4.$$

There arises the question as to what the relation between the symmetries of an equation and its factorization properties are, if any. For equations with leading derivative z_{xx} considered in the preceding theorem the answer is easy; from Proposition 4.1 it follows immediately that there are no factorizations related to the enlarged symmetry groups of cases (i), (ii) or (iii). This is different for equations with leading derivative z_{xy}; a few examples are discussed in the subsequent exercise.

D.1 Exercises

Exercise D.1. Investigate the factorization properties of the equations considered in Theorem D.1 and discuss the results.

Appendix E
ALLTYPES in the Web

Many calculations described in this monograph cannot be performed by pencil and paper because they are too voluminous. The website www.alltypes.de provides a collection of user functions for this purpose. In addition to an interactive environment for applying them, a short documentation for each function is given, including some examples. This appendix contains a list of those functions that are particularly relevant for the subject of this book; details may be found on the above website.

`Adjoint(u)`. The single arguments u is a linear ode; returns adjoint equation.

`Commutator(u,v)`. The two arguments u and v are ordinary or partial differential operators; returns commutator.

`ExactQuotient(u,v)`. The two arguments u and v are ordinary or partial differential operators; returns exact quotient or error message if it does not exist.

`FirstIntegrals(u,n)`. The first argument u represents a system of first-order ode's; returns polynomial or quasipolynomial first integrals up to order n.

`FirstOrderRightFactors(u)`. The single argument u represents an ordinary or partial differential operator in the plane; returns all first-order right factors with rational function coefficients.

`Gcrd(u,v)`. The two arguments u and v may be individual ordinary or partial differential operators in the plane, or generators of an ideal; returns generators of greatest common right divisor or sum ideal.

`IntegrabilityConditions(u)`. The single argument is a set of generators for a left ideal in the ring of differential operators or module over such a ring; returns integrability conditions for the coefficients.

`JanetBasis(u)`. The single argument u represents a list of generators of an ideal or a module of differential operators. Returns Janet basis generators in the same term ordering as applied for input.

`LaplaceDivisor(u,k)`. The first argument u is a partial differential operator in the plane, the second argument k a natural number; returns Laplace divisor of order not higher than k if it exists.

F. Schwarz, *Loewy Decomposition of Linear Differential Equations*, Texts & Monographs in Symbolic Computation, DOI 10.1007/978-3-7091-1286-1,
© Springer-Verlag/Wien 2012

LaplaceInvariants(u,k). The first argument u is a partial differential operator in the plane, the second argument k a natural number. Returns Laplace invariant of order not higher than k.

Lclm(u,v). The two arguments u and v are ordinary or partial differential operators in the plane; returns generators of least common left multiple or left intersection ideal.

LoewyDecomposition(u). The single argument u is an ordinary or partial differential operator in the plane; returns Loewy decomposition of u.

LoewyDivisor(u). The single argument u is an ordinary or partial differential operator in the plane; returns largest completely reducible right component,

RationalSolutions(u). The single argument u may be a linear ode or a linear pde in the plane.

Solve(u). The single argument u may be a linear or nonlinear ode, a first- or second-order linear pde in the plane, or a system of linear pde's of differential type zero.

List of Notation

Symbol	Meaning	Page of definition
$\prec, \preceq, \succ, \succeq$	Order relation between terms	23
$Gcrd$	Greatest common right divisor	26
$Lclm$	Least common left multiple	26
$o(z_{x^m y^n})$	Subsumes all terms not higher than argument	29
$\langle l_1, l_2, \ldots \rangle$	Ideal generated by $l_1, l_2, \ldots$	22
$\langle\!\langle l_1, l_2, \ldots \rangle\!\rangle$	Ideal generated by Janet basis $l_1, l_2, \ldots$	25
$\langle \partial_{xx}, \partial_{xy} \rangle_{LT}$	Ideal with leading derivatives ∂_{xx} and ∂_{xy}	30
$\mathscr{D}$	Ring of differential operators	22
H_I	Hilbert-Kolchin polynomial of ideal I	28
(k, j)	Differential dimension of ideal or module	29
$\mathbb{J}^{(0,k)}$	Type of ideals of differential dimension $(0, k)$	30
$\mathbb{M}^{(0,k)}$	Type of modules of differential dimension $(0, k)$	32
$\mathbb{L}_{x^m}(L)$	Laplace divisor for L of order m in x	34
$\mathfrak{l}_m$	Generator of Laplace divisor $\mathbb{L}_{x^m}(L)$	34
$\mathbb{L}_{y^n}(L)$	Laplace divisor for L of order n in y	34
$\mathfrak{k}_n$	Generator of Laplace divisor $\mathbb{L}_{y^n}(L)$	34
$\mathbb{J}_{xxx}, \mathbb{J}_{xxy}$	Generic intersection ideals for first-order operators	45
$\mathscr{E}_i(x, y)$	Shifted exponential integral for $\partial_x + a_i \partial_y + b_i$	92
$\varepsilon_i(x, y)$	Exponential integral for $\partial_x + b_i$	95
$\mathscr{L}_k^2$	Type of Loewy decomposition of second-order ode	7
$\mathscr{L}_k^3$	Type of Loewy decomposition of third-order ode	8

F. Schwarz, *Loewy Decomposition of Linear Differential Equations*, Texts & Monographs
in Symbolic Computation, DOI 10.1007/978-3-7091-1286-1,
© Springer-Verlag/Wien 2012

Symbol	Meaning	Page of definition
$\mathscr{L}_{xx}^k$	k-th decomposition type for operator with leading derivative ∂_{xx}	83
$\mathscr{L}_{xy}^k$	k-th decomposition type for operator with leading derivative ∂_{xy}	86
$\mathscr{L}_{xxx}^k$	k-th decomposition type for operator with leading derivative ∂_{xxx}	123
$\mathscr{L}_{xxy}^k$	k-th decomposition type for operator with leading derivative ∂_{xxy}	132
$\mathscr{L}_{xyy}^k$	k-th decomposition type for operator with leading derivative ∂_{xyy}	140

References

1. Abramowitz M, Stegun IA (1965) Handbook of mathematical functions. Dover, New York
2. Adams WW, Loustaunau P (1994) An introduction to Gröbner bases. American Mathematical Society, Providence
3. Beke E (1894) Die Irreduzibilität der homogenen Differentialgleichungen. Math Ann 45: 278–294
4. Bluman GW, Kumei S (1990) Symmetries of differential equations. Springer, Berlin
5. Blumberg H (1912) Über algebraische Eigenschaften von linearen homogenen Differentialausdrücken. Inaugural-Dissertation, Göttingen
6. Bronstein M (1994) An improved algorithm for factoring linear ordinary differential operators. In: Proceedings of the ISSAC'94. ACM, New York, pp 336–340
7. Buchberger B (1970) Ein algorithmisches Kriterium für die Lösbarkeit eines algebraischen Gleichungssystems. Aequ Math 4:374–383
8. Buium A, Cassidy Ph (1999) Differential algebraic geometry and differential algebraic groups: from algebraic differential equations to diophantine geometry. In: Bass H, Buium A, Cassidy Ph (eds) Selected works of Ellis Kolchin. AMS, Providence
9. Castro-Jiménez FJ, Moreno-Frías MA (2001) An introduction to Janet bases and Gröbner bases. In: Lecture notes in pure and applied mathematics, vol 221. Marcel Dekker, New York, pp 133–145
10. Chen G, Ma Y (2005) Algorithmic reduction and rational general solutions of first-order algebraic differential equations. In: Wang D, Zhen Z (eds) Differential equations with symbolic computation. Birkhäuser, Basel
11. Cohn PM (2006) Free ideal rings and localization in general rings. Cambridge University Press, Cambridge/New York
12. Coutinho SC (1995) A primer of algebraic D-modules. London mathematical society student texts, vol 33. Cambridge University Press, Cambridge
13. Cox D, Little J, O'Shea D (1991/1998) Ideals, varieties and algorithms. Springer, New York; Using algebraic geometry. Springer, New York
14. Darboux E (1972) Leçons sur la théorie générale des surfaces, vol II. Chelsea Publishing, New York
15. Davey BA, Priestley HA (2002) Introduction to lattices and order. Cambridge University Press, Cambridge/New York
16. Eremenko A (1998) Rational solutions of first-order differential equations. Ann Acad Scient FennicæMath 23L:181–190
17. Forsyth AR (1906) Theory of differential equations, vols I–VI. Cambridge University Press, Cambridge

F. Schwarz, *Loewy Decomposition of Linear Differential Equations*, Texts & Monographs in Symbolic Computation, DOI 10.1007/978-3-7091-1286-1,
© Springer-Verlag/Wien 2012

18. Goursat E (1898) Leçon sur l'intégration des équation aux dérivées partielles, vol I and II. A. Hermann, Paris
19. Grätzer G (1998) General lattice theory. Birkhäuser, Basel/Boston
20. Greuel GM, Pfister G (2002) A singular introduction to commutative algebra. Springer, Berlin/New York
21. Grigoriev D (1990) Complexity of factoring and calculating the GCD of linear ordinary differential operators. J Symb Comput 7:7–37
22. Grigoriev D, Schwarz F (2004) Factoring and solving linear partial differential equations. Computing 73:179–197
23. Grigoriev D, Schwarz F (2005) Generalized Loewy decomposition of D-modules. In: Kauers M (ed) Proceedings of the ISSAC'05. ACM, New York, pp 163–170
24. Grigoriev D, Schwarz F (2008) Loewy decomposition of third-order linear PDE's in the plane. In: Gonzales-Vega L (ed) Proceedings of the ISSAC 2008, Linz. ACM, New York, pp 277–286
25. Grigoriev D, Schwarz F (2010) Absolute factoring of non-holonomic ideals in the plane. In: Watt SM (ed) Proceedings of the ISSAC 2010, Munich. ACM, New York
26. Hillebrand A, Schmale W (2001) Towards an effective version of a theorem due to stafford. J Symb Comput 32L:699–716
27. Hubert E (2003) Notes on triangular sets and triangulation-decomposition algorithms; Part I: polynomial systems; Part II: differential systems. In: Winkler F, Langer U (eds) Symbolic and numerical scientific computing. LNCS 2630. Springer, Berlin/London
28. Imschenetzky VG (1872) Étude sur les méthodes d'intégration des équations aux dérivées partielles du second ordre d'une fonction de deux variables indépendantes. Grunert's Archiv 54:209–360
29. Ince EL (1926) Ordinary differential equations. Longmans (Reprint by Dover, New York, 1960].
30. Janet M (1920) Les systèmes d'équations aux dérivées partielles. J Math 83:65–123
31. Kamke E (1962) Differentialgleichungen I. Partielle Differentialgleichungen. Akademische Verlagsgesellschaft, Leipzig
32. Kamke E (1964) Differentialgleichungen I. Gewöhnliche Differentialgleichungen. Akademische Verlagsgesellschaft, Leipzig
33. Kamke E (1965) Differentialgleichungen, Lösungsmethoden und Lösungen II. Partielle Differentialgleichungen. Akademische Verlagsgesellschaft, Leipzig
34. Kamke E (1967) Differentialgleichungen, Lösungsmethoden und Lösungen I. Gewöhnliche Differentialgleichungen. Akademische Verlagsgesellschaft, Leipzig
35. Kaplansky I (1957) An introduction to differential algebra. Hermann, Paris
36. Kolchin E (1964) The notion of dimension in the theory of algebraic differential equations. Bull AMS 70:570–573
37. Kolchin E (1973) Differential algebra and algebraic groups. Academic Press, New York
38. Kondratieva M, Levin A, Mikhalev A, Pankratiev E (1999) Differential and difference dimension polynomials. Kluwer, Dordrecht/Boston
39. Landau E (1902) Ein Satz über die Zerlegung homogener linearer Differentialausdrücke in irreduzible Faktoren. J Reine Angew Math 124:115–120
40. Laplace PS (1777) Mémoires de l'Aacademie royal des sciences (See also Œuvres complètes de Laplace, vol. IX, 5–68).
41. Li Z, Schwarz F (2001) Rational solutions of riccati-like partial differential equations. J Symb Comput 31:691–716
42. Li Z, Schwarz F, Tsarev S (2002) Factoring zero-dimensional ideals of linear partial differential operators. In: Mora T (ed) Proceedings of the ISSAC'02. ACM, New York, pp 168–175,
43. Li Z, Schwarz F, Tsarev S (2003) Factoring systems of linear PDE's with finite-dimensional solution space. J Symb Comput 36:443–471
44. Lie S (1881) Über die Integration durch bestimmte Integrale von einer Klasse linear partieller Differentialgleichungen. Arch Math VI:328–368. (Reprinted in Gesammelte Abhandlungen III (1922) Teubner, Leipzig, pp 492–523)

45. Liouville R (1889) Mémoire sur les invariants de certaines équations différentielles et sur leurs applications, Journal de l'Ecole Polytechnique 59:7–76
46. Loewy A (1906) Über vollständig reduzible lineare homogene Differentialgleichungen. Math Ann 56:89–117
47. Magid A (1994) Lectures on differential Galois theory. AMS university lecture series, vol 7. AMS Press, Providence
48. Matiyasevich Y (1993) Hilbert's tenth problem. MIT, Cambridge
49. Miller FH (1932) Reducible and irreducible linear differential operators. PhD Thesis, Columbia University
50. Oaku T (1997) Some algorithmic aspects of D-module theory. In: Bony JM, Moritomo M (eds) New trends in microlocal analysis. Springer, Tokyo/New York
51. Olver P (1986) Application of Lie groups to differential equations. Springer, Berlin
52. Ore Ö (1932) Formale theorie der linearen differentialgleichungen. J Reine Angew Math 167:221–234; 168:233–257
53. Petrén L (1911) Extension de la méthode de Laplace aux équations

$$\sum_{i=0}^{n=1} A_{1i}(x, y) \frac{\partial^{i+1} z}{\partial x \partial y^i} + \sum_{i=0}^{n} A_{0i}(x, y) \frac{\partial^i z}{\partial y^i} = 0.$$

Lunds Universitets Arsskrift. N. F. Afd.2, Bd 7. Nr 3, Lund
54. Plesken W, Robertz D (2005) Janet's approach to presentations and resolutions for polynomials and linear pdes. Arch Math 84:22–37
55. Polyanin A 2002 Handbook of linear partial differential equations for engineers and scientists. Chapman and Hall/CRC, Boca Raton
56. Prelle MJ, Singer M (1983) Elementary integrals of differential equations. Trans Am Math Soc 279:215–229
57. Ritt JF (1950) Differential algebra. Dover, New York
58. Schlesinger L (1897) Handbuch der Theorie der linearen Differentialgleichungen. Teubner, Leipzig
59. Schwarz F (1989) A factorization algorithm for linear ordinary differential equations. In: Gonnet G (ed) Proceedings of the ISSAC'89. ACM, New York, pp 17–25
60. Schwarz F (1998) Janet bases for symmetry groups. Buchberger B, Winkler F (eds) In: Gröbner bases and applications. London mathematical society lecture notes series 251. Cambridge University Press, Cambridge/New York, pp 221–234
61. Schwarz F (2007) Algorithmic Lie theory for solving ordinary differential equations. Chapman and Hall/CRC, Boca Raton/London
62. Schwarz F (2008) ALLTYPES in the Web. ACM Commun Comput Algebra 42(3):185–187
63. Schwarz F (2011) Ideal intersections in rings of partial differential operators. Adv Appl Math 47:140–157
64. Shemyakova E, Winkler F (2011) Linear partial differential equations and linear partial differential operators in computer algebra In: Progress and Prospects in Numerical and Symbolic Scientific Computing, P. Paule et al. (eds), Springer, Texts & Monographs in Symbolic Computation, pp. 333–358
65. Sit W (1974) Typical differential dimension of the intersection of linear differential algebraic groups. J Algebra 32:476–487
66. Sit W (2002) The Ritt-Kolchin theory for differential polynomials. In: Guo L, et al. (eds) Differential algebra and related topics. World Scientific, Singapore/Hong Kong
67. Stafford JT (1978) Module structure of Weyl algebras. J Lond Math Soc 18(Ser II):429–442
68. Tsarev S (1999) On Darboux integrable nonlinear partial differential equations. Proc Steklov Inst Math 225:389–399
69. Tsarev S (2000) Factoring linear partial differential operators and the Darboux method for integrating nonlinear partial differential equations. Theor Math Phys 122:121–133

70. Tsarev S (2005) Generalized Laplace transformations and integration of hyperbolic systems of linear partial differential equations. In: Kauers M (ed) Proceedings of the ISSAC'05. ACM, New York, pp 163–170
71. van der Put M, Singer M 2003 Galois theory of linear differential equations. Grundlehren der mathematischen Wissenschaften, vol 328. Springer, New York
72. van Hoeij M (1997) Factorization of differential operators with rational function coefficients. J Symb Comput 24:537–561
73. Yiu-Kwong Man (1993) Computing closed form solutions of first order ODEs using the Prelle-Singer procedure. J Symb Comput 16:423–443
74. Yiu-Kwong Man, MacCallum AH (1997) A rational approach to the Prelle-Singer algorithm. J Symb Comput 24:31–43

Index

F. Schwarz, *Loewy Decomposition of Linear Differential Equations*, Texts & Monographs 229
in Symbolic Computation, DOI 10.1007/978-3-7091-1286-1,
© Springer-Verlag/Wien 2012